D1256669

DAVENPORT'S DREAM

21st CENTURY REFLECTIONS ON HEREDITY AND EUGENICS

DAVENPORT'S DREAM

21ST CENTURY REFLECTIONS ON HEREDITY AND EUGENICS

EDITED BY

JAN A. WITKOWSKI

JOHN R. INGLIS

Cold Spring Harbor Laboratory Press
Cold Spring Harbor, New York

COLD SPRING HARBOR LABORATORY PRESS
Cold Spring Harbor, New York • www.cshlpress.com

DAVENPORT'S DREAM

21ST CENTURY REFLECTIONS ON HEREDITY AND EUGENICS

Printed in the United States of America

Publisher	John Inglis
Book Development, Marketing, and Sales Director	Jan Argentine
Developmental Editor	Judy Cuddihy
Project Coordinator	Maryliz Dickerson
Production Editor	Kathleen Bubbeo
Desktop Editor	Susan Schaefer
Production Manager	Denise Weiss
Cover Designer	Ed Atkeson
Marketing Manager	Ingrid Benirschke
Sales Manager	Elizabeth Powers

Front cover: "The Man of Mystery Who Is Searching for the Secret of Life," *The World Magazine*, article on Charles Davenport (1907). American Philosophical Society image courtesy of Susan Lauter, Dolan DNA Learning Center.

Back cover: Plate I (frontispiece), Eye Colors in Man, reproduced from Davenport's *Heredity in Relation to Eugenics* (1911; Henry Holt).

Library of Congress Cataloging-in-Publication Data

Davenport's dream : 21st century reflections on heredity and eugenics /
edited by Jan A. Witkowski and John R. Inglis.
p. cm.
Includes bibliographical references and index.
ISBN 978-0-87969-756-3 (hard cover : alk. paper)
1. Human genetics--Social aspects. 2. Human genetics--History. 3.
Eugenics. 4. Davenport, Charles Benedict, 1866-1944 Heredity in relation to
eugenics. I. Witkowski, J. A. (Jan Anthony), 1947- II. Inglis, J. R. (John
R.) III. Davenport, Charles Benedict, 1866-1944 Heredity in relation to
eugenics.
QH438.7.D38 2008
599.93'5--dc22

2007050996

10 9 8 7 6 5 4 3 2 1

All Cold Spring Harbor Laboratory Press publications may be ordered directly from Cold Spring Harbor Laboratory Press, 500 Sunnyside Blvd., Woodbury, New York 11797-2924. Phone: 1-800-843-4388 in Continental U.S. and Canada. All other locations: (516) 422-4100. FAX: (516) 422-4097. E-mail: cshpress@cshl.edu. For a complete catalog of all Cold Spring Harbor Laboratory Press publications, visit our World Wide Web site http://www.cshlpress.com/.

Contents

A Reprint of *Heredity in Relation to Eugenics* by Charles Benedict Davenport follows page 192
(Published by Henry Holt and Company, New York, 1911)

Preface

It is unusual to reissue a book that has long been unavailable, with content long outdated. Charles Davenport's *Heredity in Relation to Eugenics* was published almost 100 years ago, and its subject matter—eugenic studies in the early 20th century—was consigned in the 1940s to what is now referred to as "pathological science." Why, then, revisit it?

One reason is the book's significance in the history of biology. While traits "running in families" had been recognized for centuries, it was not until the discovery of Mendel's work in 1902 that the theoretical framework became available for a scientific study of heredity, including that of human beings. Davenport, with his wife Gertrude, was one of the first to undertake a scientific analysis of human heredity in the post-Mendel era, and *Heredity in Relation to Eugenics* presents a review of the state of knowledge of human genetics in 1911. Although original copies are now quite rare, it remained in print for many years and achieved some significance as a university textbook, even though the ideas advanced in the book struck even some contemporary reviewers, like the psychologist R.M. Yerkes, writing in *The American Journal of Sociology*, as having "uncertain value."

A second, associated reason is the book's role as a manifesto for the early 20th century human eugenics movement in America. Eugenics, in principle, had the goal of improving human "stock," and therefore human existence, through careful attention to familial traits. The traits that became a source of particular interest to eugenicists, we now know, were and still are those most resistant to genetic analysis. Davenport, however, became the leading scientific advocate for eugenics, and established the Eugenics Records Office at Cold Spring Harbor, where it kept company with the high-powered and admirable genetics studies on corn and other organisms going on in the Carnegie Institution of Washington's Department of Genetics. *Heredity in Relation to Eugenics* sets out in detail the arguments for a eugenics program and proposes how such a program should be achieved. *Heredity in Relation to Eugenics* is therefore a milestone in the history of eugenics, providing apparent academic respectability to a program that at its foundation was not scientific but driven by social and political motives.

But the most compelling reason for bringing Davenport's book once again to public attention is our observation that although the eugenic plan of action advocated by Davenport and many of his contemporaries has long been rejected, the problems that they sought to ameliorate and the moral and ethical choices highlighted by the eugenics movement remain a source of public interest and cautious scientific inquiry, fueled in recent years by the sequencing of the human genome and the consequent revitalization of human genetics. So we invited a group of eminent scholars in history, genetics, psychiatry, and the law to reflect on themes from Davenport's book—human genetic variation, mental illness, nature versus nurture, and human evolution—in the context of the current debates about genetics and society, as the era of the "personal genome" becomes a reality.

These thoughtful essays show very clearly that as science reveals ever more about essential aspects of human nature, we will be faced with real ethical, legal, and moral dilemmas. This reprinted edition of *Heredity in Relation to Eugenics* provides ready access to one of the key documents in the application of genetics to human beings and offers a reminder of the imperative to approach the discoveries of science with care, respect, and caution.

JAN WITKOWSKI
JOHN INGLIS

Foreword

Charles Davenport had the best of intentions. His indefatigable collection of pedigrees, anecdotes, diagnoses, and symptoms was supposed to inform the improvement of the nation and eventually the entire human race, reducing the suffering caused by disease and deformity. Ideals do not come much higher than that. He based his program on the notion that a hereditary element contributes to almost all the defects of the human condition—not just hemophilia and polydactyly, but susceptibility to tuberculosis or propensity to criminality. Therefore, if people with these defects could be dissuaded from breeding, or at least persuaded to marry those with balancing advantages, then the defects could be largely eliminated from society to the benefit of all.

Yet this well-intentioned road led in many countries to the forced sterilization, and in some countries, to the murder, of millions of people. Whether the eugenics movement, to which so many prominent people on the left as well as the right enthusiastically subscribed, was a direct cause, a justification, or merely a reflection of the dreadful atrocities committed in its name can never fully be known. But that it bears some responsibility cannot be doubted.

What then was wrong with Davenport's dream? On the whole his mistake was not scientific. True, he made many scientific claims that would later prove false or foolish, such as his notorious belief that a seagoing family had an inherited factor for thalassophilia. But then all scientists who are prepared to break new ground make mistakes. And there is little in his method, the collection of pedigrees, that allowed him to test, rather than assert, his claim that heredity influences many human traits. But modern science suggests that he did not on the whole claim too much for heredity. Indeed, he sounds almost modern when he refuses to "regard heredity and environment as opposed" (p. 252).

He argues that most defects are the result of "the reaction of a specific sort of protoplasm to a specific stimulus" (p. 252)—precisely what we now fashionably call a gene–environment interaction. "We explain that Mr A. has gone insane from business losses or overwork. . . . It would be more accurate to say A. went insane because his nervous mechanism was not strong enough to stand the stresses to which it was put" (p. 254). The pejorative wording apart, he sounds like the modern biologist Terrie Moffitt describing how people with two short alleles of the 5HTT (serotonin transporter) gene are more likely to suffer depression than people with two long alleles—but only after three or more stressful life experiences. Different genes endow different people with different reactions to the same experiences: That is why personality shows high heritability in affluent western society.

Nor was Davenport necessarily wrong to believe in simple, single-gene causes of disease: This is an empirical matter that will vary from case to case. It has become fashionable in recent years to believe that many traits are caused by multiple genes. Many are, but the evidence suggests that in other cases, although many genes may be needed for the successful functioning of a metabolic pathway or the successful construction of an organ, it may take a mutation in only one to produce symptoms of a disorder. So for the affected individual it may be only one altered gene—usually interacting with an environmental trigger—that caused the condition. For many of the conditions Davenport was describing, such as myopia, language defects, or susceptibility to cancer, he may therefore have been right that certain pedigrees had simple mutations running down them.

Davenport was correct to argue that segregating, sterilizing, and advising people on marriage could very gradually reduce the burden of disease and in society. There is no practical reason people could not be bred like cattle, dogs, or pigeons. But to what benefit and at what price? Had we all followed Davenport and become enthusiastic eugenic marriers, testing our partners' pedigrees for defects before accepting their proposals, the benefit that society, in the United States or the world, would have gained in the century since Davenport wrote would have been genuine, but small and diffuse: slightly better economic productivity, perhaps, or slightly lower welfare costs. But the price would have been individual and terrible: Millions of people deprived of families, rejected, stigmatized, punished, incarcerated, or even killed. For the truth is that to gain even the smallest social benefits of eugenics, you would have to force people to obey Draconian rules at pain of severe punishment. You would have to trample on individual rights to get social results—which is, of course, precisely what the totalitarian societies of the 20th century did with gruesome enthusiasm. What is wrong with eugenics is the authoritarian means, not the scientific ends: the forced nationalization of breeding.

Today, it is argued, eugenics is back but in individual, not state-sponsored form. People can choose the sex of their child, or choose to abort a fetus with a certain disability, or select an embryo without a detected defect or disability. One day they may be able to choose a more intelligent, musical, athletic, or attractive genome for their child or to add or subtract genes. If so—and assuming the techniques used become safe—there are no individual victims from such practices. There is no boy, or unmusical child, that would have been born had the parents not chosen a girl or a musical child. True, society may eventually bear a cost if everybody chooses musical girls, but this is vanishingly unlikely because people have diverse preferences. In a sense we have been practicing such individual eugenics since the Stone Age—simply by avoiding inbreeding and by selecting mates on the grounds of relative beauty, intelligence, or talent—and there have been victims among those who were judged less beautiful, less intelligent, or less talented, so if anything the new techniques might be more humane. However, there is every difference in the world between the goal of individual eugenics and Davenport's goal. One aims for individual happiness with no thought to the future of the human race; the other aims to improve the race at the expense of individual happiness.

MATT RIDLEY
Newcastle-upon-Tyne

Illustrations

Other Titles of Interest

Murderous Science: Elimination by Scientific Selection of Jews, Gypsies, and Others in Germany 1933–1945

The Unfit: A History of a Bad Idea

Times of Triumph, Times of Doubt: Science and the Battle for Public Trust

A Passion for DNA: Genes, Genomes, and Society

Genes and Politics

James D. Watson
Cold Spring Harbor Laboratory

Editors' Note: The quotes below from Davenport show the duality of his desire to know the details of the mechanism of inheritance and his belief that this knowledge and its use was the domain of the State. In this essay, originally published in the 1996 Cold Spring Harbor Laboratory Annual Report, James Watson discusses why scientific justification of political policies just does not work. Dr. Watson puts forth several examples showing the dangers of the co-option of science to political aims—early U.S. immigration law, race hygiene in Germany during World War II, Larmarckism in Soviet Russia, and the more recent Science for the People movement in the United States—as well as how he headed off such a clash when he headed the government-funded Human Genome Project (HGP). From his unique perspective, he also considers issues such as genetic determinism, genetic injustice, and the ethical implications of knowing our genetic futures. He concludes that "In the last analysis, we should accept the fact that if scientific knowledge exists, individual persons or families should have the right to decide whether it will lead to their betterment."

January, 1918.

TRANSFER OF E. R. O.

On December 14th, the trustees of the Carnegie Institution of Washington accepted from Mrs. E. H. Harriman the gift of the Eugenics Record Office, with its land, buildings and records, valued at over $200,000, and a fund of $300,000 of which the income will be available for its maintenance. By this wise and generous gift the future of the Eugenics Record Office becomes established. The relations of the Station for Experimental Evolution and the Eugenics Record Office, always close, bcome still more intimate. The name of Mrs. Harriman is to be always associated with that of the Office as its Founder.

EUGENICS RESEARCH ASSOCIATION.

Since this essay was written in 1997, the HGP has been completed and many variations of this project are now under way, from sequencing the genomes of hundreds of organisms to studying the nature of variation in the human genome with single-nucleotide polymorphisms (SNPs) and haplotypes. And now the genomes of individual humans have been sequenced and analyzed at much lower cost in time and money than the HGP, opening the door for anyone to have their genome sequenced and their genetic future revealed—one culmination of Davenport's dream.

> *If one is provided with a knowledge of the methods of inheritance of unit characters it might seem to be an easy matter to state how each human trait is inherited and to show how any undesirable condition might be eliminated from the offspring and any wished for character introduced. Unfortunately, such a consummation cannot for some time be achieved.*
>
> DAVENPORT, PP. 23–24

> *The commonwealth is greater than any individual in it. Hence the rights of society over the life, the reproduction, the behavior and the traits of the individuals that compose it are, in all matters that concern the life and proper progress of society, limitless, and society may take life, may sterilize, may segregate so as to prevent marriage, may restrict liberty in a hundred ways.*
>
> DAVENPORT, P. 267

THE SCIENCE OF GENETICS AROSE TO study the transmission of physical characteristics from parents to their offspring. When closely studied, much variation exists for virtually any characteristic, say, in size or color, among the members of all species, be they flies, dogs, or ourselves, the members of the *Homo sapiens* species. The origin of this variability long fascinated the scientific world, which already in the nineteenth century asked how much of this variation is due to environmental causes (nurture) as opposed to innate hereditary factors (nature) that pass unchanged from parents to offspring. That such innate heredity exists could never be realistically debated. One need just look at how characteristics in the shape of the face pass through families. Ascribing, say, the uniqueness of the Windsor face to nurture as opposed to nature goes beyond the realm of credibility.

GENES AS THE SOURCE OF HEREDITARY VARIATION BOTH WITHIN AND BETWEEN SPECIES

The key conceptual breakthrough in understanding the nature component of variation came in the mid-1860s from the experiments of the Austrian monk and plant breeder, Gregor Mendel (1822–1884). In his monastery gardens he created, by self-breeding, strains of peas that bred true for a given character like pea color or pod shape. Then he crossed his inbred strains with each other and observed how the various traits assorted in the progeny pea plants. In his seminal scientific paper, published in 1865, Mendel showed that the origin of this hereditary variability lay in differences in discrete factors (genes) that pass unchanged from one plant generation to another.

Most importantly, he showed that each pea has two sets of these factors, one coming from the male parent, the other from the female. Some of those factors are expressed when present in only one copy (dominant genes), whereas others become expressed only when two copies, one from each parent, are present (recessive genes). Mendel's results later were used by the Danish botanist, Wilhelm Johannsen (1857–1927), to make the important distinction between the physical appearance of an individual (its phenotype) and its genetic composition (genotype). Mere examination of a plant's physical appearance need not reveal its genetic composition. Recessive genes present in only one copy can be identified only by further genetic crosses. Mendel further made the equally important observation that genes do not necessarily stay together when the male and female sex cells are formed. Instead, they often independently assort from each other, giving rise to progeny with sets of features very different from those of either parent.

Mendel's work, done before the behavior of chromosomes during cell division was understood, almost had to lay unappreciated until the turn of the century, when three plant breeders working on the European continent—Correns, de Vries, and Tschermak—independently rediscovered the basic rules for hereditary transmission, which today we call Mendel's laws. It was not until 1890 that the sex cells were found to possess only half the number of chromosomes present in adult cells. Fertilization through combining the haploid N number of chromosomes of the sperm with the haploid N number of the egg restores the $2N$ diploid chromosome number of adult plants and animals. Except for those special chromosomes that determine sex, adult cells contain two copies of each distinct chromosome, each of which is exactly duplicated prior to the cell division. With the basic facts of chromosome behavior so established for both ordinary cell division (mitosis) and sex cell formation (meiosis), the rediscovered laws of Mendel were given a chromosomal basis by the American, Walter Sutton. Perceptively, he noted in 1903 that the segregation patterns of Mendel's genes exactly parallel the behavior of chromosomes during the meiotic cell divisions that produce the male and female sex cells (the Chromosomal Theory of Heredity). During the next several decades, an ever-increasing number of genes were found to have precise locations along specific chromosomes. In essence, each chromosome came to be seen as a linear collection of genes running between its two ends.

Genes first were of interest because they were the source of the variability between the members of a species, but they soon began to be appreciated more properly as the source of information that gives an organism its unique form and function. Its collection of genes (its genome) is what gives each organism its own unique developmental pathway. A dog is a dog, a bacterium a bacterium, and so on, because of the information carried by their respective genomes. Thus, gene duplication prior to cell division must be based on a very accurate copying process. Otherwise, there would be no constancy of genetic information and of the development processes they make possible. Correspondingly, genetic variation arises when genes are not accurately copied (mutated) and give rise to changed (mutant) genes.

HEREDITARY VARIABILITY GENERATED BY CHANGES IN GENES (MUTATION) UNDERLIES EVOLUTION BY NATURAL SELECTION

As soon as the first spontaneous gene mutations became known, they were perceived as the obvious source of the new genetic variants necessary for Darwinian evolution by "survival of the fittest." Many more dysfunctional than more functional genes, however, resulted from random mistakes in the gene-copying process. Thus, the rate at which the gene-copying process makes mistakes is likely also to be under strong evolutionary pressure. If too many spontaneous mutations occur, none of the mutant-gene-bearing organisms are likely to develop and produce viable offspring. Correspondingly, too low a mutation rate will not generate sufficient gene variants to allow species to compete effectively with those species evolving faster because of their more frequent generation of biologically fitter offspring.

EUGENIC SOLUTIONS FOR HUMAN BETTERMENT

The coming together of Darwinian and Mendelian thinking immediately raised the question of the applicability of the new science of genetics to human life. To what extent was human success due to the presence in their recipients of good genes that led to useful biological traits like good health, social dependability, and high intelligence? Correspondingly, how many individuals at the bottom of the human success totem pole were there because they possessed gene variants perhaps useful for earlier stages in human evolution but now inadequate for modern urbanized life? Social Darwinian reasoning viewed the sociocultural advances marking humans' ascent from the apes as the result of continual intergroup and interpersonal strife, with such competitive situations invariably selecting for the survival of humans of ever-increasing capabilities. Social Darwinism came naturally to the monied products of the industrial revolution, a most prominent one being the talented statistician, Francis Galton (1822–1911). Early in his career, he wrote the 1869 treatise *Hereditary Genius*, later coining the term "eugenics" (from the Greek meaning wellborn) for studies that would bring about improvements of the human race through the careful selection of parents.

Clever though he was, and able to take comfort that he was Charles Darwin's (1809–1882) cousin, Galton's eugenic prescriptions offered no basic improvement on the long-attempted practice of seeing that offspring from families of attainment married into families of similar high function. In this way, supposed good germ plasm would not be diluted by inputs of putative bad heredity. But whether Galton was promoting reality, as opposed to an unjustified prejudice against the vulgarity of the lower classes, had no way of being even half-tested before the arrival of Mendelian analysis. So the eugenics movement naturally became galvanized by the new laws of Mendelian heredity. But immediately, their hopes had to be tempered by the fact that human genetics never would have the power of other forms of genetics where genetic crosses could be made as well as observed. For better or worse, the eugenicists' main research tool had to be hopefully well-collected, multigenerational pedigrees of physical and mental traits that passed through families from one generation to the next. Toward that end, Galton, then already 84, co-authored in 1906 the book *Noteworthy Families*, an index to kinships in near degrees between persons whose achievements are honorable and have been publicly recorded.

Initially, there were hopes that simple Mendelian ratios would characterize the inheritance of a broad-ranging group of human traits. But in addition to the limitations brought about through the inability to confirm genetic hypotheses through genetic crosses, many of the studied traits appeared in too few families for appropriate statistical analysis. Particularly difficult to analyze were progeny traits not present in either parent. Conceivably, individuals had inherited one copy of the same recessive gene from each parent. Such tentative conclusions became more convincing when the respective traits, like albinism, were found more often in highly inbred, isolated populations where marriages of cousins were frequent.

Easier to assign as bona fide genetic determinants were dominant-acting genes that need be inherited from only one parent for their presence to be felt. Once Mendelian thinking had appeared, the inheritance mode of Huntington's disease, the terrible neurological disease that leads to movement and cognition disorders, was quickly ascertained as a dominant gene disorder. Similar clear genetic attributions could be assigned to traits, such as red–green color blindness and hemophilia, which preferentially appear in males but which are never passed on to their own male offspring. This is the behavior of a trait caused by a gene present on the X sex chromosome, two copies of which are present in females but only one in males, whose sexuality is determined by the Y chromosome.

Important as these diseases of the body were to the individuals and families of those so afflicted, the main focus of early twentieth century eugenicists soon moved to potential genetic causations for disabilities of the mind, embracing a wide spectrum of manifestations from insanity through mental defectiveness, alcoholism, and criminality, to immorality. With poorhouses, orphanages, jails, and mental asylums all too long prominent features of the most civilized societies, eugenicists with virtually religious fervor wanted to prevent more such personal and societal tragedies in the future. They also desired to reduce the financial burdens incumbent on civilized society's need to take care of individuals unable to look after themselves. But in their evangelical assertions that genetic causations lay behind a wide variety of human mental dysfunctions, the early eugenically focused geneticists practiced sloppy, if not downright bad, science and increasingly worried their more rigorous geneticist colleagues.

AMERICAN EUGENICS: SLOPPY GENETICS FOR THE LEGITIMATION OF CLASS STRATIFICATION

The most notable American eugenicist, whose conclusions went far beyond his facts, was Charles B. Davenport (1866–1944), who parlayed his position as Director of the Genetics Laboratory at Cold Spring Harbor, New York, to establish in 1910 a Eugenics Record Office using monies provided by the widow of the railway magnate E.H. Harriman. In his 1911 book *Heredity in Relation to Eugenics*, pedigrees were illustrated for a wide-ranging group of putative hereditary afflictions ranging from bona fide genetic diseases, such as Huntington's disease and hemophilia, to behavioral traits of much less certain hereditary attribution, such as artistic ability and mechanical ability with reference to shipbuilding. With so little then known about the functioning of the human brain, Davenport's early rush to associate highly specific accomplishments of the human brain to specific genetic determinants could not automatically be dismissed as nonsense. In today's intellectual climate, however, a predilection for genes that predispose individuals to city life as opposed to rural life would not be the way to an academic career. But, even his fellow early eugenicists must have regarded as more wartime patriotism than science his 1917 claim that a dominant gene for thalassophilia predisposed its recipients to careers as naval captains.

In addition to its family pedigree assembly and archival roles, the Cold Spring Harbor Eugenics Record Office frequently counseled individuals with family backgrounds of genetic diseases, particularly when they were considering marriage to blood relatives. Many such seekers of help must have been misled by advice that never should have been given, considering that era's limited power for meaningful genetic analysis. Worries about insanity were a major concern, where manic–depressive disease was seen to move through some families as if it were a dominant trait. In contrast, schizophrenia had more aspects of a recessive disease. Yet, even with today's much more powerful human genetics methodologies, we still do not know the relative contribution of dominant versus recessive genes to these two major psychoses or any other form of mental disease.

The Eugenics Record Office's pre-World War II message was that insanity usually expressed itself only when genes predisposing it were inherited from both parents. If this were so, siblings of individuals displaying mental instability were at risk of being carriers of insanity-provoking genes. Because recessive genes for insanity would be silently passing through many families, marriage to any individual with mentally disturbed siblings was not prudent. Even more certainly, marriage should be avoided between individuals having severe mental illness in both parents. In those days, when no effective medicines existed for any form of psychiatric illness, most families bearing mental disease not surprisingly kept this knowledge as secret as possible. There must have been many couples, perhaps overworried about producing mentally disturbed offspring, who chose not to have children.

The eugenicists predictably were concerned about mentally unstable individuals marrying those with similar disturbances. Also of obsessive concern to them were individuals with feeblemindedness, where Davenport believed that recessive genes were also involved. With his certainty that all children of two feebleminded parents would be defective, he wrote of the "folly, yes the crime, of letting two such persons marry." In his mind, the inhabitants of rural poorhouses were there largely because of their feeblemindedness, and he considered one of our nation's worst dangers to be the constant generation of feebleminded individuals by the unrestrained lusts of parents of similar conditions.

It was to stop such further contaminations of the American germ plasm that Davenport, as early as 1911, saw the need for state control of the propagation of the mentally unstable or defective. Initially, he did not favor adoption of state laws allowing for their compulsory sterilization, an idea then considered wise and humane by much of that era's socially progressive elite. Clearly somewhat sexually repressed (obsessed?), he feared that with pregnancy no longer a worry, the sexual urges of the sterilized, mentally unstable impaired might cause more harm to society than even the procreation of more of their kind. Instead, he wanted mentally impaired women to be effectively segregated (imprisoned?) from the impaired of the opposite sex until after they passed the age of procreation. This prescription, however, was totally unrealistic and the American eugenics movement as a whole enthusiastically promoted the compulsory sterilization legislation that spread to 30 states by the start of World War II.

If the eugenics movement had focused its attention predominately on genetic afflictions that truly disabled its recipients, we might now be able to look back at it as a mixture of sloppy science and well-intentioned but kooky naiveté. Photos of the eugenics booths of the 1920s state farm fairs are virtually laughable. In them can be seen "fitter families" displayed near the pens at which prize cattle were shown. The thought that sights of their earnest faces would lead to preferential procreation of more of the same now stretches our credulity. In contrast, the words and actions of Harry P. Laughlin, Davenport's close associate and Superintendent of the Eugenics Record Office, today can only make our minds flinch.

Pleased that his ancestors were traceable to the American Revolution, Laughlin shared Davenport's belief that the strengths and weaknesses of national and religious groups were rooted in genetic as well as in cultural origins. While, at least in public, Davenport wrote that no individual should be refused admission to the United States on the basis of religious group or national origin, Laughlin stated as scientific fact before appropriate Congressional bodies that the new Americans from Eastern and Southern Europe were marked by unacceptable amounts of insanity, mental deficiency, and criminality. Although he lacked any solid evidence, he nonetheless promoted the belief that the newest immigrants to our shores were much more likely to be found in prisons and insane asylums than were the descendants of earlier waves of English, Irish, German, and Scandinavian settlers. Even though the then-current postwar hysteria against unrestrained immigration by itself might have led to the 1924 legislation, there is no doubt that Laughlin's testimony tilted the composition of the future immigrants to Northern Europeans.

With legislation in place, Davenport no longer had to fear that "the population of the United States will on account of the great influence of blood (genes) from South-Eastern Europe rapidly become darker in pigmentation, smaller in stature, more mercurial, more attached to music and art, given to crimes of larceny, kidnapping, assault, murder, rape, and sex immorality and less given to burglary, drunkenness, and vagrancy than were the original English settlers." Through propagating such racial and religious prejudices as scientific truths, the American eugenics movement was, in effect, an important ally of the ruling classes, many of whose privileges inevitably came through treating those less fortunate as inherently unequal.

USING THE FIRST IQ TESTS TO JUSTIFY RACIAL DISCRIMINATION WITHIN THE UNITED STATES

The emergence of intelligence measuring reinforced the belief of America's prosperous people that their wealth reflected their respective family's innate intellectual superiority. The French psychologist, Alfred Binet (1857–1911), was the first person to try to systematically measure intelligence, responding to a 1904 request from the French government to detect mentally deficient children. The resulting Binet–Simon tests crossed the Atlantic by 1908, being first deployed in the United States by Henry Goddard in New Jersey at a training school for feebleminded boys and girls. Soon afterward, he went on to test 2000 children with a broad range of mental abilities. Initially, there was considerable public opposition to the testing of "normal" individuals because of the test's first use on the feebleminded. Within only a few years, however, revised Binet–Simon tests, more appropriate for precocious children, were prepared by Lewis Terman (1877–1956) at Stanford University. These so-called IQ (intelligence quotient) tests were soon employed during World War I on hundreds of thousands of army draftees. Their main function was not to weed out mental defectives, but to assign recruits to appropriate army roles. Those administering the tests, led by the noted psychologist Robert M. Yerkes (1876–1956), claimed they were seeing native intelligence independent of the recruit's environmental history. Yet, clearly, many of the questions or arithmetic problems would be more easily answered by those with extensive schooling and possessing a broad vocabulary. Not surprisingly, the non-English-speaking recruits just off the immigration boats tested badly, allowing a test leader to privately confide to Davenport, "we are well on the right track in our contention that the germ plasm (now) coming into the country does not carry the possibilities of that arriving earlier." Such "objective test data" further convinced the eugenicist world that not only was mental deficiency genetically determined, but so was general intelligence.

Although black men from urban areas tested higher than white southern rural men, their IQ scores were significantly lower than their white equivalents from the same communities. Given today's realization that intelligence measurements virtually by necessity have cultural biases, the comparative data assembled from the army recruits had little real meaning. In many ways, it was like comparing oranges with apples. Nonetheless, the data summarized in *Psychological Examining in the United States Army* were used to justify the discriminatory segregation laws that effectively made America's black population second-class citizens. Genetic inequalities across so-called race boundaries were taken for granted, and 29 states maintained laws against black–white intermarriages, often using the argument that the superior white germ stock would be diluted with inferior genes.

Although eugenics had its origin in England, it never affected the national consciousness there as it did in the United States. With social class stratification so long a characteristic feature of British life, the ruling classes had no need of further justification for their privileged existence. To a lesser but real extent, social inequalities also were taken-for-granted features of most European countries, many of which still had royal families and their attendant aristocracy. Enthusiastic prewar eugenics movements nonetheless sprang up all over the continent, extending even to Southern America and Japan in the 1920s. Everywhere, the chief adherents were the professional middle class, naively proselytized into believing that genetic thinking could soon lead to human beings with heightened hereditary capabilities. Although the continent's eugenicists frequently used the now unacceptable term "race hygiene" for their movement, their ways for the betterment of human heredity for the most part in no way infringed upon preexisting human liberties. Offered as the future panacea was the standard package of marriage between genetically healthy individuals, with correspondingly strong disapproval of marriage for individuals bearing obviously bad genes like those leading to Huntington's disease. Only in two European countries, Germany and Sweden, was legislation enacted for obligatory sterilization of individuals thought to be the bearers of disabling genes.

NAZI EUGENICS (RACE HYGIENE): A MURDEROUS MÉNAGE À TROIS OF BAD GENETICS, RACIAL ANTHROPOLOGY, AND PSYCHIATRY AT THE BECKON OF HITLER AND HIMMLER

Although it was the Hitler-led Nazi government that quickly passed the 1933 Eugenic Sterilization Law, the broadly based German eugenics movement of the 1920s laid the groundwork. Then it was embraced by a spectrum of political thought, much of it totally respectable by the ethical standards of those days. The Germany of that time was a nation undergoing a great moral crisis brought on by its humiliating defeat in the World War. Its four awful years of trench warfare had killed a significant fraction of its better younger men and left it vulnerable to the hyperinflation that wiped out much of the savings of its professional middle class. Unlike England or France, Germany as a world power had only a fleeting existence, and the German people then saw the need to somehow reinvigorate themselves. The eugenicists' vision that human beings' futures lie in their genes struck a receptive chord in the immediate postwar German psyche. Even in the postwar chaos of its Weimar government, human genetics gained strong governmental support. Genetics quickly became a high-quality science in Germany, with Berlin becoming one of the world's leading centers for genetics. Study of supposed genetic differences between the so-called races was vigorously promoted, with it being accepted as fact that the commercial colonization of the world by major countries in Europe and the United States reflected the inherent superiority of the Nordic people's genes for intelligence and strength of moral purpose. Anthropological-based research had strong genetic components, with genes being perceived as the crucial element determining human behavior.

In total contrast, genetic explanations for human successes were not favorably received in the Soviet Union, whose communist doctrines emphasized social, as opposed to genetic, causation for the currently existing inequalities between humans. Already by the mid-1930s, eugenic thinking had become strongly inimical to Russian Communist policymakers, who increasingly favored the Lamarckian explanations (inheritance of acquired characteristics) of Trofim Lysenko (1898–1976), its homegrown agriculturist, over the foreign-originating Morgan–Mendelian analysis of heredity. Those pursuing genes within the Soviet Union soon were putting not only their careers, but also their lives at risk. The great American geneticist, Hermann J. Muller (1890–1967), whose left-wing views led him to leave Texas and go to Russia in 1933, effectively ended his Soviet career when, in 1936, he compared Lamarckian thinking to alchemy, astrology, and shamanism.

Seeking backing for the putative superiority of the Caucasian race, Adolf Hitler (1889–1945), while imprisoned in 1924 for the failed Munich putsch, read *Menschliche Erblichkeitslehre und Rassenhygiene* (The Principles of Human Heredity and Race Hygiene), a leading German genetic text of the time coauthored by E. Baur, Eugen Fischer, and Fritz Lenz. Enveloped by an uncritical eugenic perspective, it strongly reinforced Hitler's view of Germans as the *master race* that justifiably should rule the world. If, however, the Germans were indeed the master race, the Nazis had to explain their nation's humiliating defeat in the Great War and its subsequent devastating hyperinflation.

A perfidious scapegoat was needed, and here Hitler drew upon the long-existing, anti-Semitic feelings of many German people. Long segregated in rural enclaves dating back to the Middle Ages, Jews became effectively part of Germany's commercial and professional life only by the middle of the nineteenth century. Gravitating especially to the professions where their talents could more easily prevail over still-existing prejudices, Jewish importance in German commerce and professional life soon became disproportionate to their numbers, creating jealousies that inevitably fanned preexisting anti-Semitism. Clearly, much of this Jewish success reflected its religion and its respect for the intellect as opposed to oft revelatory-based opinions of their Christian equivalency. Their anti-Semitic opponents, however, saw the Jews' upward trajectory as manifestation of inherent immoralities that let them take unfair advantage of the more honest Christian Germans.

Until the arrival of Mendelian thinking, German anti-Semites never consistently decided whether their enemy was the Jews themselves or their religion. If their failure to acknowledge Christ was the problem, Jews who converted posed no further threat to Christian civilization. But if their reputed unscrupulous behavior and sexual licentiousness reflected innate hereditary qualities, their presence within Christian societies threatened their country's moral resolve, if not its very existence. Assertions by eugenicists that gene differences lay behind human behavioral differences were thus made to order for Nazi needs. From the 1933 start of their absolute rule, the Nazi propaganda machine ruthlessly portrayed Jews and Communists as the two main villains blocking the ultimate triumph of National Socialism. No words were vile enough to express their hatred for the genes that supposedly let Germany's one million Jews steal for themselves the monies and jobs of the honest Germans, or their horror of the Communists who wanted to redistribute monies from those who worked hard to those not able or willing to take care of themselves. Treated with equal contempt by the Nazis, but of less importance because of their much smaller numbers (30,000), were the Gypsies. Because of their wandering, supposed sexually unrepressed lifestyles, and their lack of respect for property, the German Gypsies were regarded by Nazi anthropologists as descendants of peoples of primitive etiological origin who had mated repeatedly with the German criminal, asocial subproletariat. So considered, the further breeding of this mixed-blood people must be stopped.

STERILIZATION OF THE MENTALLY UNFIT AS A PRELUDE TO MORE BROADLY BASED WARTIME GENOCIDE

Gypsies, however, were not specifically targeted under the 1933 Eugenics Law that mandated compulsory sterilization for schizophrenia, manic–depressive psychoses, hereditary epilepsy, Huntington's chorea (disease), hereditary blindness, hereditary deafness, severe physical deformity, and severe alcoholism. Tribunals of Hereditary Health, consisting of a judge, a government medical officer, and an "independent" physician, made the resulting decisions where the individuals concerned often knew they were at risk only when called before its members. With appeals extremely difficult, these psychiatrist-led verdicts between 1934 and 1939 led to some 400,000 compulsory sterilizations, many to noninstitutionalized persons. These reputedly hereditary-damned individuals were further subjected to a subsequent 1934 law forbidding persons with serious mental disturbances from marrying. A year later, legislation specifically affecting Jewish marriages came through the 1935 "Nuremberg Decrees" for the protection of German blood and health. They forbade not only marriages, but also sexual intercourse, between the so-called German and Jewish races.

Concomitant with these eugenic actions was the assembly of vast record collections documenting individual hereditary–biological characteristics. To the Reich Kinship Bureau were referred decisions as to the origin of individuals with potential partial Jewish blood. Many such anthropological "expert conclusions" were made using only photos of the putative fathers. However, with the census of 17 May 1939 providing supposed "confidential" information of any Jewish grandparent, the Nazis, as the war started, felt they had a firm handle on the Jewish blood within their midst. So encouraged, later that year Professor Eugen Fischer, this time responding to coal barons of the Ruhr, wrote, "When a people wants somehow or other to preserve its own nature, it must reject alien racial elements and when they have already insinuated themselves, it must suppress and then eliminate them. The Jew is such an alien... ."

With the war on, the German government, seeing no reason to waste scarce resources to keep what they considered genetically inferior peoples alive, proceeded to what it termed a "euthanasia policy of mercy killing." In a one-sentence letter postdated 1 September 1939, Hitler himself wrote, "Reichleiter Bouhler and Dr. Brandt are entrusted with responsibility of extending the rights of specifically designated physicians such that patients who are judged incurable after the most thorough review of their condition which is possible can be granted mercy killing." So authorized, 3000 mental patients in occupied Poland were summarily shot by storm troopers. In the Reich itself, where German citizens were involved, somewhat more formal procedures were used. Questionnaires were distributed to the mental hospitals, where they were completed in their capacity as experts by nine Professors of Psychiatry assisted by 39 other medical doctors. For their labors, they were paid 5 pfennigs (the cost of a cigarette) per questionnaire when they processed more than 3500 per month, but up to 10 pfennigs when fewer than 500 questionnaires per month were processed. The patients so selected for "euthanasia" had their respective questionnaires marked with a cross. Subsequently, carbon monoxide supplied by I.G. Farben was used for the elimination process. Before the killings stopped in the fall of 1941, some 94,000 mental patients had been killed. Subsequently, covert "euthanasia" by starvation, drugs, and failure to treat infectious diseases led to only 15% (40,000 persons) of Germany's prewar mental hospital population remaining alive at the war's end.

The primary reason for the supposed stopping of mental patient "mercy killings" was the need to transfer the personnel trained in killing by gas to the concentration camps, primarily in Poland (e.g., Auschwitz), to which most German Jews and gypsies had already been deported. With the decision already taken to invade the Soviet Union, a conference was held in March 1941 in Frankfurt at the Institute for the Investigation of the Jewish Question. At this conference, Dr. Gross, the head of the Race Policy Institute of the Nazi party, stated, "The definitive solution must involve the removal of Jews from Europe and he demands sterilization of quarter Jews." In an October letter to Himmler, Oberdiensleiter Brack of the Fuhrer's chancellery wrote that there are no objections to doing away (gassing) Jews who are unfit for concentration camp work. Less than a month later, Rosenberg, theoretician and minister of the occupied eastern territories, before representatives of the German press, announced the Final Solution of the Jewish Question, revealing plans, still to be kept secret, for the eventual mass murder of all European Jews, including the six million then living in the Soviet Union. The gas chambers so used were in no way restricted to Jews and Gypsies, with Soviet prisoners of war being victims of the first uses of Zyklon B (hydrocyanic acid) at Auschwitz. Anxious to give their racial extermination policies "scientific" justification, Himmler later in 1943 specified in a decree that only physicians trained in anthropology should carry out selection for killing and supervise the killings themselves in extermination camps. Some quarter Jews were to be spared, but not those with Jewish facial features who should be treated as half Jews. By the war's end, five to six million European Jews were so killed, the majority by the gassing procedures that the Nazis' co-opted human geneticists, psychiatrists, and anthropologists thought appropriate for individuals bearing genes inimical to the best interests of the German people.

With the liberation first of Poland and then Germany, the full horror of the racially based genocide policies of National Socialism quickly became known, generating even further disgust for the pseudoscientific theories of race superiority and purity that underpinned Nazi ideology. Anyone subsequently calling himself a eugenicist put his reputation as a decent moral human at risk. In fact, before the war even started, eugenics in the United States already was being perceived more as a social than a scientific movement. Already in 1930, the leaders of the Carnegie Institute of Washington had been told that its Cold Spring Harbor Eugenics Record Station practiced sloppy, if not dishonest, science. But with its founder Charles Davenport nearing retirement, it was allowed to expire more slowly than in retrospect it should have. Its doors closed only when Miloslav Demerec became director of the Department of Genetics in 1942. There thus was the embarrassment of Harry Laughlin's receipt in 1936 of an honorary degree from the University of Heidelberg in recognition of his contributions to racial hygiene. Undoubtedly pleased that eugenics, then fading in the United States, was becoming even more ascendant in Germany, Laughlin went to New York to receive his diploma from the German Diplomatic Counsel.

EUGENICS, A DIRTY WORD, AS THE SEARCH FOR THE CHEMICAL NATURE OF THE GENE BEGINS

By the time I first came to Cold Spring Harbor for the summer of 1948, accompanying my Ph.D. supervisor Salvador Luria, then a professor at Indiana University, the Eugenics Record Office had been virtually expunged from its consciousness. Only in the library was its ugly past revealed through the German journals of the 1920s and 1930s on human genetics and race hygiene. No one that summer showed any interest in human genetics as a science or toward the general question of how much of human behavior reflects nature as opposed to nurture. Instead, genetic research there focused on the fundamental nature of genes and their functioning. It was not that human genetic diseases had suddenly become unimportant to its director, Miloslav Demerec. But there was general agreement both by the year-round staff and the many summer visitors that until the chemical identity of the gene was elucidated and the general pathways by which it controlled cell structure and functioning were known, it was premature to even speculate how genes contributed to human development and behavior.

Then, much sooner than anyone expected, the gene was revealed in 1953 to be DNA. The genetic code was established by 1966, and gene expression was seen to be controlled by DNA-binding regulatory proteins between 1967 and 1969. Genetics, happily then, had no reasons to intersect politics, except in Russia, where the absurdity of its Lamarckian philosophy became painfully more clear to its intelligentsia with every new major advance in molecular genetics. These major genetic breakthroughs were largely accomplished using the simple genetic systems provided by bacteria and their viruses that go under the name phages. By 1969, phage had become so well understood genetically that it became possible to create specific phage strains cleverly engineered to carry specific bacterial genes from one bacterial strain to another. Yet, seeming more ashamed than pleased with their neat science, James Shapiro and Jonathan Beckwith of Harvard Medical School held with much fanfare a press conference to announce that their new way to isolate specific genes was on the pathway to eugenically motivated genetic engineering of human beings. Knowing of the left-wing views of their "Science for the People" group, I, like most of my colleagues in the Boston region, saw their self-denunciations as manifestations of unrepentant leftist fears that further genetic research would render inviable the Communist dogma that assigned all social inequalities to capitalistic selfishness. Shapiro then moved (temporarily) to Cuba to regain his ideological purity.

Although the phage transductional system developed by Shapiro and Beckwith proved not to be a forerunner for eventual human genetic engineering, this was not true for the much more powerful and general "recombinant DNA" methodologies that Herbert Boyer and Stanley Cohen developed 4 years later, in 1973, just 20 years after the discovery of the double helix. Their new procedures allowed the isolation (cloning) of specific genes, through their insertion into tiny chromosomes (plasmids) that could be moved from one cell to another. At roughly the same time, unexpectedly powerful new ways to determine the exact sequences of the four letters (A, G, T, and C) of genetic messages were worked out by Fred Sanger in Cambridge, England, and by Walter Gilbert and Allan Maxam at Harvard. Together, using these two new techniques, the exact structure of any gene could eventually be determined, given the appropriate facilities and resources.

The resulting recombinant DNA era, however, despite all the promises it held for major scientific advances, did not immediately take off. It initially stalled because of fears that among the many new forms of DNA created in the laboratory would be some that would pose unacceptable dangers to life as it now exists. In particular was the fear that highly pathogenic new forms of viruses and bacteria would be created. To give time to assess such potential dangers scientifically, a scientist-initiated moratorium on recombinant DNA research was declared in 1975. Effectively, it blocked virtually all recombinant DNA research for the next 2 years; and research concerned with cancer, where worries were expressed that a cancer gene might become bacterially transmitted, was held up for 2 more years.

During the moratorium, governmental committees were set up in the United States and in various European countries to assess the potential dangers from recombinant DNA experimentation in relation to its potential benefits for biology, medicine, and agriculture. No plausible scientific reasons for stopping such research emerged, and such committees, often containing public as well as scientific representatives, invariably concluded that in the absence of any quantifiable potential dangers, it would be irresponsible not to move ahead with experiments that could dramatically change the nature of biology. In retrospect, these decisions to move ahead were always the correct ones.

For example, cancer research and our knowledge of the genetic basis of the immune system would effectively be back in the scientific middle ages if the enlightenments made possible through recombinant DNA had not occurred.

To my knowledge, moreover, not one case of recombinant DNA-induced illness has since occurred. No person has been so killed, nor has even one case of serious illness been attributed to recombinant DNA, nor do we know of any case where the release into nature of any recombinant DNA-modified organism has led to any known ecological disaster. This is not to say that someday a recombinant DNA-induced disease or ecological upset will not occur. Today, however, there is certainly no logical reason for not exploiting recombinant DNA procedures as fast as possible for human betterment.

IDEOLOGICAL AND VALUE-BASED OPPOSITIONS TO RECOMBINANT DNA RESEARCH

Although there was no evidence of danger from recombinant DNA, there soon arose much visible and sometimes regretfully effective opposition to recombinant DNA research. Here the distinction should be made between objections from scientists who understand the technical issues involved and opposition from groups of public citizens who, though not understanding the science involved, nonetheless oppose much to all recombinant DNA research. Although some initial opposition arose from scientists whose own DNA research was not going well, virtually all the continuing scientific opponents at their heart had political hang-ups. As leftists, they did not want genes involved in human behavioral differences and feared that the onslaught of scientific advances that would follow from the unleashing of recombinant DNA might eventually allow genes affecting mental performance to be isolated and studied.

As a member of the Harvard Biology Faculty between 1975 and 1977, I watched in despair when "Science for the People" successfully assisted the public members of the Cambridge, Massachusetts, City Council to block recombinant DNA research at our Biological Laboratories. Later, I asked Salvador Luria, who was then at Massachusetts

Institute of Technology and who knew that his left-wing friends were putting forth scientifically dishonest statements, why he never publicly criticized them. His reply was that politics was more important than science. This remark has long haunted me, because my own career owes much to the generous way he shared his great scientific talents with me at the beginning of my scientific career. But as a Jew who had to flee first his native Italy and then France for the eventual safety of the United States, Luria's left-wing political affinities were understandable, and I'm lucky I never had to so choose.

Specific political ideologies, however, are not the cause of the prolonged and sometimes effective opposition to recombinant DNA from parts of the general public, particularly in German-speaking regions. With professional agitators like Jeremy Rifkin playing important roles in heightening these public fears, such leadership would never have been effective if their audiences were at emotional ease with the gene and the geneticists who study it. The concept of genetic determinism is inherently unsettling to the human psyche, which likes to believe that it has some control over its fate. No one feels comfortable with the thought that we, as humans, virtually all contain one to several "bad" genes that are likely to limit our abilities to fully enjoy our lives. Nor do we necessarily take pleasure in the prospect that we will someday have gene therapy procedures that will let scientists enrich the genetic makeups of our descendants. Instead, there has to be genuine concern as to whether our children or their governments decide what genes are good for them.

Genetics as a discipline must thus strive to be the servant of the people, as opposed to these governments, working to mitigate the genetic inequalities arising from the random mutations that generate our genetic diseases. Never again must geneticists be seen as the servants of political and social masters who need demonstrations of purported genetic inequality to justify their discriminatory social policies. On the whole, I believe that genetics still commands broad respect in the United States and in much of Europe, despite the efforts of the recombinant DNA opponents to portray the genetic manipulations underlying the biotechnology industry as money-driven actions done at the expense of the public's health and the world's environment. Unfortunately, genetics and geneticists remain much less highly respected in Germany. There even today the most benign of recombinant DNA experiments remain controversial and subject to needless regulation. Propagation of genetically engineered plants is routinely sabotaged, with the mere practice of human genetics regarded as a criminal act by extremists on the left.

This German dislike for the gene and its human-directed manipulations is easily assignable to their Nazi eugenics past. The vile actions then done in the name of the gene hover as almost permanent nightmares never erasable from their national identity. As human beings, never sure that the world is immune from further such depraved behavior, we should never let this awful past slip from our consciousness. At the same time, the whole civilized world will suffer if today's German geneticists are unfairly thought to be cut from the same material that clothed those German geneticists, anthropologists, and psychiatrists who not only assisted the Nazi eugenic efforts, but promoted them as scientific-based necessities for German progress.

Part of today's problem may lie in the postwar fate of Hitler's biological conspirators. Naively as outsiders we long assumed that they would have all been treated as potential if not real war criminals, with even those of only slight guilt losing all further opportunities for academic existence. But as the German geneticist Benno Müller-Hill courageously pointed out in his 1984 book, *Todlicel Wissenschaft* (*Murderous Science*, Cold Spring Harbor Laboratory Press, 1998), there was no attempt by the German academic community to find out what truly happened. Instead, it was academically dangerous in Germany to explore the half-truths that allowed many key practitioners of Nazi eugenics to resume important academic posts. A number of professors who early joined the Nazi Party or SS and were directly involved with its genocide programs committed suicide, but there were many Nazi-assisting scientists, successfully claiming that they were only apolitical advisors, who slid quietly back into academic prominence.

The most damning example was that of Professor Otmar von Verschuer, who actively helped the Nazis—first at the Kaiser Wilhelm Institute of Anthropology under Professor Eugen Fischer and later at his own Institute of Human Genetics in Frankfurt. Involved in distinguishing Jews and part-Jews, he later closely collaborated with his former assistant, the now notorious Josef Mengele, then doing "scientific" research at Auschwitz. After the war, he nonetheless was appointed to be Professor of Human Genetics at the University of Münster. Equally disturbing was the postwar appointment of Fritz Lenz as head of an institute for the study of human heredity at the University of Göttingen, Germany's most distinguished university. Although clearly a very competent scientist, he was a major advisor for laws on euthanasia between 1939 and 1941, as well as author of a 1940 memorandum, "Remarks on resettlement from the point of view of guarding the race."

The postwar 1949 exoneration of von Verschuer occurred despite knowledge of the 1946 article in *Die New Zeit* accusing him of studying eyes and blood samples sent to him from Auschwitz by Josef Mengele. Yet a committee of professors, including Professor Adolf Butenandt, later the head of the Max Planck Gesellschaft (the postwar name for the Kaiser Wilhelm Gesellschaft), concluded that von Verschuer, who possessed all the qualities appropriate for a scientific researcher and teacher of academic youth, should not be judged on a few isolated events of the past. I find it difficult to believe that the Butenandt committee had gone to the trouble of reading his article published in the *Volkischer Beobachter* 1-8-42. In it he wrote, "Never before in the course of history has the political significance of the Jewish question emerged so clearly as it does today. Its definitive solution as a global problem will be determined during the course of this war." Now there may be more reason to remember Professor Butenandt for his part in the von Verschuer whitewash than for his prewar Nobel Prize for research on the chemistry of the estrogen sex hormone.

GENUINE HUMAN GENETICS EMERGES FROM RECOMBINANT DNA METHODOLOGIES

Long holding back the development of human genetics as a major science was the lack of a genetic map allowing human genes to be located along the chromosomes on which they reside. As long as conventional breeding procedures remained the only route to gene mapping, the precise molecular changes underpinning most human genetic diseases seemed foreordained to remain long mysterious. The key breakthrough opening a path around this seemingly insuperable obstacle came in the late 1970s when it was discovered that the exact sequence (order of the genetic letters A, G, T, and C) of a given gene varies from one person to another. Between any two individuals, roughly 1 in 1000 bases are different, with such variations most frequently occurring within the noncoding DNA regions not involved in specifying specific amino acids. Initially most useful were base differences (polymorphisms) that affected DNA cutting by one of the many just discovered "restriction enzymes" that cut DNA molecules within very specific base sequences.

Soon after the existence of DNA polymorphisms became known, proposals were made that they could provide the genetic markers needed to put together human genetic maps. In a 1980 paper, David Botstein, Ron Davis, Mark Skolnick, and Ray White argued that human maps could be obtained through studying the pattern through which polymorphisms were inherited in the members of large multigenerational families. Those polymorphisms that stay together were likely to be located close to each other on a given chromosome. During the next 5 years, two groups, one led by Helen Donis-Keller in Massachusetts, the other led by Ray White in Utah, rose to this challenge, both using DNA from family blood samples stored at CEPH (Centre d'Étude du Polymorphisme Humain), the mapping center established in Paris by Jean Dausset. By 1985, the mutant genes responsible for Huntington's disease and cystic fibrosis (CF) had been located on chromosomes 4 and 7, respectively.

By using a large number of additional polymorphic markers in the original chromosome 7 region implicated in CF, Francis Collins' group in Ann Arbor and L.C. Tsui's group in Toronto located the DNA segment containing the responsible gene. Its DNA sequence revealed that the CF gene coded for a large membrane protein involved in the transport of chloride ions. The first CF mutant they found contained three fewer bases than its normal equivalent and led to a protein product that was nonfunctional because of its lack of a phenylalanine residue.

THE HUMAN GENOME PROJECT: RESPONDING TO THE NEED FOR EFFICIENT DISEASE GENE MAPPING AND ISOLATION

Although the genes responsible for cystic fibrosis and Huntington's disease were soon accurately mapped using only a small number of

DNA polymorphic markers, the genes behind many other important genetic diseases quickly proved to be much harder to map to a specific chromosome, much less assign to a DNA chromosomal segment short enough to generate hopes for its eventual cloning. All too obviously, the genes behind the large set of still very badly understood diseases like Alzheimer's disease, late-onset diabetes, or breast cancer would be mapped much, much sooner if several thousands more newly mapped DNA polymorphisms somehow became available. Likewise, the task of locating the chromosomal DNA segment(s) in which the desired disease genes reside would be greatly shortened if all human DNA were publicly available as sets of overlapping cloned DNA segments (contigs). And the scanning of such DNA segments to look for mutationally altered base sequences would go much faster if the complete sequence of all the human DNA were already known. However, to generate these importantly new resources for human genetics, major new sources of money would be needed. So, by early 1986, serious discussions began as to how to start, soon, the complete sequencing of the 3×10^9 base pairs that collectively make up the human genome (the Human Genome Project or HGP).

Initially, there were more scientific opponents than proponents for what necessarily would be biology's first megaproject. It would require thousands of scientists and the consumption of some $3 billion-like sums. Those disliking its prospects feared that, inevitably, it would be run by governmental bureaucrats not up to the job and would employ scientists too dull for assignment to this intellectually challenging research. Out of many protracted meetings held late in 1986 and through 1987, the argument prevailed that the potential rewards for medicine as well as for biological research itself would more than compensate for the monies the Human Genome Project would consume during the 15 years then thought needed to complete it. Moreover, completion of each of the two stages—the collection of many more mapped DNA markers and the subsequent ordering of cloned DNA segments into long overlapping sets (contids)—would by themselves greatly speed up disease gene isolation.

Always equally important to point out, the 15 years projected to complete the Human Genome Project meant that its annual cost of

$200 million at most would represent only 1–2% of the money spent yearly for fundamental biomedical research over the world. There was also the realization that some 100,000 human genes believed sited along their chromosomes would be much easier to find and functionally understand if genome sequences were first established for the much smaller, well-studied model organisms such as *Escherichia coli, Saccharomyces cerevisiae, Caenorhabditis elegans*, and *Drosophila melanogaster*. Thus, the biologists who worked with these organisms realized that their own research would be speeded up if the Human Genome Project went ahead.

The American public, as represented by their congressional members, proved initially to be much more enthusiastic about the objectives of the Human Genome Project than most supposedly knowledgeable biologists, with their parochial concerns for how federal monies for biology would be divided up. The first congressionally mandated monies for the Human Genome Project became available late in 1987, when many intelligent molecular geneticists still were sitting on the fence as to whether it made sense. In contrast, Congress, being told that big medical advances would virtually automatically flow out of genome knowledge, saw no reason not to move fast. In doing so, they temporarily set aside the question of what human life would be like when the bad genes behind so many of our major diseases were found. Correctly, to my mind, their overwhelming concern was the current horror of diseases like Alzheimer's, not seeing the need then to, perhaps prematurely, worry about the dilemmas arising when individuals are genetically shown at risk for specific diseases years before they show any symptoms.

GENOME ETHICS: PROGRAMS TO FIND WAYS TO AMELIORATE GENETIC INJUSTICE

The moment I began in October 1988 my almost 4-year period of helping lead the Human Genome Project, I stated that 3% of the NIH-funded component should support research and discussion on the ethical, legal, and social implications (ELSI) of the new resulting genetic knowledge. A lower percentage might be seen as tokenism, while I then could not see wise use of a larger sum. Under my 3% proposal, some $6 million (3% of $200 million) would eventually be so available, a much larger sum than ever before provided by our government for the ethical implications of biological research.

In putting ethics so soon into the genome agenda, I was responding to my own personal fear that all too soon critics of the Genome Project would point out that I was a representative of the Cold Spring Harbor Laboratory that once housed the controversial Eugenics Record Office. My not forming a genome ethics program quickly might be falsely used as evidence that I was a closet eugenicist, having as my real long-term purpose the unambiguous identification of genes that lead to social and occupational stratification as well as to genes justifying racial discrimination. So I saw the need to be proactive in making ELSI's major purpose clear from its start—to devise better ways to combat the social injustice that has at its roots bad draws of the genetic dice. Its programs should not be turned into public forums for debating whether genetic inequalities exist. With imperfect gene copying always the evolutionary imperative, there necessarily will always be a constant generation of the new gene disease variants and consequential genetic injustice.

The issues soon considered for ELSI monies were far-ranging. For example, how can we ensure that the results of genetic diagnosis are not misused by prospective employers or insurers? How should we try to see that individuals know what they are committing themselves to when they allow their DNA to be used for genetic analyzing? What concrete steps should be taken to ensure the accuracy of genetic testing? And when a fetus is found to possess genes that will not allow it to develop into a functional human being, who, if anyone, should have the right to terminate the pregnancy?

From their beginnings, our ELSI programs had to reflect primarily the needs of individuals at risk of the often tragic consequences of genetic disabilities. Only long-term harm would result in the perception of genetics as an honest science if ELSI-type decisions were perceived to be dominated either by the scientists who provided the genetic knowledge or by the government bodies that funded such research. And because women are even in the distant future likely to disproportionately share the burden of caring for the genetically disabled, they should lead the discussions of how more genetic knowledge is to come into our lives.

HUMAN HESITATIONS IN LEARNING THEIR OWN GENETIC FATE

With the initial distribution of American genome monies and the building and equipping of the resulting genome centers taking 2 years, the Human Genome Project in its megaphase did not effectively start until the fall of 1990. Decisions to go ahead by funding bodies in the United States helped lead to the subsequent inspired creation of Généthon outside Paris by the French genetic disease charity, Association Française contre les Myopathies (AFM), as well as the building of the now immense Sanger Centre, just south of Cambridge, England, by the British medically oriented charity, the Wellcome Trust. Now effectively 7 years into its projected 15-year life, the Human Genome Project has more than lived up to its role in speeding up genetic disease mapping and subsequent gene cloning. It quickly made successful the search for the gene behind the Fragile X syndrome that leads to severe mental retardation in boys preferentially affected by this sex-linked genetic affliction. The molecular defect found was an expansion of preexisting three-base repetitive sequences that most excitingly increase in length from one generation to the next. The long mysterious phenomenon of anticipation, in which the severity of a disease grows through subsequent generations, was thus given a molecular explanation. Then at long last, in 1994, the gene for Huntington's disease was found. Its cause was likewise soon found to be the expansion of a repetitive gene sequence.

While the mapping to a chromosome per se of any disease gene remains an important achievement, the cloning of the disease gene itself is a bigger milestone. Thus, the 1990 finding by Mary-Claire King that much hereditary breast cancer is due to a gene on chromosome 17 set off a big gene-cloning race. With that gene in hand, there was a chance that its DNA sequence would reveal the normal function of the protein it codes for. In any case, it gives its possessors the opportunity to examine directly the DNA from individuals known to be at risk for a disease to see whether they had the unwanted gene. Thus, when in 1993 the chromosome 17 breast cancer gene (*BRCA1*) was isolated by Myriad, the Utah disease gene-finding company, it could inform women so tested for *BRCA1* whether or not they had the feared gene.

Initially, concerns were voiced that unbridled commercialization of this capability would all too easily give women knowledge they would not be psychologically prepared to handle. If so, the ethical way to prevent such emotional setbacks might be to regulate both how the tests were given and who should be allowed to be tested. I fear, however, that a major reason behind many such calls for regulation of genetic testing is the hidden agenda of wanting to effectively stop widespread genetic testing by making it so difficult to obtain. Now, however, calls for governmental regulation may fall on increasingly deaf ears. To Myriad's great disappointment, it appears that the great majority of women at 50% risk of being breast cancer gene carriers don't want to be tested. Rather than receive the wrong verdict, they seem to prefer living with uncertainty. Likewise, a very large majority of the individuals at risk for Huntington's disease are also psychologically predisposed against putting themselves at risk of possibly knowing of their genetic damnation.

Although we are certain to learn in the future of many individuals regretting that they subjected themselves to genetic tests and wishing they had been more forewarned of the potential perils of such knowledge, I do not see how the state can effectively enter into such decisions. Committees of well-intentioned outsiders will never have the intimate knowledge to assess a given individual's psychological need, or not, for a particular piece of scientific or medical knowledge. In the last analysis, we should accept the fact that if scientific knowledge exists, individual persons or families should have the right to decide whether it will lead to their betterment.

INARGUABLE EXISTENCE OF GENES PREDISPOSING HUMANS TO BEHAVIORAL DISORDERS

The extraordinarily negative connotations that the term eugenics now conveys are indelibly identified with its past practitioners' unjustified statements that behavioral differences, whether between individuals, families, or the so-called races, largely had their origins in gene differences. Given the primitive power of human genetics, there was no way for such broad-ranging assertions to have been legitimatized by the then-current methods of science. Even the eugenically minded psychiatrists' claims that defective genes were invariably at the root of their mental patients' symptoms were no more than hunches. Yet, it was by their imputed genetic imperfection that the mentally ill were first sterilized and then, being of no value to the wartime Third Reich, released from their lives by subsequent "mercy killings."

But past eugenic horrors in no way justify the "not in our genes" politically correct outlook of many left-wing academics. They still spread the unwarranted message that only our bodies, not our minds, have genetic origins. Essentially protecting the ideology that all our troubles have capitalistic exploitative origins, they are particularly uncomfortable with the thought that genes have any influence on intellectual abilities or that unsocial criminal behavior might owe its origins to other than class or racially motivated oppression. However, whether these scientists on the left actually believe, say, that the incidence of schizophrenia would seriously lessen if class struggles ended is not worth finding out.

Instead, we should employ, as fast as we can, the powerful new techniques of human genetics to find soon the actual schizophrenia predisposing genes. The much higher concordance of schizophrenia in identical versus nonidentical twins unambiguously tells us that they are there to find. Such twin analysis, however, reveals that genetics cannot be the whole picture. Because the concordance rates for schizophrenia, as well as for manic–depressive disease, are more like 60%, not 100%, environmental predisposing factors must exist and, conceivably, viral infections that affect the brain are sometimes involved.

Unfortunately, still today, the newer statistical tricks for analyzing polymorphic inheritance patterns have not yet led to the unambiguous mapping of even one major schizophrenic gene to a defined chromosomal site. The only convincing data involve only the 1% of schizophrenics whose psychoses seemingly are caused by the small chromosome 22 deletions responsible also for the so-called DiGeorge facial syndrome. Manic–depressive disease also has been more than hard to understand genetically. Only last year did solid evidence emerge for a major predisposing gene on the long arm of chromosome 18. This evidence looks convincing enough for real hopes that the actual gene involved will be isolated over the next several years.

Given that over half the human genes are thought to be involved in human brain development and functioning, we must expect that many other behavioral differences between individuals will also have genetic origins. Recently, there have been claims that both "reckless personalities" and "unipolar depressions" associate with specific polymorphic forms of genes coding for the membrane receptors involved in the transmission of signals between nerve cells. Neither claim now appears to be reproducible, but we should not be surprised to find some subsequent associations to hold water. Now anathematic to left-wing ideologues is the highly convincing report of a Dutch family, many of whose male members display particularly violent behavior. Most excitingly, all of the affected males possess a mutant gene coding for an inactive form of the enzyme monoamine oxidase. Conceivably having too little of this enzyme, which breaks down neurotransmitters, leads to the persistence of destructive thoughts and the consequential aggressive patterns. Subsequent attempts to detect in other violent individuals this same mutant gene have so far failed. We must expect someday, however, to find that other mutant genes that lead to altered brain chemistry also lead to asocial activities. Their existence, however, in no way should be taken to mean that gene variants are the major cause of violence. Nonetheless, continued denials by the scientific left that genes have a role in how people interact with each other will inevitably further diminish their already tainted credibility.

KEEPING GOVERNMENTS OUT OF GENETIC DECISIONS

No rational person should have doubts whether genetic knowledge properly used has the capacity to improve the human condition. Through discovering those genes whose bad variants make us unhealthy or in some other way unable to function effectively, we can fight back in several different ways. For example, knowing what is wrong at the molecular level should let us sometimes develop drugs that will effectively neutralize the harm generated by certain bad genes. Other genetic disabilities should effectively be neutralized by so-called gene therapy procedures restoring normal cell functioning by adding good copies of the missing normal genes. Although gene therapy enthusiasts have promised too much for the near future, it is difficult to imagine that they will not with time cure some genetic conditions.

For the time being, however, we should place most of our hopes for genetics on the use of antenatal diagnostic procedures, which increasingly will let us know whether a fetus is carrying a mutant gene that will seriously proscribe its eventual development into a functional human being. By terminating such pregnancies, the threat of horrific disease genes continuing to blight many families' prospects for future success can be erased. But even among individuals who firmly place themselves on the pro-choice side and do not want to limit women's rights for abortion, opinions frequently are voiced that decisions obviously good for individual persons or families may not be appropriate for the societies in which we live. For example, by not wanting to have a physically or mentally handicapped child or one who would have to fight all its life against possible death from cystic fibrosis, are we not reinforcing the second-rate status of such handicapped individuals? And what would be the consequences of isolating genes that give rise to the various forms of dyslexia, opening up the possibility that women will take antenatal tests to see if their prospective child is likely to have a bad reading disorder? Is it not conceivable that such tests would lead to our devoting less resources to the currently reading-handicapped children whom now we accept as an inevitable feature of human life?

That such conundrums may never be truly answerable, however, should not concern us too much. The truly relevant question for most families is whether an obvious good to them will come from having a child with a major handicap. Is it more likely for such children to fall behind in society or will they through such affliction develop the strengths of character and fortitude that lead, like Jeffrey Tate, the noted British conductor, to the head of their packs? Here I'm afraid that the word handicap cannot escape its true definition—being placed at a disadvantage. From this perspective, seeing the bright side of being handicapped is like praising the virtues of extreme poverty. To be sure, there are many individuals who rise out of its inherently degrading states. But we perhaps most realistically should see it as the major origin of asocial behavior that has among its many bad consequences the breeding of criminal violence.

Thus, only harm, I fear, will come from any form of society-based restriction on individual genetic decisions. Decisions from committees of well-intentioned individuals will all too often emerge as vehicles for seeming to do good as opposed to doing good. Moreover, we should necessarily worry that once we let governments tell their citizens what they cannot do genetically, we must fear they also have power to tell us what we must do. But for us as individuals to feel comfortable making decisions that affect the genetic makeups of our children, we correspondingly have to become genetically literate. In the future, we must necessarily question any government that does not see this as its responsibility. Thus, will it not act because it wants to keep such powers for itself?

THE MISUSE OF GENETICS BY HITLER SHOULD NOT DENY ITS USE TODAY

Those of us who venture forth into the public arena to explain what genetics can or cannot do for society seemingly inevitably come up against individuals who feel that we are somehow the modern equivalents of Hitler. Here we must not fall into the absurd trap of being against everything Hitler was for. It was in no way evil for Hitler to regard mental disease as a scourge on society. Almost everyone then, as still true today, was made uncomfortable by psychotic individuals. It is how Hitler treated German mental patients that still outrages civilized societies and lets us call him immoral. Genetics per se can never be evil. It is only when we use or misuse it that morality comes in. That we want to find ways to lessen the impact of mental illness is inherently good. The killing by the Nazis of the German mental patients for reasons of supposed genetic inferiority, however, was barbarianism at its worst.

Because of Hitler's use of the term *master race*, we should not feel the need to say that we never want to use genetics to make humans more capable than they are today. The idea that genetics could or should be used to give humans power that they do not now possess, however, strongly upsets many individuals first exposed to the notion. I suspect that such fears in some ways are similar to concerns now expressed about the genetically handicapped of today. If more intelligent human beings might someday be created, would we not think less well about ourselves as we exist today? Yet anyone who proclaims that we are now perfect as humans has to be a silly crank. If we could honestly promise young couples that we knew how to give them offspring with superior character, why should we assume they would decline? Those at the top of today's societies might not see the need. But if your life is going nowhere, shouldn't you seize the chance of jump-starting your children's future?

Common sense tells us that if scientists find ways to greatly improve human capabilities, there will be no stopping the public from happily seizing them.

Charles Benedict Davenport, 1866–1944

Jan A. Witkowski
Cold Spring Harbor Laboratory

Editors' Note: Charles Davenport was a complex character—he was one of the earliest human geneticists, who was quick to pick up on and accept Mendelism and who recognized the relationship between genetics and evolution. Yet, his original biographer E. Carleton MacDowell described Davenport as having "a deep lack of confidence," and stated that he had a "defensive attitude which led to exaggerated emphasis and dulled objective thinking"—weaknesses that are reflected in his role in the U.S. eugenics movement. Dr. Witkowski describes the arc of Davenport's life and puts into context his strengths and weaknesses, as well as his legacy at Cold Spring Harbor.

I am what the determiners in my two fused germplasms have developed into under the culture which they have experienced during their development. I am not responsible for my early culture nor for the reactions determined by it; but that culture is partly determined by my makeup, as when I find pleasure in the society of bad companions, and partly is imposed by the formal "good influences" that society has organized.

DAVENPORT, P. 265

"ONE OF THE MOST INFLUENTIAL PERSONALITIES in the biology of his day was Charles B. Davenport. Few in this field have wielded the power he won, primarily by virtue of the infectious quality of his enthusiasm." These are the first sentences of a remarkable biography of Charles Davenport, written by E. Carleton MacDowell, Davenport's colleague for almost 30 years at Cold Spring Harbor (MacDowell 1946). MacDowell had attempted, he wrote, "an intimate interpretation" of Davenport's life, although such a biography "of a man so recently in our midst" was not usually attempted. (It was written within 2 years of Davenport's death.) It is a most unusual biography because MacDowell not only relates the dry sticks of Davenport's professional career but also makes a frank assessment of Davenport's personality. A few lines later, MacDowell says that "although he himself [Davenport] in his most candid self-analysis failed to recognize it, his life was dominated by a turmoil of conflicting loyalties, and his actions gave the impression of a paradoxical mixture of qualities." MacDowell's portrait of Davenport is the only biography of Davenport written by someone who had an extensive personal knowledge of him, and I shall draw heavily on it for this essay; all quotations will be taken from it, unless specified otherwise (see also Riddle 1944).

DAVENPORT'S BEGINNINGS

Davenport was born on June 1, 1866, the eighth and last child of Amzi and Jane Davenport. His birthplace was Davenport Ridge, near Stamford, Connecticut, the farm home of five generations of Davenports. His father was a successful real estate agent and they lived for most of the year in Brooklyn Heights, the family staying at Davenport Ridge during the summer months. It does not seem that Davenport's childhood was a happy one. MacDowell writes that Davenport's father accepted the "stern repressive teachings of his Puritan ancestors...with reactionary enthusiasm," and he regarded his children (11 including those of a second marriage) as "the source of great displeasure." Davenport's mother provided the "only tender love" in the family and inspired his interest in biology; his first book, *Experimental Morphology* was dedicated to her (Davenport 1897, 1899a).

Davenport was homeschooled until he was 13 years old. He must have learned to read early, as he was reputed to have read the entire New Testament between the ages seven and eight. But by 1878, he was also working in his father's office, as a general office boy and janitor, for the niggardly sum of 25 cents per week, surely not a large amount of money even then. He performed similar functions at home, acting as handyman and stable boy. MacDowell gives one anecdote of Davenport waking too late, at 6:30 a.m., for prayers on December 26, and his father confining him to his room for 2 hours as punishment. It is not surprising that his father's absences provided much-needed relief. It was not all dark, however. Davenport did join local natural history societies, but these seem to have been only as successful as might be expected of societies run by young boys. More serious interactions with other naturalists came through sending bird and weather reports to the American Ornithological Union and the U.S. Signal Corps, respectively.

Davenport began school on November 26, 1879, attending the Brooklyn Collegiate and Polytechnic Institute and continuing to work before school in his father's office. He also worked hard in school, trying to make up for having fallen behind because of his homeschooling. By the end of high school, he was ranked at the top of his class. During the summers he worked on the Davenport Ridge farm, but in his last summer at college, when he was 19, he wrote a 20-page letter to his father, describing in detail the course of work he wanted to do during that summer. He had, he told his father, decided that his studies—geology, zoology, and farm surveying—had "a direct bearing upon the

Science of Agriculture which I should like to study as my life work, if you approve of it." His father did not: "[Y]ou are too theoretical" and "the question of prime importance is how much money can you make for yourself and for me." So, instead of pursuing his proposed course of studies that summer, Davenport surveyed the family farm before returning to the Brooklyn Collegiate and Polytechnic Institute to complete his B.S. degree in civil engineering, graduating in 1886.

FINDING A CAREER

Jobs in science were hard to find and the 20-year-old Davenport wrote a plaintive letter to *Science* describing his plight, many elements of which remain true to the present day (Davenport 1886)— "I thought myself fitted for investigation in scientific fields, particularly as I love it above all else. In every case I received answer, 'Places all full.' I have begun to doubt if investigators and workers are needed in the natural or experimental sciences." With no career in natural history available to him (and no doubt with his father's views in mind), Davenport took a job as a surveyor on the Duluth, South Shore and Atlantic Railroad, incorporated in December, 1886. (Davenport may have surveyed the line of the railroad from Sault Ste. Marie to Soo Junction, Michigan, which was built in 1887 [DuLong 1997].) It was not to his taste. MacDowell speculates that it was this experience that gave Davenport the determination to resist his father's wishes—a "first and final revolt against his father's domination"—and instead to follow what he felt was best for himself. Davenport abandoned the railroad and, following the example of a friend, enrolled at Harvard. He took a part-time job doing quality control for a water commission to support himself and took as many natural history courses as he could.

It is clear that Davenport had found his métier. In addition to the course work, he spent summers at Woods Hole and, in 1890, at the Biological Laboratory at Cold Spring Harbor. He was graduated from Harvard in 1889 and began work for a Ph.D. on reproduction in Bryozoa, small aquatic colonial animals; the thesis was submitted in 1892. In 1891, he was awarded a monitorship and successively promoted until he became an instructor in 1899. Teaching was at least as important as research. He seems to have entered into teaching wholeheartedly and in 1893 devised a new course, Experimental Morphology, which included statistical and experimental study of variation in populations, individual variation and sports (mutations), normal inheritance and crossing, and selection. The course was successful, and Davenport wrote a two-volume text of the same name based on it.

The pressing biological issue at the end of the 19th century, following from Darwin's work, was, in the words of William Bateson, "to solve the problem of the forms of living things...How have living things become what they are, and what are the laws which govern their forms?" (Bateson 1894). "[T]he problem of Species," as Bateson put it, was largely a question of understanding the types and origins of the variations on which natural selection worked. This was also the time when increasing numbers of young scientists in the fields of physiology, embryology, and evolution came to embrace what William Coleman called the "experimental ideal" (Coleman 1971). It is not surprising that Davenport as one of the young turks—including contemporaries such as E.G. Conklin, R.G. Harrison, H.S. Jennings, R. Lillie, T.H. Morgan, and E.B. Wilson—turned to these new approaches. He concluded a lecture given at Woods Hole in 1899 with "in the application of combined experimental and statistical methods to genetic problems, zoology will reach its highest development" (Davenport 1899b). In the same year, he published *Statistical Methods: With Special Reference to Biological Variation,* arguably his most successful book, going through four editions, with the last in 1936 (Davenport 1899c).

These were the sorts of studies, known as biometry, that were occupying others, most notably William Bateson, Karl Pearson, and W.F.R. Weldon in England, all of whom carried out quantitative measurements of the variability of organisms and analyzed the results statistically using methods developed by Pearson. Weldon, for example, measured 11 features of the shore crab, *Carcinus maenas*—1000 samples from Plymouth and 1000 from Naples (Weldon 1893). Davenport's contributions included a series of papers reporting the quantitative measurements of the variability of scallop shells (Davenport 1900). By 1901, Davenport was the American editor of the journal *Biometrika,* founded by Pearson and Weldon, with the enthusiastic support of Francis Galton, to promote biometry. Pearson was a supporter of Davenport's later efforts to found an institute at Cold Spring Harbor for the study of evolution, but by 1904 Davenport was fast becoming an advocate of Mendelian genetics, which was anathema to the biometricians. They believed that natural selection worked on the small, continuous variations that they measured, and that they were the true followers of Darwin; the "large" discontinuous variation (mutations) studied by the Mendelian geneticists were, they thought, irrelevant to evolutionary processes.

Davenport was not a direct participant in the extraordinarily bitter dispute between Pearson and Weldon on the one hand and Bateson on the other, but as early as June 1903, he was arguing with Pearson who wanted Davenport to omit a speculative section from a paper Davenport had submitted to *Biometrika* (Davenport 1903a). It illustrates the divergence between the two camps, where Davenport wrote: "You say that I practically throw Darwin out & adopt a Mutation Theory." And a further cooling is evident where Davenport objected to the way in which Pearson used *Biometrika* to push biometry:

> Well, I have written a long letter in my plea that *Biometrika* be not run in the interest of one idea but that everyone who has used the method of numerical precision properly and will adopt the motto: "Ignoramus, in hoc signo laboremus" may have free entry to it to express the truth as he sees it.

The quotation was a not-so-subtle rebuke of Pearson. The first issue of *Biometrika* had included a picture of a statue of Darwin accompanied by the quotation: "We are ignorant; so let us work." Pearson's response cannot have been satisfactory. On July 7, Davenport wrote again, indignantly protesting that while he was not in the "ultra mutationist" camp of de Vries and Bateson, he had found no evidence against "sports" (mutations) and pointed out that Darwin accepted "sports" as a source of new races (Davenport 1903b). He was not going to remove the speculative section and withdrew the paper. It was not all he withdrew:

> I gather from your last letter that you feel that a mistake may have been made in inviting me to be a co-editor of *Biometrika*—considering my heretical views. For my past I have been proud of the association with you on the Journal and have been glad to work for it. But I trust that you

> will feel quite free to drop my name whenever you think the welfare of the Journal *Biometrika* or other considerations make it seem best. To relieve you from any embarrassment I hereby resign.

It was not long after this that Davenport became director of an institution that was a hotbed for Mendelian studies of flies, mice, chickens, ducks, canaries, sheep, rabbits, and corn.

While teaching at Harvard, Davenport met a young woman, Gertrude Crotty, who was a zoology student in the "Society for the Collegiate Instruction of Women, by Professors and other Instructors of Harvard College," later more euphoniously called Radcliffe College. She was a competent biologist, carrying out a study of the primitive streak and notochordal canal in *Chelonia* under the direction of Edward Laurens Mark, Hersey Professor of Zoology at Harvard. (Mark was the founder of the Bermuda Biological Station for Research and its director from 1902 to 1931.) After a brief courtship, they married in 1894. MacDowell remarks that Davenport needed a "level-headed person to hold him steady and take charge of mundane affairs." They appear to have been a devoted couple, having two girls and a boy, Charlie, who died of polio at the age of 5 in the 1916 poliomyelitis epidemic. Not only did Gertrude take charge of mundane affairs, she took part in Davenport's research, publishing as senior author several papers on human genetics with him (Rosenberg 1961). Together they wrote a successful elementary textbook, *Introduction to Zoology* (Davenport and Davenport 1900). Marriage, however, brought added responsibilities and Davenport began to look for a new position with Gertrude, according to MacDowell, searching the obituary columns of *Science* for newly vacated posts.

DAVENPORT COMES TO COLD SPRING HARBOR

The most significant event of this period—although it might not have seemed so at the time—was Davenport's appointment in 1898 as Director of the Biological Laboratory at Cold Spring Harbor. Established in 1890 by the Brooklyn Institute of Arts and Sciences (just 2 years later than the Marine Biological Laboratory at Woods Hole), it offered summer courses in biology for teachers. The harbor provided a plentiful source of organisms for study, and students explored the slopes and woods of the valley that opened up at its northern end to form the harbor. Davenport published a short article in *Science* in November of 1898—intended, no doubt, to attract students to the summer courses—in which he described the geology and flora and fauna of Cold Spring Harbor (Davenport 1898). Davenport availed himself of the biological opportunities provided by Cold Spring Harbor only once, when some years later, he published a lengthy study of "The Animal Ecology of the Cold Spring Sand Spit, with Remarks on the Theory of Adaptation." He argued that to a degree an organism selects the environment in which it can survive as much as the environment selects the organisms that survive in it (Davenport 1903c). He was continuing to publish papers on biometry, as well as studies of a more classical descriptive type. It is probable that the teaching and organizational aspects of the job appealed to Charles and Gertrude and "the couple made a strong team and their devotion to the Biological Laboratory grew as they gave more and more of themselves to it."

Nevertheless, it was clear that Davenport's professional aspirations were not likely to be satisfied by a laboratory that was only a summer school, and over the next few years he pursued other openings. One came from Charles Whitman, Director of the Marine Biological Laboratory in Woods Hole and Head of the Department of Zoology at the University of Chicago. Whitman had grand plans for a new institute, to be based close to Woods Hole, that would promote studies of how organisms evolved, particularly through examining variation and heredity. Davenport had similar plans but turned Whitman down—Cold Spring Harbor was a cheaper place to live! Not to be outdone, Whitman appointed Davenport an assistant professor at the University of Chicago in 1899 and advanced him to an associate professorship in 1901. At the same time, in 1901, there appeared to be an opportunity to transform the Biological Laboratory at Cold Spring Harbor into a full-time research institute. It seemed that Columbia University might take over management of the Laboratory, and if that happened, then Davenport might find himself its full-time director. In the event, nothing came of these rumors but Davenport soon found a new avenue to explore.

PHILANTHROPY TO THE RESCUE

Andrew Carnegie was one of the great entrepreneurs who combined great business sense with a ruthlessness to succeed. In 1901, he sold his Carnegie Steel Company for $250 million and retired. He turned his skills to philanthropy and over the years invested some $350 million in worthy causes. Carnegie had considered establishing a university in Washington, D.C., but his advisors persuaded him instead to create a private research institute which would, as he put it, "in the broadest and most liberal manner encourage investigation, research, and discovery [and] show the application of knowledge to the improvement of mankind." The topics for study were to be drawn from biology, astronomy, and the earth sciences. The Carnegie Institution of Washington (CIW) was incorporated on January, 4 1902 and ratified by an Act of Congress in 1904.

The first Departments of the new Institution—Plant Biology and Historical Research—were created in 1903, and Davenport lost no time in approaching the CIW for funding. With an urgency that seemed to characterize all his dealings with the Carnegie, his first letter to the Institution's secretary was dated January 16, 1902, just 8 days after the CIW had been incorporated. It was "the opening move of a two-year campaign...the lengths to which it was carried out would be unimaginable without the original documents in the archives...the incandescence of his enthusiasm distorted his judgment and permitted exaggeration that bordered closely upon misrepresentation." Davenport's January letter was followed by further communications as well as repeated visits to John Billings in New York, one of the CIW's most influential trustees. One of Davenport's letters to Billings, six pages long, extolled the virtues of Cold Spring Harbor as the site for a new Carnegie Institute and of himself as director (Allen 2004)!

What would this new department do? It was to follow the lines that Davenport and Whitman had described in 1898, and, indeed, Whitman and Woods Hole were also applying to the CIW for funds. The proposed Department would carry out experimental investigations of evolution, combining studies of variation with studies of heredity. Mendel's work had been discovered in 1900 by de Vries, Tschermak, and Correns in the course of their own studies, and, as early as 1901, Davenport published a paper on "Mendel's Law of Dichotomy in Hybrids," a brief discussion of dominance (Davenport 1901). He was also clear about the relationship of heredity and evolution. In 1907, Morgan reviewed Davenport's *Inheritance in Poultry* (Morgan 1907) and remarked:

> It would seem that studies dealing with the hybridization of unit-characters [genes] relate to the field of heredity rather than to that of evolution. In fact, many evolutionists of the old school deny the applicability of such results to the theory of evolution. There seems to be here a difference of opinion, whether well founded or not the future will decide. The opening paragraph of Professor Davenport's paper leaves, however, no room for doubt as to his attitude in the matter. "Evolution proceeds by steps. These steps are measured by the characters of organisms... . Since the character is the unit of evolution it deserves careful study.

However, the road to Davenport's hoped-for institute continued to be rocky. He did all he could to smooth the way by enlisting the support of eminent scientists. Gertrude and he went on a trip to Europe in the latter part of 1902, collecting materials for study and visiting many of Europe's biological stations, gathering information. He visited London and persuaded Pearson and Galton to write letters of support. Neither were overly enthusiastic about Davenport's abilities. Pearson replied "Personally he seems stronger than his published work," and although Galton was enthusiastic about Davenport's goals, he declined to provide a testimonial given their short acquaintance. (Galton's letter is quoted in full in MacDowell [1946].)

Davenport's chances for success with CIW were bolstered by the failure of the Woods Hole campaign. The Woods Hole Corporation had initially been very enthusiastic about a close association with the Carnegie voting 60 to 3 to transfer the whole establishment to the Carnegie! (Davenport's was one of the dissenting votes.) However, the terms exacted by Carnegie were not acceptable and the proposal collapsed. Nevertheless, things must have seemed pretty bleak to Davenport; there was a dearth of progress despite all his efforts. But, as MacDowell put it, "Idling was one thing Davenport could not tolerate" and so he turned his attention to Whitman's original idea of an experimental farm associated with the University of Chicago where, of course, Davenport was still an associate professor. Davenport went so far as to find a suitable 700 acres, which was examined by the University and pronounced satisfactory. He selected five acres for his own house and drew up a program of research to be carried out at "The Field." Davenport was prepared to resign from his post of director of Biological Laboratory at Cold Spring Harbor if he was appointed to direct "The Field" and if the University paid him a reasonable salary.

Negotiations with the CIW continued to drag on but at the end of 1903, Davenport received word that Billings was going to recommend Davenport's proposal to the Carnegie Trustees. And so, in December 1903, almost 3 years since his first communication with the CIW, Davenport learned that the Carnegie was going to finance *two* biological research stations, one at Cold Spring Harbor and the other, the Department of Marine Biology, at Dry Tortugas, a coral reef formation about 70 miles west of Key West in Florida. Davenport's equivalent for the latter laboratory was the Harvard-trained marine biologist Alfred Goldsborough Mayer. Throughout 1903, Mayer had conducted a campaign in the pages of *Science* in support of Dry Tortugas, canvassing support from zoologists, including Davenport, Conklin, Jennings, and Morgan (Mayer 1903). Three other Departments opened in 1904, including a short-lived one for Economics and Sociology and two that continue to the present day—the Departments of Terrestrial Magnetism and the Observatories of the CIW. The Department of Embryology opened in 1914 and it continues as a leading research institute.

THE STATION FOR EXPERIMENTAL EVOLUTION OPENS

Davenport resigned his Assistant Professor position at the University of Chicago in 1904 and was appointed director of the Station for Experimental Evolution at Cold Spring Harbor that same year. (It was not made a Department of the CIW until 1921.) There was a grand opening ceremony on June 11, with a special train bringing guests from New York City. The honored guest was none other than Hugo de Vries, one of the discoverers of Mendel's studies. But it seems that Davenport had spent so much time on his campaign that he had not devoted enough effort in developing a detailed program of research for the Station. He had included a list of eight experiments as part of his submission to the CIW, but these were described in hardly more than a single sentence each and show little evidence that Davenport had thought deeply about them. MacDowell comments that while Davenport refined this list, the differences in the successive lists "did not represent progressive critical thought."

Even if Davenport's program was ill defined, he showed aptitude over the years in selecting the staff of the CIW. The first full-time staff included himself, Frank E. and Ann Lutz, who discovered the wing vein and dwarf mutants of *Drosophila*, and George H. Shull, who developed hybrid corn. All were concerned with studies of variation, initially searching for suitable organisms to study. Interestingly, Davenport's project was to determine how feasible it would be to maintain and cross wild animals in captivity, presumably reflecting a concern that animals like mice, selected for life in the laboratory, might not be representative of wild populations. In addition, a small number of Associates came to the Station for short visits; Raymond Pearl, who studied the biology of aging, and the cell biologist Edmund Beecher Wilson were among the first. During Davenport's tenure as director, the roster of scientists included several who made significant contributions to biology either at Cold Spring Harbor or later in their careers. Shull developed hybrid corn; MacDowell and Clarence Little studied cancer genetics in mice, Little going on to found the Jackson Laboratory at Bar Harbor in Maine; Albert Blakeslee and John Belling made important studies of the plant *Datura,* correlating triploids with particular phenotypic traits; Milislav Demerec (later director of the Biological Laboratory and the Department of Genetics at Cold Spring Harbor) made important contributions to *Drosophila* and bacterial genetics, as well as developing a high-yielding strain of *Penicillium;* Oscar Riddle discovered prolactin; and Gertrude and Charles published four quite respectable papers on the inheritance of eye color, hair color, hair forms, and skin color (Davenport and Davenport 1907, 1908, 1909, 1910).

EUGENICS ENTERS THE PICTURE

However, as if Davenport's duties as director of both the Biological Laboratory and the Station for Experimental Evolution were not enough, he was soon looking for ways to develop his new passion for the study of human traits. He wrote in his 1909 Annual Report that "the necessity of applying new knowledge to human affairs has been too evident to permit us to overlook it," and at the end of the 1907 paper on eye color, Davenport discussed the "practical applications of these results to human marriage." It is clear from other papers of this period—for example, "Influence of Heredity on Human Society" (Davenport 1909)—that he believed that genetic knowledge should be applied to bettering the human race. This was not a new idea and can be traced to Francis Galton who coined the term "eugenics" in 1883 for this new discipline. The Eugenics Education Society had already been formed in England (1908) to promote eugenic thinking. It boasted such luminaries of British society as Leonard Darwin (Charles Darwin's son) and Julian Huxley, as well as many other noteworthies, drawn as much from the left wing of British political as the right (Paul 1984). However, the British eugenics movement, generally regarded as "positive eugenics," never achieved the power and political influence of the American movement, generally regarded as "negative eugenics," whose members were able to capitalize on a long-standing American concern over immigration and fears that the founding stock of America was becoming tainted by degenerates.

The links between better breeding of human stock and better breeding of livestock and crops was explicit from the very beginning of the American eugenics movement. The American Breeders' Association was established in 1903 with the purpose of promoting the use of Mendelian genetics for improving livestock and plants (Paul and Kimmelman 1988). Three years later, at the Association's second meeting, a Committee of Eugenics was set up with the remit to "investigate and report on heredity in the human race" and to "emphasize the value of superior blood and the menace to society of inferior blood." (The close connection between the ways in which breeders selected their best animals and plants for breeding and the goals of the eugenics movement was emphasized repeatedly.) Luther Burbank wrote a book, *The Training of the Human Plant,* in which he drew parallels between the entire process of breeding and growing crops and the bringing up of children (Burbank 1919). The Committee included David Starr Jordan (President of Stanford University) as President, Davenport as secretary, Roswell Johnson (co-author of *Applied Eugenics*), Luther Burbank (plant breeder), Alexander Graham Bell (inventor), Vernon Kellogg (zoologist), and Henry Fairfield Osborn (President of American Museum of Natural History). (Osborn and the Museum played a very active part in the eugenics movement, hosting the Second and Third International Congresses of Eugenics in 1921 and 1922, respectively.) The program of the Committee was set out in a 1909 report, reprinted in 1910 as the second part of Davenport's small book *Eugenics: The Science of Human Improvement by Better Breeding* (Davenport 1910). (The first part on "Fit and Unfit Matings" was based on a talk he gave to the American Academy of Medicine at Yale University on November 12, 1909.) The Committee was first to investigate how human traits were inherited, and to this end subcommittees on Feeble-mindedness and Insanity were set up. It was expected that other subcommittees would be set up to examine the heredity of criminality and pauperism, cancer, muscular strength, and intellectual traits. Another line of investigation would examine the characteristics of "normal" American families by carrying out surveys and by using the vast amounts of data already in the records of hospitals, prisons, and other institutions (Haller 1963; Kevles 1985).

A BENEFACTOR STEPS FORWARD AND THE EUGENICS RECORD OFFICE IS BORN

Davenport, as always eager to get on with the task at hand, conceived the idea of setting up an institute at Cold Spring Harbor for eugenics. The second part of *Eugenics* ended with a call for help:

> One cannot fail to wonder that, where tens of millions have been given to bolster up the weak and alleviate the suffering of the sick, no important means have been provided to enable us to learn how the stream of weak and susceptible protoplasm may be checked. Vastly more effective than ten million dollars to "charity" would be ten millions to Eugenics. He who, by such a gift, should redeem mankind from vice, imbecility and suffering would be the world's wisest philanthropist.

In that same year, he found the looked-for benefactor in his own neighborhood. Mrs. E.H. Harriman was the widow of Edwin H. Harriman, the railroad magnate who had owned, among others, the Illinois Central, Union Pacific, the Central Pacific, and the Southern Pacific railroads. He had died in 1909. Their daughter, Mary, had been a student in the Biological Laboratory at Cold Spring Harbor and it was through her that Davenport met her mother. On February 16, 1910, Mrs. Harriman agreed to help and provided the funds to purchase a large house and its associated estate of 75 acres just up the road from the main Cold Spring Harbor campus. It was, Davenport wrote in his journal "A Red Letter Day for humanity!" (Mrs. Harriman's gifts were substantial, eventually amounting to some $500,000.) The house became the Eugenics Records Office (ERO) until a new house had been constructed.

Davenport appointed Harry Laughlin as Superintendent of the ERO in 1910 and for the next 25 years, inspired by Davenport and under his tutelage, Laughlin was the key figure in American eugenics (Hassencahl 1970; Wilson 2002). Laughlin was born in Oskaloosa, Iowa, in 1880. He began teaching agriculture at the North Missouri State Normal School (later Truman University) in 1900. Laughlin became an enthusiast for applying Mendelian principles to plant and animal breeding, and he and his students carried out studies breeding poultry. This was one of Davenport's interests—he had published a monograph on "Inheritance in Poultry" in 1906—and Laughlin contacted him for information. As a consequence, Laughlin came to the Biological Laboratory in 1907 to take the genetics course, a visit that seems to have induced a sense of hero worship on Laughlin's part, no doubt much welcomed by Davenport. They met again in 1909 at the American Breeders Association meeting in Columbia, Missouri. Although Laughlin had studied eye color in human beings with his students, he was primarily interested in animal breeding. But it was at this meeting, apparently, that Davenport persuaded him to concentrate on human traits.

Laughlin moved to Cold Spring Harbor in 1910 and the ERO rapidly became the scientific center of eugenics in the United States, implementing the program outlined by the Committee on Eugenics and making a survey of the hereditary characteristics of American families. To this end, Davenport listed the traits that should be studied in

The Trait Book (Davenport 1912), and the ERO ran courses training field workers who went out to interview families and fill out survey forms based on *The Trait Book.* All the data were entered on index cards, which numbered several hundred thousand by the end of the project. In addition, the ERO sponsored studies on specific inherited characteristics, racial interbreeding, and studies of degeneracy in families and in institutions. Davenport, for example, published monographs on "The Hill Folk: Report on a Rural Community of Hereditary Defectives" (Danielson and Davenport 1912), "Body-Build: Its Development and Inheritance" (Davenport 1924), and "Race Crossing in Jamaica" (Davenport and Steggerda 1929). One of the more extraordinary of Davenport's studies was "Naval Officers: Their Heredity and Development" in which he postulated that there was hereditary tendency—"thalassophilia"—to account for his finding that the sons of naval officers often went to sea (Davenport and Scudder 1919). But he also studied traits of clinical importance such as Huntington's disease, epilepsy, albinism, and neurofibromatosis.

During the First World War, Davenport put his organizing and analytical skills at the disposal of the Army, who assigned him to the Surgeon General's Office. Here, he and Albert G. Love set about reducing the measurements made on over 1,000,000 recruits to a manageable size (Love and Davenport 1919). However, the most significant information to emerge from the studies of army recruits came not from Davenport's anthropometry but from IQ testing (Gould 1981). Alfred Binet in France had devised the first IQ tests in 1905, intending that they should be used in trying to determine the learning deficits of children who had already been recognized as learning disabled. In that way, he believed, such children could receive specific "mental orthopedics" to improve their skills, attention spans, concentration, and so on. It was Henry Goddard of the Vineland Training School for Feeble-minded Girls and Boys in New Jersey who introduced Binet's IQ tests in the United States and advocated their use as a measure of a single attribute, IQ. But the widespread use of IQ tests was popularized by Lewis Terman at Stanford University. Terman's intent was eugenic (Terman 1916):

> It is safe to predict that in the near future intelligence tests will bring tens of thousands of these high-grade defectives under the surveillance

> and protection of society. This will ultimately result in curtailing the reproduction of feeble-mindedness and in the elimination of an enormous amount of crime, pauperism, and industrial inefficiency.

Although the IQ tests devised by Robert Yerkes and Terman were inappropriate for Army recruits, and sloppily administered, there was widespread consternation and uproar at the results (Gould 1981). An analysis of 160,000 individuals showed that the average mental age of the young white recruit was 13, only a fraction above what was considered moron-level. Recruits of European origin did even worse. The implications were clear to some: the United States was becoming a nation of morons and this could be controlled in part by blocking the immigration of Poles, Russians, Italians, and other "low-grade" nationalities.

These findings provided a significant boost to those who had been extolling immigration control and the abolition of unrestricted access to the United States. Eugenical arguments became an important part of the debate, and Albert Johnson, chairman of the House Committee on Immigration and Naturalization, appointed Harry Laughlin as the Committee's "expert eugenics agent." Laughlin prepared several huge reports purporting to show, for example, that there were more immigrants in mental institutions than would be expected on the basis of their proportion in the population. Laughlin's fact-finding surveys, extensive tables and charts, and presentations before the Committee were persuasive and the 68th Congress passed the National Origins Act of 1924. This severely restricted immigration from countries deemed to have higher percentages of defectives, notably Eastern and Southern Europe and the Slavic area, including the Balkans.

Davenport does not seem to have played a public role in the immigration debate, unlike Laughlin who basked in the spotlight shone by the House Committee on Immigration. Laughlin's activities drew unwelcome attention to the ERO and the extreme claims that were being made for strong genetic components in, for example, pauperism and "nomadism." Eminent geneticists such as T.H. Morgan and H.S. Jennings began to speak out against eugenics (Morgan 1925; Jennings 1930). Raymond Pearl, for example, in a talk given at the Fifth International Congress of Genetics in Berlin, 1927, described the literature on eugenics as largely (Pearl 1928)

> a mingled mess of ill-grounded and uncritical sociology, economics, anthropology and politics, full of emotional appeals to class and race prejudices, solemnly put forth as *science*, and unfortunately accepted as such by the general public.

Nor surprisingly, the CIW became concerned and instigated two evaluations of the ERO. Following the second review in 1935, the ERO gradually began to fade away. This was accelerated because of the injuries Laughlin sustained in a car accident, ironically initiated by an epileptic seizure, one of those conditions the eugenicists regarded as the mark of a "defective." Vannevar Bush became President of the CIW in 1938, and in 1939 he forced Laughlin's retirement. The ERO was quietly shut down at the end of that year and Laughlin retired to Missouri where he died in 1943. Davenport wrote Laughlin's obituary for *Science* (Davenport 1943) and continued to stand behind the man he had chosen 33 years previously to lead his eugenics project—"Some of Laughlin's conclusions and their applications in legislation were opposed by those committed to a different social philosophy, founded on a less thorough analysis of facts. One can not but feel that a generation or two hence Laughlin's work, in helping bring about restricted immigration and thus the preservation of our country from the clash of opposing ideals and instincts found in the more diverse racial or geographical groups, will be the more widely appreciated as our population tends toward greater homogeneity."

THE BIOLOGICAL LABORATORY IS REORGANIZED

The Biological Laboratory was also undergoing changes in this period. The Brooklyn Institute Trustees evidently felt that they could not exercise sufficient control over an institute so far from their base and offered the Laboratory to two universities. Both declined the offer, even though the Biological Laboratory had managed to raise an endowment that totaled $27,500 by 1917. Davenport seems, by 1921, to have felt that he had too much do, running three institutions, one of which, the Biological Laboratory, continued to struggle financially. Davenport recommended that an association of the Laboratory's wealthy neighbors should be formed that would take over governance of the Laboratory and raise funds. He also laid out a scientific program for the Laboratory—it should begin research in biophysics. It is not clear why he thought this, other than that William T. Bovie might be enticed to Cold Spring Harbor. Bovie was a biophysicist at Harvard who achieved lasting fame by developing electrosurgery in collaboration with Harvey Cushing. (Bovie received only $1 for his patent nor was his career helped; on being refused tenure at Harvard he became Professor of Biophysics at Northwestern University.) Davenport also advocated the purchase of the land surrounding the Biological Laboratory and the appointment of a resident director of the Laboratory while he concentrated on the CIW institutions.

This strategy was successful: Three years later the Long Island Biological Association (LIBA) was incorporated on February 18, 1924, and, on March 12, the Brooklyn Institute transferred its property to LIBA. By then, Davenport had relinquished the post of director of the Biological Laboratory (he remained head of the Department of Genetics until 1934), and Reginald Harris became the first full-time director of the Biological Laboratory (Cold Spring Harbor Laboratory 1982). Then 26 years old and married to the Davenports' daughter Jane, Harris set about revitalizing the Biological Laboratory, transforming it from a summer teaching institution to a research institute. He made biophysics the new focus of research and began the Cold Spring Harbor Laboratory Symposia on Quantitative Biology. These meetings have repeatedly played a key role in keeping the Laboratory afloat financially and continue to be one of the intellectual highlights of Cold Spring Harbor. They were especially important during the emergence of molecular genetics during 1949–1965, when many of the key findings were first presented at these summer meetings. Harris died in 1936, but his appointment is arguably the single most important event of Davenport's career at Cold Spring Harbor.

DAVENPORT, THE ORGANIZER

Throughout his career, Davenport was an indefatigable organizer of people and words. MacDowell tallied Davenport's membership of societies at 64, of which 24 were scientific, 7 eugenic, 5 anthropology, 5 medicine, 10 civic, 5 social, and 8 "miscellaneous." Of these, he was vice president or president of ten. He was early elected to the National Academy of Sciences, in 1912. As for organizing words, at one time or another, he served on the editorial boards of eight journals.

When he retired in 1934, Davenport was 68 years old. He became an Associate of the CIW and continued to work as hard as when he was director. He continued to write, publishing 47 papers, a book, and the fourth edition of *Statistical Methods* in the last 10 years of his life. His relentless pursuit of work led to his death. Davenport was Curator and Director of the Whaling Museum in Cold Spring Harbor village and was as passionate a player in this small museum as he had been on the international scientific stage. So, when a killer whale was beached on the end of Long Island in January, 1944, Davenport set out to procure its skull for the museum. He did so by boiling the head in a large cauldron in an outdoor shed; MacDowell relates that Davenport stank so of boiled whale meat that at the last staff meeting he attended, he sat by himself, away from everyone else. January was not a good time for a man of 78 to be outdoors and Davenport died of pneumonia, contracted from this exposure, on February 18, 1944.

It seems that Davenport's skills as a director did not come from having overarching, long-term plans for the three institutes that he headed. In the Biological Laboratory and the Station of Experimental Evolution, he provided laboratories and support for a rather eclectic group of scientists who were able, and perhaps wanted, to work alone, pursuing their individual interests. Even the Eugenics Records Office fits this pattern. Although eugenics was clearly a passion for Davenport and he maintained a long-term interest in it, Harry Laughlin was the driving force behind the ERO and its public face. MacDowell wrote that "to the end of his days he [Davenport] remained a lone man, living a life of his own in the midst of others, and feeling out of place in almost any crowd." However, Davenport's skill at giving a home to scientists who, unlike himself, could tackle a problem and work it out to completion, gave rise to his greatest legacy—Cold Spring Harbor and its various institutions. None of his scientific contributions can match those of the scientists he brought to that otherwise undistinguished bay on the North Shore of Long Island (Table 1). It is not surprising

Table 1. *Significant discoveries at Cold Spring Harbor during Davenport's administration*

An early example of ecological study

Davenport C.B. 1903. The animal ecology of the Cold Spring sand spit, with remarks on the theory of adaptation. *Decennial Publications U. Chicago* **X:** 157–176.

An early example of human genetics

Davenport G.C. and Davenport C.B. 1907. Heredity of eye color in man. *Science* **26:** 589–592.

Two studies that led to the development and widespread use of hybrid corn

Shull G. 1908. The composition of a field of maize. *Am. Breeders Assoc. Rep.* **4:** 296–301.

Shull G. 1909. A pure line method of corn breeding. *Am. Breeders Assoc. Rep.* **5:** 51–59.

Studies of the inheritance of cancer

Little C.C. 1920. The heredity of susceptibility to a transplantable sarcoma (J. W. B.) of the Japanese waltzing mouse. *Science* **51:** 467–468.

Richter M.N. and MacDowell E.C. 1930. Studies on leukemia in mice: I: The experimental transmission of leukemia. *J. Exp. Med.* **51:** 659–673.

Association between specific chromosomes and inherited traits

Belling J. and Blakeslee A.F. 1924. The configurations and sizes of the chromosomes in the trivalents of 25-chromosome *Daturas.*" *Proc. Natl. Acad. Sci.* **10:** 116–120.

A mutation later shown to be caused by transposable elements is discovered

Demerec M. 1926. Reddish–a frequently "mutating" character in *Drosophila virilis. Proc. Natl. Acad. Sci.* **12:** 11–16.

Discovery of a new hormone

Rowntree L.G., Greene C.H., Swingle W.W., and Pfiffner J.J. 1930. The treatment of patients with Addison's Disease with the "cortical hormone" of Swingle and Pfiffner. *Science* **7:** 482–483.

Genetic studies of the ability to taste phenylthiocarbamide (PTC)

Blakeslee A.F. 1932. Genetics of sensory thresholds: Taste for phenylthiocarbamide. *Proc. Natl. Acad. Sci.* **18:** 120–130.

Discovery of prolactin

Riddle O., Bates R.W., and Dykshorn S.W. 1932. A new hormone of the anterior pituitary. *Proc. Soc. Exp. Biol. Med.* **29:** 1211–1212.

that the resolution passed by the Long Island Biological Association Executive Committee (1944) read (in part):

> *Be it resolved,* That the directors of the Long Island Biological Association record with a sense of irreparable loss the death, on February 18, 1944, of Dr. Charles Benedict Davenport.... To a greater extent than any other individual, he was, indeed, the founder of both these institutions [the Biological Laboratory and the Carnegie Department of Genetics] .... Throughout periods of discouraging outlook, of disappointment and deep personal sorrow, no less than during the happier years, he held faith in the importance and assured success of our common aim.

The same theme of battling discouragement for a greater purpose are echoed in MacDowell's last words on his colleague (MacDowell 1946):

> Only an extraordinarily resilient constitution could have sustained so long the continuous pressure Charles Davenport put upon himself in his struggle to overcome a deeply implanted sense of inferiority [from his father]. Instead of destructively turning this in upon himself, he turned it outward. Instead of seeking immediate physical gratification and displaying the bombastic self-assurance typical of many ambitious men, he identified himself with great concepts—Science, Experimental Morphology, Advancement of Knowledge, Improvement of Mankind. As a result, he lived a full life and played a constructive part that gained him wide recognition. Such an interpretation gives meaning to many of the strange inconsistencies that appeared and opens the door to unqualified admiration of the valiant way that he faced his great problem. It was a grand struggle.

REFERENCES

Allen G.E. 2004. Heredity, development and evolution at the Carnegie Institution of Washington. In *Centennial History of the Carnegie Institution of Washington, Vol. V: The Department of Embryology* (ed J. Maienschein et al.), pp. 145–171. Cambridge University Press, Cambridge.

Cold Spring Harbor Laboratory. 1982. Reginald Harris and the Biological Laboratory. *The Reginald Harris Building Dedication Ceremony, May 27, 1982.* Cold Spring Harbor Laboratory, Cold Spring Harbor, New York. *

*These papers are reprinted in Witkowski 1999.

Bateson W. 1894. *Materials for the study of variation treated with especial regard to discontinuity in the Origin of Species.* Macmillan and Co., London.

Burbank L. 1919. *The training of the human plant.* The Century Co., New York.

Coleman W. 1971. *Biology in the nineteenth century. Problems of form, function and transformation.* Cambridge University Press, Cambridge.

Danielson F.H. and Davenport C.B. 1912. The hill folk. A report on a rural community of defectives. *Eugenics Record Office Mem.* 1.

Davenport C.B. 1886. Science for a livelihood. *Science* **8:** 236.

———. 1897. *Experimental morphology. Part I. Effect of chemical and physical agents upon protoplasm.* The Macmillan Company, New York.

———. 1898, The fauna and flora about Coldspring Harbor, L.I. *Science* **8:** 685–689.

———. 1899a. *Experimental morphology. Part II. Effect of chemical and physical agents upon growth.* The Macmillan Company, New York.

———. 1899b. The aims of the quantitative study of variation. *Biological Lectures,* Marine Biological Laboratory, pp. 267–272.

———. 1899c. *Statistical methods: With special reference to biological variation.* John Wiley & Sons, New York.

———. 1900. On the variation of the shell of *Pecten irradians* Lamarck from Long Island. *Am. Nat.* **34:** 863–877.

———. 1901. Mendel's Law of Dichotomy in Hybrids. *Biol. Bull.* **2:** 307–310.

———. 1903a. Letter to Karl Pearson, June 5, 1903. University College, London Special Collections. Karl Pearson papers. KP, 674/1.

———. 1903b. Letter to Karl Pearson, July 7, 1903. University College, London Special Collections. Karl Pearson papers. KP, 674/1.

———. 1903c. The animal ecology of the Cold Spring sand spit, with remarks on the theory of adaptation. *Decennial Publications U. Chicago* **X:** 157–176. *

———. 1906. Inheritance in poultry. *Carnegie Inst. Wash. Pub.* 52.

———. 1909. The influence of heredity on society. *Ann. Am. Acad. Pol. Soc. Sci.* **34:** 16–21.

———. 1910. *Eugenics: The science of human improvement by better breeding.* Henry Holt & Co., New York.

———. 1912. The Trait Book. *Eugenics Record Office Bull.* No. 6.

———. 1924. Body-build and its inheritance. *Carnegie Inst. Wash. Pub.* 329.

———. 1943. Harry Hamilton Laughlin. *Science* **97:** 194–195.

Davenport C.B. and Davenport G.C. 1900. *Introduction to zoology.* Macmillan Publishing Co., New York.

Davenport C.B. and Scudder M.T. 1919. Naval officers: Their heredity and development. *Carnegie Inst. Wash. Pub.* 259.

Davenport C.B. and Steggerda M. 1929. Race crossing in Jamaica. *Carnegie Inst. Wash. Pub.* 395.

Davenport G.C. and Davenport C.B. 1907. Heredity of eye color in man. *Science* **26:** 589–592. *

———. 1908. Heredity of hair-form in man. *Am. Nat.* **42:** 341–349.

———. 1909. Heredity of hair color in man. *Am. Nat.* **43:** 193–211.

———. 1910. Heredity of skin pigment in man. *Am. Nat.* **44:** 641–731.

DuLong J.P. 1997. Duluth, South Shore, and Atlantic Railway. http://dssa.habitant.org/chrono.htm.

Gould S.J. 1981. *The mismeasure of man.* W.W. Norton & Company, New York.

Haller M.H. 1963. *Eugenics: Hereditarian attitudes in American thought.* Rutgers University Press, New Brunswick, New Jersey.

Hassencahl F. 1970. *Harry H. Laughlin, "Expert Eugenics Agent" for the House Committee on Immigration and Naturalization, 1921–1931.* Ph.D. dissertation, Case Western Reserve University, Cleveland.

Jennings H.S. 1930. *The biological basis of human nature.* W.W. Norton and Company, Inc., New York.

Kevles D. 1985. *In the name of eugenics: Genetics and the uses of human heredity.* Alfred A. Knopf, New York.

Long Island Biological Association Executive Committee. 1944. In memory of Charles Benedict Davenport. *Science* **99:** 195.

Love A.G. and Davenport C.B. 1919. Physical examination of the first million draft recruits: Methods and results. (Compiled under direction of the Surgeon General, War Department.) *Office of the Surgeon General, Bull.* No. 11.

MacDowell E.C. 1946. Charles Benedict Davenport 1866–1944: A study of conflicting influences. *Bios* **17:** 1–50.

Mayer A.G. 1903. A tropical marine laboratory for research. *Science* **17:** 655-660. (Davenport's letter is on page 657.)

Morgan T.H. 1907. Review of *Inheritance in Poultry* by C. B. Davenport. *Science* **25:** 464–466.

———. 1925. *Evolution and genetics*, pp. 200–207. Princeton University Press, Princeton, New Jersey.

Paul D.B. 1984. Eugenics and the Left. *J. Hist. Ideas* **45:** 567–590.

Paul D.B. and Kimmelman B.A. 1988. Mendel in America: Theory and practice, 1900–1919. In *The American development of biology* (ed. R. Rainger et al.), pp. 281–310. University of Pennsylvania Press, Philadelphia.

Pearl R. 1928. *The present status of eugenics.* The Sociological Press, Hanover, New Hampshire.

Riddle O. 1944. Charles Benedict Davenport. *Science* **99:** 441-442.

Rosenberg C.E. 1961. Charles Benedict Davenport and the beginning of human genetics. *Bull. Hist. Med.* **35:** 266–276.

Terman L.M. 1916. Excerpt from Chapter 1: The uses of intelligence tests. In *The measurement of intelligence.* Houghton Mifflin, Boston; full text can be found at http://psychclassics.yorku.ca/Terman/terman1.htm.

Weldon W.F.R. 1893. On certain correlated variations in *Carcinus maenas. Proc. R. Soc. Lond.* **54:** 318–329.

Wilson P.K. 2002. Harry Laughlin's eugenic crusade to control the "socially inadequate" in Progressive Era America. *Inst. Jewish Policy Res.* **36:** 49–67.

Witkowski J. A. 1999. *Illuminating life; Selected papers from Cold Spring Harbor* (1903-1968). Cold Spring Harbor Laboratory Press, Cold Spring Harbor, New York.

The Eugenic World of Charles Benedict Davenport

Elof A. Carlson

Stony Brook University, Stony Brook, New York

Editors' Note: Eugenic ideas have a long history, being expressed in Plato's *Republic*, Aristotle, ancient Sparta, the Roman Law of Twelve Tables, as well as in the utopias of the Renaissance, such as Thomas More's *Utopia* and Tommaso Campanella's *City of the Sun*. In early 20th century America, a potent mix of new and increasing waves of immigration, new and great wealth, racial issues, the developing social role of the state, and the birth control movement confronted the new sciences of genetics and evolution. To many, here was a way to address the problems of society—"better breeding" solutions that were "justified" with the latest science. They even had the support of former President Theodore Roosevelt: "I wish very much that the wrong people could be prevented entirely from breeding, and when the evil nature of these people is sufficiently flagrant, this should be done. Criminals should be sterilized, and feebleminded persons forbidden to leave offspring behind them. ...The emphasis should be laid on getting desirable people to breed" (*Outlook*, Jan 3, 1914).

In such an atmosphere, Charles Davenport's Eugenic Records Office enterprise thrived. Beyond his scientific discourse in *Heredity in Relation to Eugenics*, his ideas entered the popular realm in the best-selling book by his friend Madison Grant, *The Passing of the Great Race: or The Racial Basis of European History* (1916), a book that used Davenport's ideas about genetics to create a master plan for ending crime and poverty and supporting racial hygiene.

Elof Carlson describes the context of Charles Davenport's times—the science of the early 20th century that made it possible for him to put forth his ideas regarding eugenics and initially gain wide respect for them. Today, we find many of his views distasteful, with no consideration for the will and wishes of the individual and with an emphasis on decision-making by the state. However, Dr. Carlson, in his discussion of *Heredity in Relation to Eugenics*, points out that although Davenport may have been wrong both morally and scientifically in the area of eugenics, he was often right, even prescient, in the area of human genetics and can be considered one of the early founders of this discipline.

> *The commonwealth is greater than any individual in it. Hence the rights of society over the life, the reproduction, the behavior and the traits of the individuals that compose it are, in all matters that concern the life and proper progress of society, limitless, and society may take life, may sterilize, may segregate so as to prevent marriage, may restrict liberty in a hundred ways.*
>
> DAVENPORT, P. 267

DAVENPORT'S BOOK, *Heredity in Relation to Eugenics,* appeared in 1911. He wrote it, I believe, to launch the study of eugenics at Cold Spring Harbor, where he had become the founding Director of the Station for Experimental Evolution of the Carnegie Institution of Washington in 1904. Davenport's interests in eugenics predate the publication of his book by at least 8 years. He met Francis Galton and Karl Pearson on a visit to London in 1902, a visit that strengthened his faith in Mendelian genetics and quantitative approaches to biology. Davenport was one of the first to endorse the rediscovery of Mendelism in 1900, and he studied chickens to apply Mendelism to animals. But Davenport was motivated by higher ambitions. He wanted to serve humanity by applying the new science of Mendelism to human beings and so better the world. This idealism permeated the eugenics movement that had begun, through Francis Galton's efforts, in Great Britain in the 1880s. Galton coined the term "eugenics" and saw it as a moral and philosophical movement. Both Galton and Davenport shared a belief that humanity should take human evolution into its own hands and transform humanity through the preservation and enhancement of its best hereditary features.

THE HISTORICAL VIEW OF DAVENPORT AND EUGENICS IS UNFLATTERING

In the verdict of history, Davenport is seen in a very different way, not as an idealist but as an opinionated mischief maker whose eugenic movement (often called by historians of science "the American eugenics movement") fed the prejudices of middle-class America and led to a dubious compulsory sterilization movement and a blatant discriminatory immigration policy in the 1920s that effectively barred eastern and southern Europeans from entering the United States. The launching of the American eugenics movement had its antecedents in 19th century American intellectual thought (David Starr Jordan, President of Stanford University, and Alexander Graham Bell among them), but Davenport was recognized early in the 20th century as the American leader of that movement. It led to the founding of the Eugenics Record Office (ERO) in 1913, one of the Cold Spring Harbor institutions directed by Davenport, and to the appointment of his pupil, Harry Laughlin, as Superintendent of the ERO. It also led to the adoption of eugenics as an accepted scientific discipline taught in more than 40 leading U.S. universities.

Laughlin is universally perceived as the Iago of the American eugenics movement. He was early the point man to lobby legislatures for passage of compulsory sterilization laws, and he helped design the "model eugenic law" (Laughlin 1922) that became the testing instrument for the U.S. Supreme Court's 1927 emphatic Buck v. Bell 8–1 decision upholding the right of the state to sterilize its least desirable citizens (Lombardo 1985). Later, he was the advocate who went to Congress as an "expert witness" to testify about the inferiority of ethnic groups who had emigrated to the United States. This testimony supported the passage of the 1924 National Origins Act, which remained in force until 1952. Laughlin worshipped Davenport and bolstered Davenport's insecurities by being his strongest supporter. I do not doubt that Davenport shared with Laughlin a common outlook on the core value of negative eugenics that the state had a responsibility to isolate the alleged genetic harm residing within those classes and individuals designated as "the unfit," a view that is clearly stated by Davenport in the opening quote to this essay. Had Davenport fundamentally disagreed with Laughlin, he would have silenced him and Laughlin would have followed Davenport's wishes.

Historians of science, as the 21st century begins, disagree on the interpretation of the eugenics movement. One school, initiated by Daniel Kevles in his book, *In the Name of Eugenics*, portrays the American eugenics movement as largely a middle- and upper-class undertaking based on a distaste for social failures who are a financial drain on taxpayers and a threat to the smooth working of society (Kevles 1985). Those social failures included the poor (or classes of them known as paupers, beggars, and vagrants), the psychotic (then known as lunatics or the insane), the mentally retarded (then known as the feebleminded), and recidivist or sexual criminals. In its more extreme form, this interpretation, exemplified by Edwin Black in his book, *War against the Weak* (2003) sees (almost certainly incorrectly) the American eugenics movement as a conspiracy launched at the beginning of the 20th century by the wealthy (Harriman, Carnegie, Rockefeller, and Kellogg families) to establish a master race of Aryans, and the export of that doctrine to Europe, especially Germany, culminating in the Holocaust (Black 2003).

Another school, which includes William Schneider in his *Quality and Quantity: The Quest for Biological Regeneration in Twentieth-Century France* (Schneider 1990) and myself in *The Unfit: A History of a Bad Idea* (Carlson 2001), portrays eugenics as a mixture of sometimes contradictory beliefs and movements that used the term eugenics as a broad umbrella, as did the American Democratic and Republican parties throughout most of the 20th century. It included positive eugenicists (who advocated encouraging those with the traits of eminent men and women to breed) and negative eugenicists (who wanted to stop those with undesirable traits from breeding), Malthusian and anti-Malthusian advocates, birth control enthusiasts, race suicide pessimists, and equally ardent environmentalists and hereditarians. Part of the confusion in these interpretations is the tendency to equate bigotry against others (religion, class, ethnicity, or race going back to biblical times) and fear or loathing of social failures as essential aspects of eugenics. Those prejudices are far older than the eugenics movement. What distinguished the eugenic movements were the attempts by geneticists and those who admired the work of geneticists to interpret human behavior largely, if not exclusively, as an outcome of heredity and to enforce those beliefs by law—especially the sterilization laws passed by many states, as well as the federal immigration quotas. One should also recognize that throughout the history of the eugenics movement (roughly 1880–1950) geneticists themselves were diverse in their support or rejection of eugenic applications by the state or advocacy by different components of the eugenics movement.

DAVENPORT CHAMPIONS MENDELISM

Witkowski (pp. 35–58 of this volume) describes the arc of Davenport's life. Here I will discuss those aspects of his career that shaped Davenport's attitudes to the applications of Mendelian genetics to human beings. Davenport's training as a naturalist was in the tradition founded by Louis and Alexander Agassiz at Harvard. Although Louis Agassiz was not a supporter of the theory of evolution, he recognized the antiquity of fossil animals and argued for a series of catastrophes and replacements by separate acts of creation. Davenport, like most of his fellow students, was an ardent Darwinian, and this group sought new ways to study evolution, especially by experimental means. Also permeating the education of Davenport and his generation was the conceptual revolution in thinking about heredity brought about by the ideas of August Weismann. Weismann successfully demolished the views of Jean-Baptiste Lamarck that the environment directly alters heredity. Weismann demonstrated by experimentation—cutting off the tails of mice—that no changes were evident even after six generations of crossing mutilated mice. He also cited the practices of circumcision, earlobe piercing, foot binding, and other practices that were carried out for centuries or millennia in different cultures or religions. In each case the newborn child shows no evidence of the treatment of previous generations having had an effect on the latest generation. Instead, Weismann argued that reproductive tissue (the germ plasm) was isolated early from the other tissues of the embryo or adult (the soma) and environmental changes are found in the tissues of the soma but not the germ plasm. Thus muscles can enlarge or skin can become calloused through hard labor, and musical skills can be acquired through practice, but they are not transmitted to the offspring. Each child must learn anew to carry out the physical labor or practice the musical instrument to attain proficiency.

It should be no surprise that these two themes—experimental evolution and the theory of the germ plasm—were formative in the 1890s education that Davenport received and that they would dominate his thinking as a scientist. In 1900 when Mendelism was rediscovered and confirmed in Europe, it rapidly found its support in the United States where a large network of federal agricultural stations at state universities were ready to apply Mendelism to agriculture. It was not eugenics that initiated Davenport's interest in natural history; it was evolution and the opportunity to do experiments that led to new knowledge that inspired the young Davenport. Many of these ideas are expressed in Davenport's 1901 *Science* paper "Zoology of the Twentieth Century," in which he advocated the study of evolution with "comparative observation, experimentation, and a quantitative study of results." He predicted that new problems would be studied with new methods, including comparative physiology, animal behavioral studies, control of biological processes, and advances in animal ecology. He also made a plea for better financial support of science in the new century (Davenport 1901).

Davenport was an enthusiast for teaching what he knew and he made a reputation by reaching out to high school teachers and bringing to them the excitement of Mendelian experimentation. Davenport was also driven by ambition to be successful to live up to his father's expectations, and his colleagues saw him as a humorless, energetic, and efficient scientist and administrator. Davenport sought funding for his interests and cultivated the attention of Jordan and other senior scientists and scholars. He discovered another talent, that of fund-raising, and he combined his engineering and naturalist backgrounds to advantage in designing and overseeing the building of Cold Spring Harbor with funding from Harriman and Carnegie family philanthropic gifts. He was an efficient administrator and recruited a staff of independent scientists to the Station for Experimental Evolution at Cold Spring Harbor. (The Station later became the Department of Genetics of the Carnegie Institution.) The Laboratory quickly became respected nationally and internationally for the quality of that research, and with justification Davenport should be recognized as a significant contributor to the rapid development of classical genetics (Witkowski 2000).

Unfortunately Davenport was insecure and could not bear criticism of his scientific, social, or administrative activities. As his biographer Carleton MacDowell noted, "Effective operation of a group depends on mutual confidence, but this is not inspired by a leader whose own deep lack of confidence is covered by assuming a role of great independent and surpassing ability" (MacDowell 1946, p. 33). He liked to exert his authority from the top down and surrounded himself with a staff (for the nonresearch components of the complex he built) who looked up to him and admired his work and who would not think of criticizing his views. This character flaw was self-defeating (MacDowell 1946). He alienated other geneticists by his imperious style and his rigidity. Because he refused to take criticism of his own work, he continued to do (or he accepted) shoddy scientific work in the field of eugenics. He engaged in polemic retorts to criticism of British eugenicists who ridiculed his ideas that traits such as carpentry, seafaring, wanderlust, and other social traits were fundamentally genetic ones. His friends, like Thomas Hunt Morgan, who were embarrassed by his views and who knew how touchy he was to criticism, resigned from his board of advisors on eugenics matters, claiming they were too busy or had other interests that dominated their time. There were also a significant number of geneticists (and other scientists) in the first half of the 20th century who had an "ivory tower" view of university research. Scientists, they felt, should seclude themselves from society, and devote themselves only to "pure" science and making discoveries that add to our general knowledge rather than be directed to applications. In their mind, Davenport was too close to politics and the corrupting effects of motivations (especially power and the pursuit of money) that guided the applied sciences.

Despite these tensions over his administrative style and his root beliefs about the applications of genetics to eugenics, Davenport had considerable support because of the good done by the basic science coming from Cold Spring Harbor, which was heavily funded by the Carnegie Institution. It included spectacular findings like quantitative inheritance and the nature of hybrid corn by George Harrison Shull. It included the cytogenetic studies of jimsonweeds (*Datura*) by Albert Blakeslee and John Belling that contributed to an understanding of plant evolution. Davenport was too busy after 1910 as an administrator to carry out major experiments in the laboratory, but he realized that he could do work in eugenics because it was not an experimental science and it relied chiefly on pedigree analysis, an activity that could be taken up whenever he had the time. As a consequence, Davenport shifted in activity and reputation from 1910 to 1940 from one of the founders of Mendelism in America to become the chief advocate of the American eugenics movement. During the years 1910–1930, Davenport represented a popular enthusiasm for using the new science of genetics for solving failed social efforts to address some of the most serious problems in American society: the submerged ten percent of the population who constituted the socially inadequate individuals and classes known as the "unfit." The promotion of compulsory sterilization and restrictive immigration seemed the right approach in the 1920s, for everyone from Presidents of the United States down to average voters.

Things changed with the 1930s. The Great Depression vastly increased the homeless, the unemployed, and the vagrant population of America. America was revolted by the excesses of hate groups like the Ku Klux Klan, whose leaders fell from grace in Indiana and other northern states where they had had a popular appeal after Griffith's film *The Birth of a Nation* (1915; also known as *The Clansman*) celebrated their family values and patriotism. The American eugenics movement seemed dated, flawed, and ineffective to geneticists who recognized the complexity of gene interaction to the characters they regulated or shaped. The rise of virulent racism and anti-Semitism in

Germany through the Nazi Party made the eugenics movement seem indistinguishable from Nazi race hygiene (which Davenport and Laughlin initially admired). To the embarrassment of the Cold Spring Harbor staff, Harry Laughlin was awarded an honorary medical degree in 1936 by the University of Heidelberg, then under Nazi control, for his work in eugenics. Laughlin was told by the State Department not to go to Germany to pick it up and he instead received his honor at the German embassy in Rockefeller Center in New York. His model eugenical sterilization law had been adopted by the Nuremberg racial hygiene codes of 1933. Intellectual support for the eugenics movement began to wither, and Davenport was under pressure from the Carnegie Institution, which in turn was being challenged by critics who did not want the Eugenics Record Office to promote what they felt was a questionable science, if not a pseudoscience, of eugenics. I have no doubt that the ever-insecure Davenport felt deeply wounded in his last years (he retired in 1934) at Cold Spring Harbor as he faced mounting hostility and ridicule for his eugenics empire. Davenport had entered a political world when he embraced the tenets of a eugenics that sought government action, and he paid the price that politicians do when their views fall out of favor. He became the Herbert Hoover of eugenics.

HOW SHOULD WE ASSESS DAVENPORT'S *HEREDITY IN RELATION TO EUGENICS?*

The title of Davenport's book in 1911 reflected his dual interests. In the 19th century when eugenics was named, there was no science of heredity that made sense of many contradictory observations. Some traits, like height and weight, were quantitative—they varied continuously and smoothly; other traits were "sports" or suddenly arising changes (mutations), like the short-legged Ancon sheep. Some traits were called atavisms or "throwbacks" to alleged ancestral forms. Some traits, such as color blindness, were associated with sex—they were usually seen in males and rarely in females. The birth of Mendelism in 1900 followed by the chromosome theory of heredity (1902–1903) and the discovery and interpretation of sex chromosomes (1905–1910) suddenly shifted heredity to its new name, genetics (introduced by

British geneticist William Bateson in 1906) and classical genetics began to emerge. Davenport was in the thick of this activity when he wrote *Heredity in Relation to Eugenics.* He had many insights into human genetics and he should be acknowledged as a founder of that field. His book reads like the first text in human genetics, but Davenport's title for his book is significant. He recognizes that if eugenics has a role in society, it has to have a scientific foundation. That foundation is the field of heredity and its newest findings. It is worth noting that approximately 204 pages of the book are devoted to genetics and about 103 pages are specifically given to eugenics.

In his introduction we learn of Davenport's values. New scientific knowledge should be applied to social problems. What are those problems? They are "the problems of the unsocial classes, of immigration, of population, of effectiveness, of health and vigor" (p. iii). He asserts that modern medicine "has forgotten the fundamental fact that all men are created *bound* by their protoplasmic makeup and *unequal* in their powers and responsibilities" (p. iv; Davenport's emphasis). These are not original insights. They are the views he has adopted from David Starr Jordan and Alexander Graham Bell, and their phrasing ("protoplasmic makeup") reveals their 19th century roots. He promotes the new Eugenics Record Office and its mission, especially its need to amass pedigrees from his readers. "We do not appeal primarily to physicians for this information but to the thousands of intelligent Americans who love the truth and want to see its interests advanced.... Thus, every one can share in the eugenics movement" (p. v). He seeks information on a variety of traits: "short stature, tallness, corpulency, special talents in music, art, literature, mechanics, invention and mathematics, rheumatism, multiple sclerosis, hereditary ataxy [ataxia], Ménière's disease, chorea of all forms, eye defects of all forms, otosclerosis, peculiarities of hair, skin, and nails (especially red hair), albinism, harelip and cleft palate, peculiarities of the teeth, cancer, Thomsen's disease, hemophilia, exophthalmic goiter, diabetes, alkaptonuria, gout, peculiarities of the hand and feet and other parts of the skeleton" (pp. iv–v).

It is a remarkable list. The absence of paupers, vagrants, psychotics, and the feebleminded can be explained by the audience he addresses. He cannot ask the "socially inadequate" to respond, and he does not expect many (if any) of those social failures in the families of those reading his book. Instead he focuses on the readers, their talents, and individual and relatively rare behavioral abnormalities (ataxia, Ménière's disease). Davenport looks at visible traits; very few physiological traits were known (he cites those—diabetes and alkaptonuria) and biochemical genetics was not developed until the late 1930s. Most of these traits we encounter today in human genetics. They are not social constructs like pauperism and vagrancy. There really are conditions with genetic determination—for example, albinism, cleft lip and palate, alkaptonuria, and a variety of skeletal malformations. We can look them up today in Victor McKusick's catalog *Mendelian Inheritance in Man* and learn of the latest scientific literature for diagnosis and treatment, as well as the genetic or molecular basis of the mutations involved.

We find strange bedfellows in the list. Why red hair? Davenport tells us later that so many redheaded people associate it with their temper. We also wonder about his choice of talents for analysis. Are these occupations and interests familial traits because children like to imitate their parents? Are these familial traits because a special gene or cluster of genes makes them more likely to be expressed among a disproportionate number of the descendents? But as this book came out in 1911, was it wrong to speculate that these familial traits might be inherited? What if their pedigrees revealed a simple Mendelian recessive or dominant mode of inheritance? We cannot fault Davenport for wanting to know. Today we would include a large number of birth defects or congenital malformations largely undiagnosed in Davenport's day because the infants died shortly after birth or were of a biochemical nature whose diagnosis was then impossible. In 1911 they would have been attributed collectively to a more general phrase such as "constitutional weakness." It was also an era that was just beginning to emerge from the bleak expectation that half of all infants born failed to survive to the first year of their birth. Physicians were far more interested in infectious diseases and the damages of malnutrition than they were in exotic birth defects. And they were right—infectious diseases and malnutrition might be corrected. Who among the readers of Davenport's book (half of them being vulnerable to such an early death had they been born about 1900) would see their survival primarily as the result of a "weak constitution" that got by, rather than a triumph of public health that kept milk, water, and food from teeming with pathogenic microbes?

In his introductory chapter, Davenport praises eugenics as "the science of the improvement of the human race by better breeding" (p. 1). It is Galton's definition he uses. And note that he calls it a science and does not distinguish between pure and applied science. That is not a bias on his part but part of the widespread use of the term "science" for a variety of organized and skilled activities; for example, we refer to both the science and art of navigation. At the time of its birth eugenics was considered a science because it was seeking to determine the beneficial and the detrimental factors in human inheritance. We can criticize eugenicists for being careless in that effort, and certainly they often were. Davenport falls into that trap when he makes a sweeping generalization that "we have become so used to crime, disease and degeneracy that we take them as necessary evils. That they were so in the world's ignorance is granted; that they must remain so is denied" (p. 4). The literary elegance is powerful in that assessment, but it hides his values. Davenport sees social problems as complex clouds of causes that can be reduced or analyzed by science. This is not necessarily wrong, but a century later we would be more cautious in our enthusiasm that everything can yield to a simple scientific analysis. It would be more accurate if Davenport had recognized or tried to disprove that a considerable amount of injustice—cronyisms, nepotism, sexism—and other forces are at work in making crime such a popular activity among those who feel cheated, ignored, or victimized. This does not justify the criminal acts, but it may certainly point to different reasons for its occurrence than an alleged defective protoplasm for the criminal's motivation.

We cannot equate Davenport's support for limiting the spread of alleged (and real) genetic defects in the human population with the murderous applications of race hygiene under Nazi Germany. He is quite specific in the values he rejects. One of them is the destruction of the unfit before or after birth (p. 4). He supports instead the state control of the unfit by isolation in sexually segregated asylums. He is ambivalent about sterilization because he feared the vasectomized males or tubally ligated females would become sexually promiscuous without the worry of children to support and that this would spread sexually transmitted diseases and promote vice. (He shifted to favor sterilization in the 1920s.)

DAVENPORT USES MENDELIAN GENETICS INAPPROPRIATELY

Despite his knowledge of Mendelism and X-linked inheritance, Davenport seems muddled in the way he expresses his human traits. He states, "[W]hen both parents have low grades of a trait-complex the children will have low grades of that complex" (p. 25). I think what Davenport meant was that if the parents are both homozygous for a recessive trait, their progeny will also be homozygous recessive. But Davenport does not use these genetic terms (they were already in use for at least a half dozen years by both European and American scientists). Throughout the book he uses terms like "low grades," "negative" traits (by which he means recessive traits), and "positive" traits (by which he means dominant traits). He also does not distinguish simple traits from complex traits with a few exceptions (not many were known in 1911). When vague terms like "low grade" are used, they can lead to misconceptions or errors. Later in the book, Davenport gives one hundred or more pedigrees of human conditions that vary from behavioral traits to physical traits. His use of "low grade" makes him favor an interpretation biased to single-gene conditions and to a simple dominant and recessive inheritance for the overwhelming number of these pedigrees. He also does a sloppy job of analyzing them. Many flatly contradict a simple recessive or dominant mode of inheritance. His use of those vaguer terms makes him fall back on his 19th century training, and he speaks of defective germ plasm as a synonym for the more precise terms he fails to use or deliberately avoids using. As MacDowell noted regarding the preparation of *Heredity in Relation to Eugenics* (MacDowell 1946, pp. 30–31), "its usefulness was reduced by its hasty preparation and the lack of critical judgment in lumping together indiscriminately cases with ample and with insignificant evidence."

Reading the pedigrees that form the bulk of his book is an eye-opener for both the historian of science and the present-day geneticist. We are shocked by these naïve pedigrees for behavioral traits that most of us would associate with family traditions and mimicry rather than innate factors. Is it really genetic that business people tend to have offspring who go into businesses? Is it really warrior genes rather than admiring a soldier's life that leads the children of a career soldier to follow the example of their parent? Do physicians pass on genes for healing or do they manifest a dedicated lifestyle that appeals to their children?

What surprises me in looking at these pedigrees is the degree to which Davenport's bias colors his interpretation of pedigrees. For example, many pedigrees show a preponderance of males, but the obvious male-to-male transmission in these pedigrees is not Y-linked and excludes X-linked inheritance. (This excess of males is hardly surprising in an era that limited the occupations of women.) However, Davenport does not see these contradictions. I can only imagine that he was so impressed by the high proportion of members of the family following in the same talent or profession that he ignored these conflicts with Mendelian and sex-linked inheritance.

DAVENPORT'S EUGENICS INCLUDE SOME LEGITIMATE TRAITS

If those were the only pedigrees that Davenport presented, he could easily be dismissed as naïve or uncritical. But as we follow up his human genetic inventory we see pedigrees for albinism (p. 38), hereditary ataxia (p. 99), Ménière's disease (p. 101), Huntington's disease (p. 103; see also his reliable article with Elizabeth C. Muncey on that disease [Davenport and Muncey 1916]), and many other traits that are unambiguous in their autosomal dominant or autosomal recessive inheritance as well as the hemophilia, color blindness, and muscular dystrophy that follow X-linked inheritance. Here Davenport is critical and his analysis is no different than that of a genetic counselor today. There is also in Davenport's lengthy presentation of pedigrees of the feebleminded, epileptic, and other socially unfit families a bias to like-for-like inheritance with glaring contradictions to Mendelism. Why do both sides of the family show the dominant trait? Why are the horrible living conditions dismissed as a cause of the depravity of mental deficiency and instead assigned as consequences of an alleged hereditary insufficiency?

The medical pedigrees (rather than the social failure pedigrees) stand out as a compendium of human and medical genetics as it existed in 1911. A good many of the citations come from France, Germany, and Great Britain, but quite a few are from American physicians who took the time to compile family histories and send them on to Davenport. We like to think that human genetics as a science is a field that had its origins after World War II and the establishment of the American Society for Human Genetics, but a careful reading of Davenport's book would have to acknowledge him as a founding contributor to that science. This is brought home in even greater impact by Davenport's analysis of the geographical distribution of traits. Davenport is clearly aware of the founder effect—a loss of genetic variation and change in genotype frequencies when a small number of individuals form a new population—before it was named, and he cites example after example of the constraints of isolation on islands and peninsulas, in secluded valleys, and through religious, racial, or other cultural barriers to the free flow of genes between populations forcing small isolates into a high degree of consanguinity.

Davenport's analysis is masterful in these cases. He notes an excess of deaf-mutism and hermaphroditism in Martha's Vineyard (p. 188) and of dwarfism in the outlying regions of the Chesapeake Bay Peninsula and states "the result is determined by the specific defect in the germ plasm of the common ancestor" (p. 188). He rejects the idea of a pure European or ethnic stock and cites the assimilation of white-, yellow-, black-, and brown-skinned slaves brought back, freed, and marrying with the Roman people throughout the Roman Empire. He sees wanderlust not as a vice but as a virtue that brought immigrants from the Old World to the New World and enriched America through its many strains of largely healthy and well-motivated travelers who sought a better life in America.

EUGENICS REMAINS AN ISSUE

Davenport's book is still worth reading today because it reveals how complex were the issues raised by eugenics and how varied were the motivations of those who looked to eugenics as a solution to social problems. It is also worth reading because we end up appreciating much of the work that Davenport did even if we reject what is naïve and clearly biased by his class prejudices. We also appreciate Davenport for recognizing that human and medical genetics were appropriate topics for geneticists to explore. At the same time we feel disappointed that Davenport does not have the benefit of our hindsight, that he was not more sympathetic to those who were victims of society's indifference and selfishness. Eugenics was largely bad in its outcomes. It did lead to a poorly conceived compulsory sterilization movement. It did lead to immigration laws that were based on ethnic bias. It did lead to a "blame the victim" mentality for avoiding tough social issues. It targeted social failures more often than it targeted proven single-gene defects in humans. It targeted those who had a low probability of reproducing their defects and avoided addressing the larger population (virtually all of humanity) who harbored the heterozygous recessive genes that would appear in virtually all future generations. It also failed to inspire the eminent to have more children, and it settled for promotion of 4H fair "fitter families" who were a mere notch above mediocrity, similar to Sinclair Lewis's portrayal of the aspiring middle class in his novel *Babbitt* (1922).

At the same time eugenics asks questions that most geneticists and the general public are afraid to ask today and even more afraid to investigate. We do have spontaneous mutations occurring and there are good estimates that at least one in ten of our sperm or eggs carries a new mutation that arose spontaneously as a consequence of DNA replication, or, less frequently, agents in our environment, internal or external, that induce mutations. We also know that most of these mutational changes are harmful rather than beneficial or neutral, and that the altered gene products lead to altered function. Add to this our belief in the Golden Rule that guides most ethical systems; we want to cure or ameliorate (if at all possible) a harmful condition when it occurs in our children or us. We gladly wear corrective lenses, use hearing aids, slip in false teeth, and take a variety of medications for our ailments to give us some semblance of normalcy consistent with our desire to function as close to our ideals as possible. Those who were (and are) idealistic in their eugenic thinking, worry about these "corrections." New mutations in plants and animals, if harmful, are selected out by natural selection. But human beings have escaped such natural elimination and the detrimental effects accumulate in the population. In the long run, the concerns about our mutations will have to be addressed because the technology for genetic services and reproductive decisions has increased enormously since the 1980s. It will continue to provide more options and more decision-making as the Human Genome Project reveals a wealth of information on the disorders in McKusick's catalog and about a host of chronic conditions in which gene interaction (including environmental effects) is the most likely hereditary component of complex traits.

We know that we cannot keep adding prosthetic components to our bodies (and eventually our brains) without creating enormous burdens to future generations and individuals that would have to include these health costs in their budgets. Eventually society will have to adopt some form of individually chosen artificial means—whether germinal choice, gene replacement with the germ line, or in vitro fertilization/embryonic cell screening—if assortative mating (nonrandom conscious or unconscious choice of a partner) is insufficient to address this worry. But that is why research is so important. Eugenics failed because it was not scientific enough. It knew too little and it plunged into social policy with a self-assurance that was unwarranted. We still do not know how much of a "genetic load" we can tolerate without an overwhelming financial and social burden to patch ourselves up. We do not know the molecular biology of the most important behavioral traits. We properly fear tinkering with the good to make it better when that concept of the better is based on traits whose genetics or molecular biology is virtually nonexistent. We cannot and should not make social policy for eugenics based on hopes rather than on proven science.

If Davenport (and the American eugenics movement) failed and our reading of Davenport's book has some merit outside the history of science itself, it may well be the caution it calls for when dealing with human and medical genetics. We need the knowledge; in the long run it is more beneficial to know rather than to remain ignorant. Research in human genetics is the only way to acquire knowledge of how our genes work and how our traits are shaped. That research has to be guided by ethical principles (e.g., we cannot breed people out of curiosity the way we breed fruit flies or mice—or tell who to marry whom). The values of our generation respect the autonomy of the individual in making reproductive choices and in seeking the information they need to make those choices. If there is a role of eugenics in our time and in the immediate future as the human genome analysis reveals an avalanche of new knowledge, it is in maximizing that information and its availability to those who need it and minimizing the temptation to use the State as the means of enforcing eugenic ideals.

REFERENCES

Black E. 2003. *War against the weak: Eugenics and American's campaign to create a master race.* Four Walls, Eight Windows, New York.

Carlson E.A. 2001. *The unfit: A history of a bad idea.* Cold Spring Harbor Laboratory Press, Cold Spring Harbor, New York.

Davenport C.B. 1901. Zoology of the twentieth century. *Science* **14:** 315–324.

Davenport C.B. and Muncey E.B. 1916. Huntington's chorea in relation to heredity and eugenics. *Am. J. Insanity* **73:** 195–222.

Kevles D.J. 1985. *In the name of eugenics: Genetics and the uses of human heredity.* Alfred A. Knopf, Inc., New York.

Laughlin H.H. 1922. Model eugenical sterilization law. In *Eugenical sterilization in the United States*, pp. 446–452. Psychopathic Laboratory of the Municipal Court of Chicago.

Lombardo P. 1985. Three generations, no imbeciles: New light on Buck v Bell. *New York University Law Review* **60:** 31–62.

MacDowell E.C. 1946. Charles Benedict Davenport 1866–1944: A study of conflicting influences. *Bios* **17:** 1–50.

Schneider W.H. 1990. *Quality and quantity: The quest for biological regeneration in twentieth-century France.* Cambridge University Press, New York.

Witkowski J.A. 2000. *Illuminating life: Selected papers from Cold Spring Harbor (1903–1969).* Cold Spring Harbor Laboratory Press, Cold Spring Harbor, New York.

Davenport's Dream

Maynard V. Olson
Department of Medicine and Genome Sciences,
University of Washington, Seattle

Editors' Note: Eugenics is basically about classifying humans, but how can we classify ourselves objectively? Early 20th century eugenics schemes were often culturally laden. Willet Hays, in his 1912 "Constructive Eugenics" paper, proposed assigning unique 11-digit "number names" to each person on earth to express "the individual value of the general efficiency of the person" and provide his or her genealogical lineage. Hays posited that this system would facilitate "mating with those of equal genetic excellence, and the more rapid multiplication of their numbers." Davenport's classification scheme listed categories such as class, criminality, seafaring ability, and many other qualities that had nothing to do with heredity.

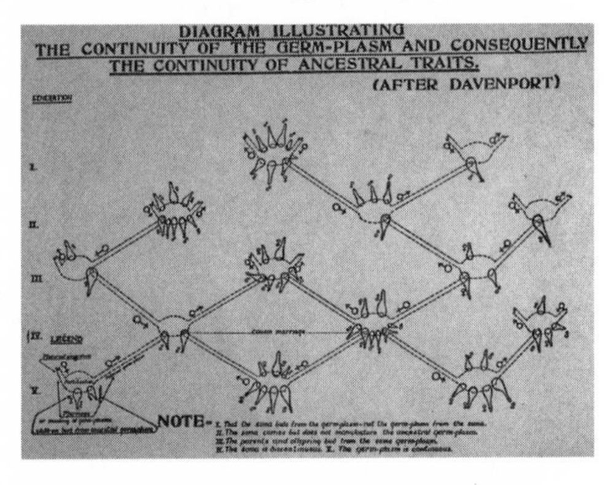

By mid-century, the idea that information theory might be a way to study human populations gained some credence, with Claude Shannon (the founder of information theory) even spending the summer of 1939 at Cold Spring Harbor Laboratory to learn genetics. Discovery of the structure of DNA in 1953 and the "digital" nature of the genetic code made it possible for human genetic variation to be studied in a more biologically basic and objective way. This approach culminated in the Human Genome Project, which produced the complete letter-by-letter DNA sequence of the human genome, and the International HapMap Project, which aims to "identify and catalog genetic similarities and differences in human beings" (http://www.hapmap.org/thehapmap.html.en).

As Dr. Olson notes in this essay, Charles Davenport's ideas about heredity were "solidly rooted in the principles of Mendelian inheritance." But he was a man of his time when it came to the cultural issues that intersected with the new science of heredity, resulting in views about human classification that we find repugnant today. However, almost 100 years later, these issues have not gone away. This essay emphasizes the importance of realizing Davenport's mistakes (the "red flags" in *Heredity in Relation to Eugenics*), focusing on obtaining objective knowledge about human genetics, and participating in open, critical discourse about the power of this knowledge and its application in improving human health. For, as Dr. Olson points out, "the real debate about eugenics still lies ahead."

> *If one is provided with a knowledge of the methods of inheritance of unit characters it might seem to be an easy matter to state how each human trait is inherited and to show how any undesirable condition might be eliminated from the offspring and any wished for character introduced. Unfortunately, such a consummation cannot for some time be achieved.*
>
> DAVENPORT, PP. 23–24

THE HUMAN GENOME SEQUENCE AND dbSNP—the rapidly growing database of natural sequence variants—fill once-gaping holes in Davenport's dream. Here is the raw material for a real science of human genetic perfection. There is still much to learn about how to sculpt these materials into ideal human forms, but we now have the tools with which to pursue this forbidden knowledge. Uninformed as we still are about the complexities of human inheritance, once scientists bring big questions into focus and develop needed tools, ignorance is often swept away with remarkable speed. Even if it takes a generation or two to work out the details of human genotype–phenotype relationships, eugenicists are patient people. They need to be in order to envision breeding a long-lived species of slight reproductive potential to biological perfection.

Certainly, the next 50 years will see a sweeping expansion of our knowledge of human biology and attendant ability to alter its course. Inevitably, there will be a revival of eugenic ambitions. Hence, now is a good time to revisit the lessons offered by Davenport's early 20th century burst of eugenic enthusiasm. In particular, I find it instructive to identify, albeit with the benefit of 100 years of hindsight, the warning flags in Davenport's book that foreshadowed disaster.

IN DAVENPORT'S BOOK, THREE RED FLAGS STAND OUT

In my reading of *Heredity in Relation to Eugenics*, three red flags stand out. Each remains relevant today because all three still flutter over contemporary discussions of genetics and society. The first red flag is that Davenport deeply commingles science and social values. This commingling occurs on many scales—the book, the chapter, the sentence, and even the choice of words. The second, and perhaps most ubiquitous, red flag is that Davenport's thinking is often hopelessly muddled. This muddling is independent of the validity of his facts and our sympathy, or lack thereof, for his goals. Finally, the third red flag is that Davenport overpromises what genetics could achieve even if it were vigorously applied to societal problems.

Commingling of Science and Social Values

With respect to the first flag—the commingling of science and social prejudices—race is the touchstone. The subject first arises when Davenport discusses the inheritance of skin color. In this section, he points out correctly that, although the skin color of a child with a black and a white parent is predictably intermediate, the apparent blending of traits is illusory: the "determiners of unit characteristics" (i.e., allelic variants of genes; Davenport had not yet adopted the word "gene," which had been introduced by Wilhelm Johannsen just 2 years before publication of *Heredity in Relation to Eugenics*) remain discrete, and, hence, the intermediate phenotype does not breed true. Subsequent intermarriages between mixed-race children are likely, in just two or three generations, to produce some children with skin that is almost fully "white" and others whose skin is almost fully "black." So far, so good. But then the trouble begins: "So far as skin color goes," these light-skinned progeny of a mixed-race lineage "are as truly white as their greatgrandparent and it is quite conceivable that they might have mental and moral qualities as good and typically Caucasian as he had" (pp. 37–38). This sentence requires multiple levels of deconstruction. In the second half of the sentence, Davenport lends the authority of scientific analysis to three propositions. The first of these is blatantly racist: Davenport associates good "mental and moral qualities" with Caucasians, a perspective widely shared in early 20th century America and hardly vanquished today.

The second proposition is that a strong form of genetic determinism applies to an extraordinary range of human characteristics. Of course, this belief is a core dogma of the eugenics movement. I distinguish Davenport's genetic determinism from his racism because, quite apart from race, he implies that desirable mental and moral characteristics, in whatever individual or socially defined group they might occur, are heritable—indeed as simply heritable as traits such as skin color. This belief is of particular interest today because genetic determinism is undergoing a revival in the genomic era. The lead article in the December 18, 1994, *The New York Times* "News of the Week in Review" was entitled "The Curse of Living Within One's Genes." The article is accompanied by a cartoon showing a bewildered looking everyman, one of whose ankles is connected to an enormous black ball by a double-helical chain. We are, in the cartoonist's view, prisoners of our genes. The same theme reappears on the April 21, 1997, cover of *U.S. News and World Report*: Printed in bold white letters over a photograph of an endearingly cute infant, dressed in the traditional black-and-white-striped garb of a prison inmate, is the question "Born Bad?" Renewed interest in genetic determinism, which represents a remarkable swing of the social-opinion pendulum from the nurturist views of the 1950s, is undoubtedly a consequence of the DNA revolution that followed the discovery of the double helix in 1953. That the scientific advances of recent decades, extraordinary as they have been, have contributed little to the nature–nurture debate appears not to matter. The public has embraced the idea that we become obese, alcoholic, or antisocial because of our genes rather than because of defective parenting or Calvinist predestination.

The third proposition implied by the sentence quoted above is the most subtle. If we abstract the sentence's scientific message away from Davenport's racism and genetic determinism, we uncover a claim that can be paraphrased roughly as follows: If most members of two human groups with different ancestries differ in some visible trait, that trait is a good marker for all the invisible ways in which those groups differ. Hence, after intermarriage between representatives of the two groups, descendents of the initial intermarriage who look more like typical members of one group than the other will also resemble the members of that group in invisible traits of potentially greater importance. This claim is simply wrong, regardless of the definition of the groups or the nature of the visible or invisible traits to which it applies.

Skin color is actually a good illustration of this point. Many studies that roughly follow Davenport's design (i.e., depend on analysis of variations in skin color in the generations that follow intermarriage between two individuals with highly divergent pigmentation) have indicated that skin color is largely controlled by a small number of genes, something on the order of five. The most important genetic variant imparting light pigmentation to individuals of European ancestry was only discovered recently (Lamason et al. 2005): This single gene accounts for roughly a third of the average difference in skin color between individuals of European and sub-Saharan African ancestry, with most of the rest of the difference still unexplained. Nonetheless, Davenport's claim is basically correct that after two or three generations of intermarriage between individuals who trace half their recent ancestry to an African lineage and half to a European lineage, the children produced would display nearly the full European–African spectrum of skin colors. Where Davenport is wrong is in generalizing these results from skin color to *any* other trait—not just the ill-defined, race-baiting ones that he mentions. To the extent that there are any heritable traits—for example, patterns of genetic resistance or susceptibility to various diseases—that occur at different frequencies in individuals of European and African descent, there would be no correlation between the skin color of the children in his hypothetical pedigree and the extent to which they possessed the more typically European or more typically African versions of these other traits. We are all a hodgepodge of the genetic variants present in our ancestors, and selection in some line of descent for a particular ancestral trait leads, at most, to individuals with that trait. Other ancestrally variable traits will fall where they may in each individual.

Hence, in one sentence about the consequences of racial intermarriage, Davenport raises all three of the red flags that permeate *Heredity in Relation to Eugenics*: He commingles science and social values, indeed particularly noxious ones; his thinking about both the science and its implications is badly muddled; and he overpromises what genetics could accomplish even if it were applied in pursuit of a biologically feasible and socially acceptable goal. However, the big flag flying over this sentence relates to the commingling of science and social values. My scientific exegesis is not intended to dull the impact of Davenport's racism: Eugenics and racism have been natural bedfellows ever since Galton, a point we should never lose sight of.

A Thicket of Contradictions

If we look beyond Davenport's frequent commingling of science and social prejudice, we quickly encounter the most pervasive red flag in *Heredity in Relation to Eugenics*: Regardless of topic, Davenport is an astonishingly muddled thinker. He is actually most reliable on purely scientific points. In contrast, on the frequent occasions when he wanders into social policy or ethics, Davenport invariably leads us into a thicket of contradictions from which there is no escape. Consider his comments about classifying humans—not just with respect to race but with respect to any traits. Because eugenics is basically about classifying humans, this topic is of central interest to anyone seeking to understand Davenport's thinking. Early in *Heredity in Relation to Eugenics*, Davenport sounds a promising note (p. 8):

> The theory of independent unit characters has an important bearing upon our classifications of human beings and shows how essentially vague and even false in conception these classifications are. A large part of the time and expense of maintaining the courts is due to this antiquated classification with its tacit assumption that each class stands as a type of men. Note the extended discussion in courts as to whether A

belongs to the white race or to the black race, or whether B is feeble-minded or not.

Davenport follows up this admirable summary of the difficulty of classifying humans with a perceptive comment that remains as valid now as it was then: "if experts be called in to give a definition the situation is rendered only worse" (p. 9). However, anyone who expects Davenport to build coherently on this promising start will soon be disappointed. Once Davenport hits his stride, we learn that "a text-book of Human Anatomy must be rewritten for the Chinese, for the Ethiopians, and for the Eskimos, so must it be rewritten for the Rumanian, for the North Italian, for the Norwegian and for the Spaniard" (pp. 246–247). Evidently classification is back in favor. So too are experts—who else would consider cataloging the anatomical distinctions between North Italians and Spaniards?

A similar mental fog envelops the core of Davenport's program. Early in the book, Davenport takes pains to distance himself from negative eugenics (i.e., restricting the reproduction of the "unfit"): "The general program of the eugenist ... is to improve the race by inducing young people to make a more reasonable selection of marriage mates; to fall in love intelligently" (p. 4). Although acknowledging that the program "also includes the control by the state of the propagation of the mentally incompetent," he hastens to reassure us that "[i]t does not imply destruction of the unfit either before or after birth" (p. 4).

Interestingly, as discussed below, destruction of the "unfit" before birth is the one form of eugenics that has gained widespread acceptance 100 years after Davenport wrote. The form of negative eugenics that Davenport advocated, preventing the reproduction of the "unfit," has proven almost entirely unacceptable in modern, liberal societies. This form of negative eugenics comes to dominate *Heredity in Relation to Eugenics* after its upbeat start. As the book progresses, Davenport has less and less to say about inducing young people "to fall in love intelligently," and his real obsession emerges. This obsession, which dominated 20th century eugenics throughout its course, is the social burden of supporting a population that includes "about half a million insane, feeble-minded, epileptic, blind and deaf, 80,000 prisoners and 100,000 paupers" (p. 4). Each entry on this astonishingly heterogeneous list poses an imposing classification challenge. Davenport is well aware of this difficulty, even if he appears oblivious to its implications (p. 257):

> ...laws against the marriage of the feeble-minded are unscientific because they attempt no definition of the class. If feeble-mindedness were always as clearly distinct from normality as polydactylism [the occurrence of extra fingers or toes] then there would be no objection to the law on this score. But this is by no means the case.

Given these difficulties, he affirms that only a fool would attempt to enact eugenic legislation (p. 258):

> Shall we sterilize or forbid marriage to all children whose mental development is retarded as much as one year? That would include 38 per cent of all children, and one of yours, O legislator! Shall the limit be two years of retardation? That would include 18 per cent of the children. ... Is it not reckless to pass such serious legislation in such loose terms?

It would be reckless, indeed, as 20th century experience quickly confirmed. However, Davenport's qualms about state-mandated mechanisms for controlling the reproduction of the unfit pass quickly in *Heredity in Relation to Eugenics*. On the very next page, after warning of the recklessness of eugenic legislation, Davenport proposes the following alternative to sterilization or establishing a genetic-fitness test for marriage: "Such then, would seem to be the proper program for the elimination of the unfit—segregation of the feeble-minded, epileptic, insane, hereditary criminals and prostitutes throughout the reproductive period" (p. 259). Because "the reproductive period" lasts until death for males and spans at least three decades for females, Davenport's proposed remedy actually goes far beyond "forbidding marriage": It envisions lifetime incarceration of many groups whose reliable classification would challenge the most exuberant expert. I emphasize the inconsistencies in Davenport's prescriptions because they have received too little attention. Incoherent pronouncements on public policy are a sign of trouble regardless of content.

Promising the Moon

It is perhaps unfair to accuse Davenport of overpromising. He wrote in 1911, at the dawn of the "genetic century" and his role was that of pied piper rather than technocrat. Davenport would surely have admitted that there remained major uncertainties about what could be practically achieved through eugenics. We have fewer excuses today for our extravagant claims about imminent practical benefits of expanding genetic knowledge. When it comes to "promising the moon," modern geneticists live in a glass house.

Davenport overpromised in the relatively benign style of Progressive Era social optimists. At the time he was writing *Heredity in Relation to Eugenics*, technocrats were coming into fashion, and early 20th century progressives believed that all manner of social ills could be attenuated by rational application of scientific principles. To Davenport, the prospect of human improvement through genetics simply mirrored those that were opening up in agriculture, as breeding programs began to acquire rigorous scientific underpinnings: "The experience of animal and plant breeders who have been able by appropriate crosses to increase the vigor and productivity of their stock and crops should lead us to see that proper matings are the greatest means of permanently improving the human race—of saving it from imbecility, poverty, disease and immorality" (p. 260). Davenport's confidence that the new science of genetics would lead to spectacular improvements in food crops proved well founded. However, the extrapolation to humans is entirely disingenuous. Davenport wanted to emphasize the ability to concentrate positive "determiners of unit characteristics" in particular lineages through selective matings. He did not want to emphasize the extent to which the success of artificial breeding depends on an inextricable link between positive and negative eugenics. Humans did not turn wolves into greyhounds by preserving all progeny along the way who were perfectly good dogs but not particularly fleet of foot.

More generally, the opportunities for applications of positive-eugenic principles that Davenport saw all about him would be unlikely to deliver the promised results even if they were widely adopted. Assortative mating is already ubiquitous in human societies. Indeed, it is difficult to see how eugenicists could improve upon current practice in societies that have developed merit-based systems for segregating talented young adults in elite educational institutions. Even farther out of touch with mate-selection practices in the real world is Davenport's idea that individuals with deleterious recessive traits could have "normal" children by choosing genetically "normal" spouses. Although this practice is an effective means of attenuating the burden of some rare genetic diseases, it is hardly a plausible approach to minimizing the transmission of common traits that are judged undesirable. Would Davenport have wanted his child to marry someone with major heritable deficiencies even if he were confident that the relevant traits were recessive? Davenport's overpromising involved promoting a world in which great benefits could be had at little cost to the elements of society he deemed worthy. That world is not only unjust but illusory.

DO DAVENPORT'S RED FLAGS REMAIN TODAY?

At great human cost, we have learned the most blatant of Davenport's lessons during the 20th century. However, the basic warning signs that mark *Heredity in Relation to Eugenics* as the siren song of a dangerous thinker permeate contemporary discussions of human genetics. I will focus on the first and third of Davenport's red flags: the commingling of science with social values and the overpromising of imminent benefits from expanding genetic knowledge. Muddled thinking is ubiquitous in all eras, thus there is little to be gained by documenting that it continues to flourish today.

The commingling of science and social values now takes a more subtle form than it did a century ago. Indeed, quite perniciously, this commingling has come to be seen as less of a sin than it was formerly judged to be. Mainstream social values have evolved in liberal directions since Davenport's time; hence, many academics now appear confident that the commingling of science and social values can only have beneficent effects. In the "critical studies" movement, the view has even arisen that all claims of scientific objectivity are hollow.

Racial classifications provide a good illustration of contemporary thinking on university campuses, which are awash with the view that there is "no biological basis for race" (Petit 1998). Scientific support for this view comes from the finding—first based on electrophoretic variants of proteins and amply supported by subsequent studies of DNA polymorphisms—that a high proportion of the worldwide variance in human genotypic diversity can be found within every human population. The weakness of this argument is that the residual "between-group variance," although numerically small, is ample enough to allow nearly absolute classification of humans into the same five geographic-ancestry clusters that have long been evident: Africa, East Asia, Oceania, America, and Europe. Of course, this system fails, as expected, when applied to many smaller culturally and geographically defined populations that have extensive recent histories of admixture (Rosenberg et al. 2002).

In reality, we know almost nothing about how important group differences in average genetic makeup are in influencing biologically important characteristics, nor are we likely to learn much anytime soon, given extreme sensitivities about studying the question. The National Human Genome Research Institute (NHGRI) took the unusual measure of stripping group identifiers off the samples in the "Polymorphism Discovery Resource" that was constructed to facilitate early studies of DNA-sequence variation in the human genome (Collins et al. 1998). As discussed above, the patterns of variation in the major geographic-ancestry groups are sufficiently distinctive that these identifiers could actually be easily rediscovered by any investigator. However, to prevent this possibility, the NHGRI required, as a condition for acquiring the DNA samples, that investigators sign an affidavit agreeing "not to attempt to identify ethnicity when they obtain samples and when they access the data derived from the DNA Polymorphism Discovery Resource" (Collins et al. 1998). This radical prohibition against even analyzing data in particular ways generated surprisingly little reaction, given its implications for the core values of science. Presumably the church authorities, too, would have been happy to support development of the telescope as long as Galileo had signed an affidavit promising not to turn it in certain directions.

That human geneticists have become "gun-shy" about race is indisputable. However, many modern geneticists appear equally fearful of what they might learn if they studied everyday genetic differences within any group. Intelligence, athletic potential, physical beauty, social behavior—the very traits of particular interest to Davenport—remain almost wholly out of bounds for serious genetic investigation. Meanwhile, in the absence of significant supporting evidence, we teach our students that the human traits on which society places a high value are too complex to manipulate effectively by old-fashioned selective breeding or its high-tech successors such as in vitro fertilization coupled to pre-implantation diagnosis. Most of these traits, we are told, are heavily modulated by the environment. If not, they are influenced by variation in large numbers of genes that interact in idiosyncratic and unpredictable ways. Indeed, the very definitions of many of the traits that Davenport highlights—intelligence, for example—are commonly said to be so culturally laden as to preclude their biological manipulation.

There is a real risk that these infusions of social values into scientific discussions of human genetic variation have lulled contemporary geneticists into overestimating the extent to which science itself will rescue us from future eugenic temptations. This lulling effect is particularly dangerous because of the tautology on which it is based: We first infuse social values into our science and then look to the science for support of these social values. In this echo chamber, we see good science as an inevitable ally of progressive social policies. Stated another way, if Davenport was wrong about the science, we are spared the need to develop other bulwarks against abuses of genetics.

I see little but danger in attributing the misdeeds of 20th century eugenics to bad science. If one disentangles Davenport's social agenda from his scientific claims and simply ignores his cheerleading, Davenport's science was not so bad. *Heredity in Relation to Eugenics* is solidly rooted in the principles of Mendelian inheritance, a remarkable accomplishment for a book written only a decade after the rediscovery of Mendel's laws. Davenport understood that there is no "blending" from one generation to the next of the "determiners of unit characteristics." He similarly understood the profound implications for genetic counseling of the distinction between dominant and recessive inheritance. Indeed, Davenport's discussion of traits such as Huntington's disease and color blindness—the latter informed by the newly discovered chromosomal mechanism of sex determination—differ little from those in contemporary textbooks. He also basically understood the reasons why many traits that cluster in families fail to display simple Mendelian inheritance. Consider his discussion of body weight (pp. 43–44):

> Adult weight (assuming density to be constant) depends upon stature and circumference. It is, therefore, still more complicated than stature and still further removed from any semblance of a unit character. Moreover, it is much more dependent upon conditions of life, for, as is well known, a sedentary life with overfeeding and drinking tends, *in persons so disposed*, to increase weight, even as strenuous activity and dieting favor the reduction, within certain limits, of weight.

We can do little better today. Indeed, subsequent research has vindicated Davenport's optimism that "Despite [the] dependence of weight on environment we may attempt to learn if it shows any trace of heredity" (p. 44). By straightforward extensions of Davenport's methods, we now know that adult body weight shows more than a "trace of heredity": indeed, it is largely genetically determined, at least when assessed in environments that typify contemporary affluent societies.

As discussed above, the ways in which we infuse social values into contemporary science have become more subtle than they were in Davenport's day. In contrast, our style of overpromising has become more blatant. We live in a culture dominated by advertising and "sound bites" rather than reasoned discourse or even basic candor. For all its flaws, I read *Heredity in Relation to Eugenics* with a certain nostalgia. However muddled, misguided, and dangerous Davenport's thinking may have been, his book represents a serious effort to explore the practical implications of the new science of genetics. Mixed in with the book's hype are clear indications of what he actually thought. In contrast, most of what I and my colleagues think about these issues remains unspoken, at least in any public forum.

Contrast *Heredity in Relation to Eugenics*, which is at least worth reading for its negative lessons, with modern promotions of better living through genetics. Brightly colored advertisements from Agilent, a biotechnology company, show a double helix winding upward into the mist. "At the top of this ladder," the advertisements promise, "is a world without disease" (Agilent Technologies 1999). In December of 2000, I used this advertisement as an example of the absurdity of current claims about the potential of genetic research at a biotechnology conference; I was startled after my talk when one of the attendees asked me to defend my view that biotechnology was inherently incapable of producing "a world without disease." J. Craig Venter, a genomic entrepreneur, opened his Congressional testimony in April of 2000 by emphasizing the importance of speed in sequencing the human genome because of the extreme—and, one presumes, imminent—importance of the data to curing cancer and increasing drug safety (Subcommittee on Energy and Environment of the Committee on Science, U.S. House of Representatives 2000):

> At Celera we have adopted the motto "Speed matters" because "Discovery can't wait." Since the Congress began funding the human genome effort over 5 million Americans have died of cancer and over a million people have died because of adverse reactions to drugs.

One Celera advertisement went so far as to make a similar case with respect to childhood malnutrition in poor countries (Olson 2002). Although there are several problems with this style of advocacy, the most pernicious is the implicit claim of a tight causal and temporal link between sequencing the human genome and dramatic improvements in human health.

Some prominent advocates of increased support for biomedical research have gone so far as to set fanciful time lines for realizing the full medical benefits of genetic research. In 2002, Andrew C. von Eschenbach, then Director of the National Cancer Institute (NCI) of the National Institutes of Health, cited rapid progress in understanding the genetic mechanisms of cancer and revolutionary improvements in DNA-based diagnostics as enabling the NCI to set the goal of "eliminating death and suffering from cancer by 2015." Although one might have imagined that Congress would have held hearings to explore the feasibility of achieving this fantastic goal, political discussion soon focused instead on whether or not the goal could be stepped up to 2010 (von Eschenbach 2005). When *Science* magazine sought cancer researchers who were willing to say publicly that NCI planning had lost all touch with reality, few were willing to speak for attribution (Kaiser 2003). Cancer experts have evidently become so dependent on research grants from the NCI that they are reluctant to describe dangerous nonsense for what it is. To be sure, many scientists find the hype surrounding genetic research unsavory; however, most regard it as an unavoidable cost of doing business in a media-driven world.

WHAT IS THE STATE OF EUGENICS TODAY?

It is perhaps a vain hope that the current republication of *Heredity in Relation to Eugenics* will stimulate a new interest in rational discourse about eugenics. Vain or not, stimulation of such a dialog should be our goal. As a start, we might ask how eugenic principles are presently faring in real-world applications. I will confine my attention to affluent countries with modern medical systems, and simply note that the disastrous practice of prenatal sex selection, still widespread in Asia, is undoubtedly the most widespread application of eugenic principles, albeit of a peculiarly self-limiting sort.

Like sex selection, contemporary eugenics in mainstream medicine depends largely on the destruction of the "unfit" before birth, an approach that Davenport took off the table on the fourth page of *Heredity in Relation to Eugenics.* Direct DNA testing and chromosomal karyotyping now allow reliable detection of many genetic anomalies through the testing of fetal cells. This practice, coupled to selective termination of pregnancy, has won substantial acceptance in countries with advanced medical systems. The most dramatic effects have been achieved in relatively small, culturally cohesive groups that have a high prevalence of particular genetic diseases with severe effects in children. For example, the incidence of β-thalassemia in Sardinia has dropped by more than 90% since the widespread introduction of prenatal testing (Cao et al. 2002). Similar results have been achieved for Tay–Sachs disease in a worldwide sampling of Jewish populations (Kaback 2000). In both instances, there is some contribution from mate selection following carrier testing of young adults, but this contribution is thought to be small compared to that of prenatal testing and selective abortion. In the general population, the major impact of the new eugenics has been on Down syndrome. Current estimates in the United States are that there are fewer than 2000 children born each year who are affected with Down syndrome, whereas approximately 7000 would be expected in the absence of prenatal testing (Egan et al. 2004).

The inroads that negative eugenics has made into mainstream medicine have relied on technical means that Davenport did not envision. More compellingly, their broad—although certainly not universal—social acceptance has been aided by the evolution of an ethical framework for genetic counseling that would have been entirely foreign to Davenport. At least in the United States, Europe, and other countries with similar cultural mores, nondirective counseling has become the basic dogma of genetic services. As articulated by Rowley (1984), "the aim of genetic screening programs and prenatal cytogenetic diagnosis should be to maximize the options available to families rather than to reduce the prevalence of genetic diseases." Davenport could scarcely have conceived of separating these aims. In his classically rationalist view of human culture, there would have been little room for the idea that the most productive way to reduce disease prevalence would be to renounce this end point as a goal. Current attitudes may or may not survive as advances in biology create an expanding range of potential interventions into the human life cycle. However, the story of how negative eugenics has found a place in mainstream medicine illustrates the cultural complexity of evolving human values.

The main refinement on the crude process of prenatal testing followed, when patients so choose, by selective abortion in the first or second trimester depends on the rapidly evolving technologies of assisted reproduction. In particular, creation of embryos by mixing eggs and sperm in culture plates is now sometimes followed by genetic testing of a single dispensable cell that is detached from the early embryo by micromanipulation. Embryos in which genetic defects are detected are destroyed, while a sampling of those that pass genetic testing are introduced into the uterus of hormonally prepared females for gestation to term. This clear-cut form of negative eugenics has considerable potential to grow as the range of genetic tests that can be carried out and, particularly importantly, the number of tests that can be reliably performed on a single cell expand.

THE REAL DEBATE ABOUT EUGENICS LIES AHEAD

As we gain knowledge of the genetic basis of traits such as intelligence, physical appearance, and social behavior, a "boutique eugenics" industry will surely arise based on in vitro fertilization and pre-implantation diagnosis. How effective these methods will prove to be is a matter of speculation, given our present ignorance of the underlying science. A major biological obstacle to pre-implantation enrichment for desired traits is that the supply of eggs from any given donor is sharply limited; hence, current technology would only support selection for traits influenced by a small number of genes, and the number of traits accessible to selection within any given batch of embryos would also be small. Intrinsic limits on our ability to establish genotype–phenotype correlations may also emerge as a fundamental barrier to positive eugenics. Many heritable traits may display such extreme genetic heterogeneity that it will be difficult to establish generally applicable cause-and-effect relationships. Although the heterogeneity within individual families is less than that in the general population, there may often be too few individuals in a family to establish the statistically significant guidance that "personalized eugenics" would require. However, biological obstacles such as the limited supply of eggs and the effects of genetic heterogeneity are likely to yield, over time, to advancing knowledge of human biology, even if we cannot presently envision exactly how this will happen. We should not assume that biological limitations will save us from ourselves. As we acquire a detailed knowledge of human genotype–phenotype correlations, this knowledge—just like the capability of determining the sex of a fetus—will undoubtedly be used in ways that are morally dubious and socially deleterious. Indeed, the comfortable idea that Davenport's misguided world has been left behind us, albeit at great human cost, is undoubtedly wrong. The real debate about eugenics still lies ahead as the science and technology of human improvement finally catch up with Davenport's dream.

WE NEED A RETURN TO ENLIGHTENMENT VALUES

What can we learn from *Heredity in Relation to Eugenics* that will help prepare us for this debate? As my reading of Davenport suggests, I think we need more vigilance when we see the red flags in his arguments reappear today. The commingling of social values and science was dangerous in 1911 and remains dangerous today. We need a robust counterattack on the increasingly popular argument that such commingling is inevitable and even desirable. As argued below, acquisition of objective knowledge about human genetic variation and genotype–phenotype relationships must be the goal of the basic science of human genetics and will surely provide the key to humane and effective applications. We inherited this belief that objective knowledge is achievable and inherently beneficial to society from the European Enlightenment. Although it has been much battered during the past 250 years, there is no other acceptable foundation for value-laden discussions of how society should function.

The only antidote to the second red flag in *Heredity in Relation to Eugenics*, Davenport's frequent lapses into muddled thinking, lies in open, critical discourse—another Enlightenment value. In the politically correct environment of modern academia, the range of permissible discussion about eugenics is dangerously narrow. Therein lies the risk that ideas will move from theory to practice without the robust criticism that is the best bulwark an open society can construct against misguided social and political enthusiasms. Finally, we need to bring a similar culture of criticism to our reliance on hype for public support. Overpromising by scientists has become such a prominent feature of the science–society dialog that this echo of Davenport's dream may prove the most difficult to silence.

SELF-IMPOSED IGNORANCE IS RISKY

In our eagerness to escape the long shadow of eugenics, geneticists have retreated into a narrow utilitarianism, whose actual utility is largely unproven. In the process, we have largely abandoned Davenport's one appealing characteristic, which was a genuine curiosity about why we are the way we are (p. 6):

> When we look among our acquaintances we are struck by their diversity in physical, mental, and moral traits. Some of them have black hair, others brown, yellow, flaxen, or red. The eyes may be either blue, green, or brown; the hair straight or curly; noses long, short, narrow, broad, straight, aquiline, or pug. They may be liable to colds or resistant; with weak digestion or strong. The hearing may be quick or dull, sight keen or poor, mathematical ability great or small. The disposition may be cheerful or melancholic; they may be selfish or altruistic, conscientious or liable to shirk.

Given the importance of genetic variation to our understanding of human biology—and the spectacular progress in basic genetics during the past century—one might suppose that we now know a great deal about the genes, and the types of variation in these genes, that underlie heritable components of the traits Davenport mentions. Actually, we know almost nothing, and there is little momentum toward acquiring this knowledge. Why?

The standard answer is a hybrid of two propositions. The first is that the key to success in human genetics lies in narrowly defining phenotypes and pushing phenotypic definitions as close to biochemical processes as possible. Hence, human genetics first gained scientific respectability by retreating from its early 20th century fascination with gross physical and behavioral characteristics. Blood groups and inborn errors of metabolism led the way. Certainly human genetics benefited scientifically from this movement, which spanned much of the 20th century, toward the analysis of simpler phenotypes than the full "diversity of physical, mental, and moral traits" that we observe when "we look among our acquaintances." However, there is little indication that 21st century geneticists are any more eager than their forebears to tackle the broader dimensions of human genetic variation. I am hard pressed to name variable traits, other than those deemed to be "medically relevant," that have been analyzed at the molecular level. Red hair, light skin, and the ability to taste the chemical phenylthiocarbamide (PTC), an odd carryover from the 1930s, stand out as exceptions just because they are so unusual (Valverde et al. 1995; Kim et al. 2003).

Our failure even to attempt to build a robust science of human genetic variation has two related roots. First, human geneticists have succumbed to their own propaganda, whose central tenet is that the analysis of disease phenotypes and susceptibilities is the key to producing medical benefits. Geneticists cling to this comfortable tenet because it "sells well," not because of any convincing evidence that it is the only, or even the best, path toward practical benefits. Indeed, the promised benefits of genetics for mainstream medicine, while not nonexistent, remain surprisingly slight despite an enormous investment in studying the genetics of "medically relevant" traits. Second, there is little doubt that many human geneticists are afraid of what they might learn if they brought contemporary tools to bear on families of the sort Davenport diligently describes in *Heredity in Relation to Eugenics.* It is more comfortable to continue teaching our students that Davenport's science was bad than to risk finding out that many of his scientific instincts were actually right. Certainly it is quite plausible that many traits on which society places a high value—not to mention an even larger number that are held in disdain—have relatively simple genetic contributions. Particularly as the human environment becomes more uniform, these contributions may often outweigh the effects of environmental variability.

As argued above, I believe that the fear of eugenic abuses has been a major deterrent to broad applications of contemporary genetic and genomic tools to the types of traits that were of primary interest to Davenport. I can envision at least two scenarios under which humanity could suffer from this misguided effort. First, fearful science is typically bad science. If pursued too tentatively, genetics may simply fail to deliver on its promise of contributing to better, broadly affordable health care. Science has never prospered when scientists pick and choose what knowledge to acquire on the basis of social sensitivities. Our present selectivity about the phenotypes that deserve pursuit is not sustainable. History offers few examples in which scientists have been able to pluck selectively the bits of knowledge that would ultimately prove useful from a vast sea of ignorance. In confronting the core scientific questions that dominate *Heredity in Relation to Eugenics*—and, indeed, that dominate efforts to define a central role for human genetics in mainstream medicine—we are presently adrift on such a sea.

If genetics, by ignoring this basic lesson from the history of science, fails to deliver otherwise achievable practical benefits, the social consequences could be grave. In addition to the avoidable human suffering associated with missed opportunities, a failure of biomedical research to deliver promised benefits could lead to a backlash against social investment in basic science. Just because we have enjoyed the immense benefits of expanding basic knowledge about the natural world for 300 years, there is no assurance that society will stay on this path.

Second, extreme utilitarianism, which is the most plausible alternative to a continued faith in the value of basic scientific knowledge, carries its own risks. These risks may prove particularly great in medicine. Quite apart from direct uses of genetic information to guide human reproduction, opportunities to intervene in the human life cycle are certain to proliferate. Botox, anabolic steroids, and long-term life support of comatose patients are early indicators of where extreme utilitarianism will take us in biology. A headlong rush into this future would be a harsh and dehumanizing experience. In contrast, a more balanced understanding of human biology, and our place in the natural world, might direct our impulse to re-engineer ourselves into more effective and humane directions. Consider our relationship with the environment. Throughout the world, extractive industries have gone through a phase, still rampant in many places, of inflicting massive environmental destruction in pursuit of short-term gains. Even the most developed countries have yet to master these impulses, but they have surely been attenuated by an expanding knowledge of the basic principles of ecology and environmental science. Imperfect as this path may have been, it was better than one that concentrated wholly on the development of better chain saws and bulldozers. One must hope that an expanding knowledge of human biology—including the genetics of human differences—will similarly attenuate our impulse to redesign ourselves. This expanding knowledge would be accompanied by its own risks, but we should embrace them relative to the alternatives.

KNOWLEDGE SHOULD PRECEDE TEACHING

Perhaps the major lesson from a modern reading of *Heredity in Relation to Eugenics* is that, in our rush to suppress Davenport's distressing social agenda, we should avoid losing confidence in the net social benefits of curiosity-driven science. In Davenport's words, "It would seem a self evident proposition, but it is one too little regarded, that knowledge should precede teaching" (p. 26). One of Davenport's great failings is that he did not take this, his own best advice. We should take it now and interpret "teaching" in the broadest sense, encompassing all efforts—including the pursuit of medical advances—to use genetics for human betterment. Just as evolving social values protected us from the darker ambitions of the 20th century eugenicists, we need confidence that a more objective understanding of human biology, and the biological basis of human differences, will protect us from the worst effects of a revived pursuit of human perfection

REFERENCES

Agilent Technologies. 1999. Advertisement. *U.S. News World Rep.*, September 6, pp. 2–3.

Cao A., Rosatelli M.C., Monni G., and Galanello R. 2002. Screening for thalassemia—A model of success. *Obstet. Gynecol. Clin. North Am.* **29:** 305–328.

Collins F.S., Brooks L.D., and Chakravarti A. 1998. A DNA polymorphism discovery resource for research on human genetic variation. *Genome Res.* **8:** 1229–1231.

Egan J.F., Benn P.A., Zelop C.M., Bolnick A., Gianferrari E., and Borgida A.F. 2004. Down syndrome births in the United States from 1989 to 2001. *Am. J. Obstet. Gynecol.* **191:** 1044–1048.

Hays W.M. 1912. Constructive eugenics. *J. Hered.* **3:** 113–119.

Kaback M.M. 2000. Population-based genetic screening for reproductive counseling: The Tay-Sachs disease model. *Eur. J. Pediatr.* (suppl. 3) **159:** S192–S195.

Kaiser J. 2003. NCI goal aims for cancer victory by 2015. *Science* **299:** 1297–1298.

Kim U.K., Jorgenson E., Coon H., Leppert M., Risch N., and Drayna D. 2003. Positional cloning of the human quantitative trait locus underlying taste sensitivity to phenylthiocarbamide. *Science* **299:** 1221–1225.

Lamason R.L., Mohideen M.A., Mest J.R., Wong A.C., Norton H.L., et al. 2005. SLC24A5, a putative cation exchanger, affects pigmentation in zebrafish and humans. *Science* **310:** 1782–1786.

Olson M.V. 2002. The human genome project: A player's perspective. *J. Mol. Biol.* **319:** 931–942.

Petit C. 1998. No biological basis for race, scientists say. *San Francisco Chronicle,* February 23, p. 1.

Rosenberg N.A., Pritchard J.K., Weber J.L., Cann H.M., Kidd K.K., Zhivotovsky L.A., and Feldman M.W. 2002. Genetic structure of human populations. *Science* **298:** 2381–2385.

Rowley P.T. 1984. Genetic screening: Marvel or menace? *Science* **225:** 138–144.

Subcommittee on Energy and Environment of the Committee on Science, U.S. House of Representatives, 106th Congress, Second Session. 2000. *The Human Genome Project,* April 6, 2000. U.S. Government Printing Office, Washington, D.C.

Valverde P., Healy E., Jackson I., Rees J.L., and Thody A.J. 1995. Variants of the melanocyte-stimulating hormone receptor gene are associated with red hair and fair skin in humans. *Nat. Genet.* **11**, 328–330.

von Eschenbach A.C. 2005. A message from NCI Director Dr. Andrew C. von Eschenbach. *NCI Cancer Bulletin* July 19, **2:** 1–2, 8.

Genetic Determinism and Evolutionary Ethics

A Mitochondrial Perspective

Douglas C. Wallace
Center for Molecular and Mitochondrial Medicine and Genetics, University of California, Irvine

Editors' Note: How do we determine what constitutes a good versus a bad genotype? Do our genes determine our ethics? These are issues of both science and morality, and whether the facts of science can be used to determine human morality has been vigorously debated for hundreds of years. Charles Darwin recognized this problem in relation to evolution and addressed it in his 1871 book *The Descent of Man* (pp. 159–160): "It must not be forgotten that although a high standard of morality gives but a slight or no advantage to each individual man and his children over the other men of the same tribe, yet that an increase in the number of well-endowed men and an advancement in the standard of morality will certainly give an immense advantage to one tribe over another." For Darwin, morality could evolve by natural selection. Debates about evolutionary ethics have continued ever since Darwin's ideas were published and these debates remain active and sometimes contentious to this day.

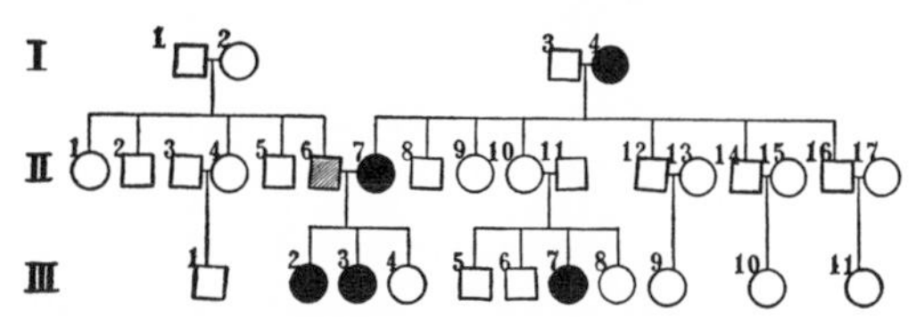

FIG. 58.—Inheritance of nervousness and brilliancy. I, 4, is subject to headaches and nervousness. Her daughter, II, 7, is similarly affected. She married a man, II, 6, who has had temporary attacks of paralysis. One of their children, II, 2, has nervous prostration and one, III, 3, is subject to headaches and nervousness. F. R.; Cla. 3.

Charles Davenport thought that geneticists could determine bad genotypes by knowing the "unit characters" of heredity; he believed that this knowledge would make it possible to breed out of the human race what he and other eugenicists perceived as bad genotypes. However, things were not so simple as Davenport had hoped, and Dr. Wallace explains why in this essay.

Mendelian genetics do not determine everything about genetic disease; the role of random factors, polygenic disease, and the interaction of multiple chromosomes are but a few issues that make genetic prediction very difficult. Mitochondrial DNA and its unique nature and role in human disease add an additional layer of complication. Dr. Wallace then draws from this current knowledge about human genetics and disease to discuss issues regarding evolutionary ethics—the evolutionary value of maintaining genetic diversity, the chance nature of each person's genotype, and the price tag that society must pay to maintain our human genetic heritage. Charles Davenport did not want society to pay this price tag, but Dr. Wallace argues that this is the price we must pay for the human species to survive.

> *If one is provided with a knowledge of the methods of inheritance of unit characters it might seem to be an easy matter to state how each human trait is inherited and to show how any undesirable condition might be eliminated from the offspring and any wished for character introduced. Unfortunately, such a consummation cannot for some time be achieved...we do not yet know all of the unit characters in man...we can hardly know in advance which of them are due to positive determiners and which to the absence of such.*
>
> DAVENPORT, PP. 23–24

IN 1911, WHEN DAVENPORT WROTE the paragraph quoted above in *Heredity in Relation to Eugenics*, concepts in medical genetics were coalescing around Mendelian genetics with its rigid rules for inheritance. As a consequence, Davenport interpreted the various clinical pedigrees that he studied based on the concepts that all familial traits were the result of inherited "unit characters" and that these characters were transmitted in a Mendelian and thus deterministic fashion. With discrete unit characters (genes) and rigid rules, a time could be envisioned when the genetic future of an individual, family, or even species might be predicted. Accurate predictions would then permit the elimination of deleterious "unit characters" from individuals and the population to the benefit of all.

Today, almost 100 years later, our understanding of medical genetics is much greater and consequently much more complex than could be envisioned in 1911. As a result, all three of Davenport's assumptions—that all "diseases" are caused by "unit characters," that all unit characters are inherited exclusively according to the rules of Mendel, and that hereditary traits that cause "disease" are inherently bad and thus should be eliminated—have been shown to be incorrect. Therefore, the proposal that selective breeding could rid humanity of "disease" can now be seen as not only untenable but also counter to the best interests of our descendants.

For Mendelian and thus chromosomal genes, the hereditary traits (mental retardation, psychiatric disorders, epilepsy, etc.) about which Davenport and his contemporaries were most concerned are now known to be influenced not just by invariant "unit characters" but also by genes that can change over successive generations via nucleotide repeat expansion, chromatid rearrangements, or transient epigenetic modifications of the genome. Hence, tracking disease traits has proven to be a far more daunting task than was previously envisioned. In addition, it now appears that many of the common traits of concern to Davenport are the result not of just one unit character but of the interaction of multiple genetic and environmental factors—so called multifactorial disease. Therefore, it is now clear that many stochastic, or random, factors act on chromosomal genes to generate a clinical phenotype, making deterministic predictions difficult.

The capacity to predict genetic outcomes is made even more complicated by the fact that the human mitochondrion has its own independent DNA-based genetic system and that the genetics of the mitochondrial DNA (mtDNA) has strong stochastic aspects. The mtDNA codes for critical genes needed for energy production and many of the phenotypes addressed by Davenport have been associated with mtDNA variation. In addition, the mtDNA is maternally inherited, present in thousands of copies per cell, and intracellular mixtures of mutant and normal mtDNAs can segregate randomly during both mitosis and meiosis. Therefore, mtDNA genetics is not only relevant to the etiology of common diseases, but its inheritance is inherently stochastic.

Although the complexities of human phenotypic traits and their inheritance make genetic prediction, and thus genetic purification, infeasible, recent studies on human origins and evolution have revealed that the very concept of "bad" traits is an oversimplification and potentially dangerous. It is now clear that for many phenotypes, the categorization of "good" or "bad" depends on the environmental context in which the genotype occurs. Modern molecular genetics has shown that every individual is genetically different, and that these differences have an important purpose. Because local and planetary environmental conditions change, every successful species must retain an array of genetic variants to permit adaptation to these changes. A phenotypic trait that is marginal in one environment might be essential in another. Therefore, genetic diversity is not bad but both good and essential for the species.

This new understanding of the origins and importance of genetic variation, plus recent advances in biotechnology, now require that we reexamine the ethics of the applications of biotechnology for influencing reproductive outcomes. Because the issues at hand address genes and genetic diversity, any new eugenic principle for evaluating the application of interventional genetic procedures should take into account the importance of the genes for individual and species health and longevity. This logic leads to two new eugenic principles. The immediate, and thus prime, genetic objective is the successful transmission of the genes to the next generation. The long-term but central genetic objective is the preservation of human genetic diversity to insure the survival of our species. These two principles provide a rational basis for making decisions about genetic intervention. They also lead to the elaboration of a new "evolutionary ethics" that may explain a number of aspects of human behavior and provide a compelling justification for honoring diversity in all members of our society and species.

GENETIC DETERMINISM IS ELUSIVE

The Human Genome Project has determined the sequence of human chromosomal DNA (Lander et al. 2001). This feat should have identified all of the Davenport "unit characters" responsible for human genetic disease. Yet, even with enormous expenditures of effort and funds, most of the genetic factors responsible for common diseases such as diabetes, obesity, mental illness, cardiovascular disease, blindness, deafness, dementias, and movement disorders have not been found. Furthermore, those unit characters that have been associated with disease cannot be eliminated. Why?

The Hardy–Weinberg equilibrium ($p^2 + 2pq + q^2 = 1$) and the dawn of human population genetics led to the realization that it would be virtually impossible to remove even well-defined deleterious chromosomal genes from the human gene pool (Cavalli-Sforza and Bodmer 1971). For diseases caused by autosomal recessive genes, such as cystic fibrosis and Tay–Sachs, both copies of the chromosomal gene must be mutant to manifest the disease. Given a population frequency of the mutant allele of q, the probability of a homozygous mutant is q^2. Thus, for rare alleles, the proportion of the mutant alleles in the expressed homozygous state is negligible when compared to the number of mutant copies retained in the population in the invisible heterozygous state ($2pq$). Therefore, even the consistent elimination of all "affected" individuals would fail to reduce significantly the frequency of the mutant allele in the general population (Cavalli-Sforza and Bodmer 1971).

The elimination of rare dominant deleterious alleles might seem more feasible, because every individual harboring a mutant gene would be affected. However, lethal dominant genes are already eliminated by natural selection in the heterozygous ($2pq$) state. Therefore, these genetic diseases are sustained in the population by new mutations (Cavalli-Sforza and Bodmer 1971). There is no way to predict who will acquire a new dominant mutation, and so dominant lethal mutations cannot be eliminated by genetic screening. Hence, the frequency of dominant lethal genes cannot be changed.

Beyond classical Mendelian disease genes, a number of clinically relevant loci become pathogenic only when they change. For example, an allele can be inactivated by expansion or contraction of a repeated sequence in the DNA from one generation to the next. This mechanism is important in the Fragile X syndrome, Huntington disease, multiple spinal cerebellar atrophies, and facioscapulohumeral muscular dystrophy, to name a few (Garber et al. 2006; Orr and Zoghbi 2007; van der Maarel et al. 2007). Deletions and rearrangements can alter regions of individual chromosomal chromatids, resulting in partial deficiencies that can now be detected by comparative genomic hybridization (CGH) (Cai 2007; Ijssel and Ylstra 2007). Such changes are beginning to be implicated in a variety of common diseases. The phenotype associated with a mutant gene can also be modified by transient genomic changes, which can be transmitted through the germ line. Such epigenetic changes have been implicated in both cancer and inherited diseases (Feinberg 2007), the latter including Prader–Willi, Angelman, and Beckwith–Wiedemann syndromes (Jiang et al. 2004). Therefore, the transmission and expression of even well-defined Mendelian "unit characters" can have a strong stochastic component.

The Mendelian inheritance of "unit characteristics" cannot explain the transmission of diabetes, obesity, psychiatric disorders, and macular degeneration. One common explanation for this dilemma is that these problems result from the inheritance and adverse interaction of a number of different Mendelian factors. This "polygenic" model has two subhypotheses: (1) The disease may result from inheritance of any one of a variety of different rare deleterious gene alleles, or (2) the phenotype results from the inheritance and interaction of several common alleles. The latter subhypothesis requires that the common alleles preexist in the population, indicating that these gene alleles were previously enriched by selection. That selection has acted on certain human genes has been confirmed by the discovery of extended regions of linkage disequilibrium surrounding various chromosomal loci (Wang et al. 2006; Sabeti et al. 2007).

Recently, evidence that multiple loci may contribute to certain common phenotypes has been sought by scanning large numbers of single-nucleotide polymorphisms (SNPs) (Frazer et al. 2007) in "whole-genome scans" (WGSs) of the chromosomes of patients versus controls. SNP alleles that are at statistically higher frequency in the patients than in the controls are taken as evidence that a disease-related locus resides in that region of the chromosome. For diabetes, in a series of 35,550 diabetes patients and controls that were screened for several hundred thousand chromosomal SNPs, nine chromosomal loci were revealed at increased frequency in diabetic patients. Unfortunately, each locus accounted for only a small fraction of the disease risk. Thus, multiple chromosomal loci must interact to result in diabetes (Saxena et al. 2006; Scott et al. 2007; Sladek et al. 2007; Willer et al. 2007; Zeggini et al. 2007). In summary, even if all adverse chromosomal loci could be known, it would still be impossible to predict who would develop a common clinical problem in the next generation. Hence, no mating strategy could be devised that would significantly reduce disease prevalence.

MITOCHONDRIAL DNA GENETICS IS STOCHASTIC

The probabilistic nature of human chromosomal genetics is further enhanced by the stochastic nature of mtDNA genetics. Life requires the interaction between structure, energy, and information, and the genes of the mtDNA are pivotal to cellular energy production. Hence, the mtDNA genes are critical to health (Wallace 2007).

One Cell–Two Organisms

The eukaryotic cell was assembled some 2 billion years ago from two different life-forms fused in a symbiotic event. One cell gave rise to the modern nuclear–cytosol organism and was probably a motile glycolytic cell. The other generated a cellular organelle—the mitochondrion—from an oxidative bacterium. At the time of the initial symbiotic event, both organisms contained the genetic machinery necessary for a free-living microorganism. However, in the subsequent 1.2 billion years, the genomes of the two organisms underwent repeated genetic exchange until the chromosomal genes of the nuclear–cytosolic organism became specialized in defining the structural components of the cell, whereas the mtDNA became specialized in encoding critical elements for energy generation by oxidative phosphorylation (OXPHOS). Today, roughly 1500 of the mitochondrial genes are located within the chromosomal nuclear DNA (nDNA), whereas 13 critical energy production genes have been retained by the mtDNA (Wallace 2007).

The mtDNA is located within the cytoplasmic mitochondrion, and so it is not inherited with the nuclear chromosomes. Rather, the mitochondria and mtDNAs are transmitted through the cytoplasm of the egg and thus are maternally inherited (Giles et al. 1980). Because the mitochondria and mtDNAs of the sperm are destroyed at fertilization, male and female mtDNAs never mix and thus cannot recombine. Consequently, the mtDNA can change only by the sequential accumulation of mutations along radiating maternal lineages. Each cell contains hundreds of mitochondria, and each mitochondrion contains multiple mtDNAs; thus, the typical human cell contains thousands of mtDNA molecules. As a result, cells can contain various mixtures of mutant and normal mtDNAs, a state known as heteroplasmy (Wallace 2005).

The nDNA encodes genes for the structure of the cell and organism, and the mtDNA encodes proteins for the electron and proton "wires" of OXPHOS (Wallace 2007). By analogy, when building a house, an architect develops the plan and blueprints for the structure of the house (shape, wings, floors, rooms, and location of the appliances and light fixtures). However, an electrical engineer must design the wiring diagram to connect the electrical components and energize the house. Likewise, the chromosomal DNA contains the structural blueprints for the cell and body, but the mtDNA contains the circuit diagram. It is possible to add structural components to the house with minimal effect on the integrity of the preexisting structure, but it is impossible to add electrical elements to a circuit without reorganizing the circuit. This is because all of the circuit elements function in series and thus must work together.

Similarly, for the cell, it is advantageous to recombine the chromosomal genes to generate novel structures. However, it would be disastrous to assort mtDNA genetic elements randomly because this could degrade the efficiency of the mitochondrial circuit. Thus, the biparental nDNA is designed to undergo recombination, whereas the uniparental mtDNA is sequestered to block recombination (Wallace 2007).

The Genetics of Energy

The human mtDNA genes encode 13 subunits of the OXPHOS enzyme complexes, plus the 12S and 16S rRNA and 22 tRNAs necessary for mitochondrial protein synthesis. The 13 polypeptides include seven

(ND1, ND2, ND3, ND4L, ND4, ND5, ND6) of the 45 polypeptides of complex I, one (cytochrome *b*, cytb) of the 11 polypeptides of complex III, three (COI, COII, COIII) of the 13 subunits of complex IV, and two (ATP6 and ATP8) of the about 16 polypeptides of complex V. Cytochrome *b*, COI-III, and ATP6 have all been shown to be involved in ion transport. The precise function of the ND and ATP8 subunit genes is not known, but I will assume by extrapolation that they also involve OXPHOS charge transport (Wallace 2005).

The mitochondrion burns hydrogen (reducing equivalents) recovered from dietary calories with the oxygen that we breathe to generate energy in the form of adenosine triphosphate (ATP) to perform work and to generate heat to maintain our body temperature at 37°C. Dietary electrons flow down the electron transport chain (ETC) through complex I, coenzyme Q, complex III, cytochrome *c*, complex IV, and oxygen to generate water. The energy released as the electrons flow through complexes I, III, and IV is used to pump protons out through the mitochondrial inner membrane, creating an electrochemical gradient ($\Delta P = \Delta? + \Delta\mu^{H+}$). Thus, each of the 10^{17} mitochondria within the human body is a tiny charged capacitor. The potential energy stored by the mitochondrial capacitance can then be expeditiously allocated to drive many biological processes. Some of the energy stored in ΔP is used by the mitochondrial ATP synthase (complex V) to convert ADP + Pi to ATP. ATP is then exported out of the mitochondrion where the phosphate bond energy can be used to promote otherwise unfavorable reactions (Wallace 2005).

Like any furnace, combustion in the mitochondria is incomplete. This generates cellular reactive oxygen species (ROS) or oxygen radicals, the equivalent of smoke from a furnace. Mitochondrial ROS, encompassing superoxide anion, hydrogen peroxide, and hydroxyl radicals, can have two actions within the cell. At low levels, ROS serve as a signal by which the nucleus monitors the number and status of the mitochondria. Current data suggest that as the number of cellular mitochondria increases, the hydrogen peroxide released into the cytosol increases. This signals the nucleus that there are too many mitochondria. Consequently, the nucleus replicates its DNA, the cell undergoes cytokinesis, and the number of mitochondria is divided in half. Inhibition of the mitochondrial ETC redirects the electrons into increased ROS production, and the increased hydrogen peroxide can then drive uncontrolled replication leading to cancer. However, when produced at high levels, mitochondrial ROS also acts as an oxidizing agent that damages mitochondrial proteins, lipids, and DNA. This ultimately leads to energetic failure and cell death (Wallace 2005).

Because the mitochondria make most of the cellular ROS, and ROS is mutagenic, the mtDNA mutation rate is one to two orders of magnitude greater than that of the nDNA. The occurrence of a mtDNA mutation creates a heteroplasmic cell. As this cell undergoes cytokinesis, the mutant and normal mtDNAs undergo replicative segregation, ultimately resulting in homoplasmic normal cells (i.e., the cells contain only normal mtDNA) or homoplasmic mutant cells. As the percentage of mutant mtDNAs increases, the capacity of the mtDNAs to repair damaged mitochondria declines, resulting in the progressive decline in mitochondrial function and energy production. Mitochondria turn over continuously in all cells, including postmitotic cells such as neurons, cardiomyocytes, and muscle fiber; as a result, mutant mtDNAs accumulate in postmitotic cells over time. As the proportion of mutant mtDNAs increases, the capacity of the mtDNA to repair mitochondrial damage declines. This results in a progressive decline of mitochondrial function, the cellular equivalent of a metropolitan brownout. Thus, the accumulation of mtDNA mutations in postmitotic tissues creates the aging clock (Wallace 2005).

As mitochondrial and mtDNA oxidative damage increases, electron transport stalls, mitochondrial energy production declines, and mitochondrial ROS production is further stimulated. These factors impinge on the mitochondrial permeability transition pore (mtPTP), a mitochondrial self-destruct switch. When the mtPTP is activated, a channel opens in the mitochondrial inner membrane, the electrochemical gradient short-circuits, pro-apoptotic proteins are released into the cytosol, and the energy-deficient and/or precancerous cell is destroyed by apoptosis. Because the probability that the mtPTP will be activated is stochastic, some cells in postmitotic tissues will always be in the process of dying, and the tissue cell number will decline with age. When tissues and organs have lost sufficient cells that they can no longer function properly, the symptoms of age-related metabolic and degenerative disease and aging ensue (Wallace 2005).

The high mutation rate of the mtDNA means that new variants are constantly arising along the female germ line. Many of these mutations damage essential mitochondrial genes. When the percentage of mutant mtDNAs becomes high enough, energy becomes insufficient for normal cellular function and the patient develops a mitochondrial disease. If mtDNA mutations were permitted to accumulate in the female germ line over many generations, ultimately there would be too few normal mtDNAs to sustain the species. To avoid this progressive increase in mitochondrial genetic load, the primordial germ cells of the female germ line contain relatively few mtDNA segregating units, on the order of 100–200. Moreover, the primordial germ cells undergo approximately 20 cell divisions, generating several million proto-oocytes. As a result, deleterious mtDNA mutations rapidly segregate out, because of intracellular genetic drift, ultimately generating individual proto-oocytes with either high or low percentages of mutant mtDNAs (Jenuth et al. 1996). Selection then acts within the female germ line to eliminate those proto-oocytes harboring high levels of the most severely deleterious mutations (Fan et al. 2008). In this way, only those mtDNA mutations that fall within a defined mitochondrial physiological range are transmitted to offspring and thus can contribute to the genetic diversity of the species.

Although constrained in the severity of mtDNA mutations, the high mutation rate means that deleterious mtDNA mutations are very common. More than 200 pathogenic mtDNA mutations have been reported (Wallace et al. 2007), and the frequency of individuals harboring deleterious mtDNA mutations has been estimated to be approximately 1/5000 (Schaefer et al. 2004, 2007).

Mitochondrial defects preferentially affect the central nervous system, heart, muscle, renal, and endocrine systems, although any tissue can be involved. Common symptoms associated with mtDNA mutations include certain forms of blindness, deafness, dementias, movement disorders, epilepsies, cardiovascular disease, muscle weakness and degeneration, renal dysfunction, and diabetes and aspects of the metabolic syndrome (Wallace et al. 2007). Moreover, because the more severe mtDNA mutations are heteroplasmic, the proportion of mutant mtDNAs can differ among family members. As a result, maternal relatives can harbor the identical mtDNA mutation but manifest totally different symptoms. In summary, mtDNA mutations have been implicated in all of the symptoms associated with aging and age-related diseases and thus encompass most of the adverse phenotypes of concern to Davenport and his compatriots.

A possible example of a Davenport pedigree that might have harbored a heteroplasmic mtDNA is shown in Figure 58 of *Heredity in Relation to Eugenics* (see chapter-opening image or p. 97 in the Davenport book). Multiple individuals, over two generations, were affected by a variety of symptoms, including "headaches" (migraines?), "nervousness" (myoclonic epilepsy?), and "nervous prostrations" (stroke-like episodes?), all of which can be observed in pedigrees harboring mtDNA tRNA gene mutations (Wallace et al. 2007). This pedigree shows clear parent-to-child transmission as might be expected for an autosomal dominant gene mutation. However, the symptoms among family members are highly variable and skip generations (e.g., mother II10 was unaffected yet her daughter III7 was affected) and all of the affected individuals are related through the maternal lineage. An X-linked recessive mutation can be eliminated because women are preferentially affected. Therefore, this pedigree could be explained by a maternally inherited, heteroplasmic, mtDNA mutation.

Davenport would have assumed that this pedigree was autosomal dominant; however, if it were due to a mtDNA mutation, then the predictions made as to who in the family would develop clinical symptoms would be very different. According to the Mendelian hypothesis, individuals III4, III5, III6, and III8 would be unlikely to have affected children. However, according to the mtDNA hypothesis, women III4 and III8 would have a significant probability of having affected children, whereas males III5 and III6 would have no possibility of transmitting the disease. Thus, uncertainty about the mode of transmission negates the predictive power of Mendelian genetics.

In summary, the deterministic perspective of Davenport based on the concepts of "unit characters" and rigid Mendelian inheritance was an illusion. In reality, human disease can result from a variety of genetic changes and combinations of changes, and the inheritance of the resulting traits, both Mendelian and mitochondrial, is strongly stochastic. Therefore, the accurate prediction of many genetic outcomes is not feasible.

mtDNA VARIATION AND ENVIRONMENTAL MODULATION—WHY GENETIC DIVERSITY IS GOOD

Not only was the eugenic view of genetic determinism incorrect, but the concept of "good" and "bad" traits was oversimplified. It is now clear that genetic diversity is not only rampant, it is critical to the survival of our species. Therefore, the indiscriminant destruction of genetic variation would be a threat to the survival of our descendants.

Although certain autosomal recessive genes appear to be uniformly bad, such as that causing Tay–Sachs disease, other mutations that are deleterious in the homozygous state may be beneficial in the heterozygous state. The classic example of this heterozygous advantage is the resistance to malaria imparted by heterozygous β-globin gene mutations. A β-globin mutant is deleterious to the occasional homozygous individual, but the presence of the mutant allele in the more common heterozygous state protects individuals from a potential lethal infection by malaria (Cavalli-Sforza and Bodmer 1971; Akide-Ndunge et al. 2003; Williams et al. 2005).

Genetic variation that results in other traits, such as diabetes, obesity, hyperactivity, bipolar disorder, or reduced athletic ability capacity, might also be deleterious in one environment but beneficial in another. In fact, the high prevalence of the common genetic variants that combine to generate common clinical problems implies that these gene alleles may have been beneficial and enriched in certain ancient populations.

Mitochondrial variation provides an excellent example of how environmental context can modulate the relative advantage or disadvantage of individual genetic variants. The inner mitochondrial membrane potential (ΔP) is maintained by burning dietary calories with oxygen via the ETC. However, the efficiency by which ΔP is converted to ATP (the coupling efficiency) is not constant. An individual whose mitochondria are tightly coupled will generate the maximum ATP per calorie burned and produce little excess heat. By contrast, an individual whose mitochondria are loosely coupled will need to burn more calories to generate the same amount of ATP, and the differential will be dissipated as heat. Variation in coupling efficiency could then have obvious adaptive advantages in different climatic conditions. Individuals living in the tropics need to optimize physical performance and minimize heat production. Hence, they would benefit from more tightly coupled mitochondria. Alternatively, individuals living in the Arctic need to generate much more heat to survive extreme cold. Hence, they would benefit from less coupled mitochondria and would need to eat a higher-calorie diet to produce enough ATP to function normally (Mishmar et al. 2003; Wallace et al. 2003; Ruiz-Pesini et al. 2004; Ruiz-Pesini and Wallace 2006).

Because all of the mtDNA-encoded proteins contribute to the generation, maintenance, and use of ΔP, mutations in the mtDNA can affect the coupling efficiency and thus may facilitate climatic adaptation. As humans moved into new environments, mtDNA mutations arose along radiating maternal lineages. Those mutations that changed the coupling efficiency to be more optimal for the new environment enhanced individual survival and increased in frequency. Consequently, new mtDNA variants came to predominate in the new environments. This phenomenon resulted in discrete branches of the mtDNA tree being associated with the indigenous peoples of a particular region. These regional groups of related mtDNA haplotypes are known as haplogroups (Ruiz-Pesini et al. 2007; Wallace 2007).

By reconstructing the human mtDNA mutational tree and overlaying the tree on the geographic distribution of indigenous populations, it has been possible to reconstruct the ancient migrations of women. The human mtDNA tree originated in Africa about 200,000 years before present (YBP) (Johnson et al. 1983; Cann et al. 1987; Merriwether et al. 1991) and radiated into four African-specific mtDNA lineages (L0, L1, L2, and L3). About 65,000 YBP, in northeastern Africa, two new mtDNA lineages (M and N) arose from L3, and only women harboring M and N succeeded in colonizing all of Eurasia. Lineage N gave rise to the European-specific lineages H, I, J, K, T, U, V, W, and X, whereas both N and M each gave rise to a plethora of Asian-specific mtDNA lineages including A and B for N and C and D for M. About 40,000 YBP Asian-specific lineages A, C, and D came to predominate in northeastern Siberia, and these were the mtDNAs in position to cross into the Americas when the Bering land bridge appeared about 20,000 YBP. About 15,000 YBP, individuals carrying the predominantly European lineage X entered the Great Lakes region of North America, and about 12,000 YBP, lineage B, found along the central and southeastern Asian coast, bypassed Siberia and settled in the temperate and tropical regions of the Americas. At 7,000–9,500 YBP another Siberian migration brought modified haplogroup A lineages to the Americans to found the Na-Dene of northwestern North America. Finally, migrations carrying haplogroup A and D mtDNAs around the Arctic are associated with the Eskimos and Aleuts (Wallace et al. 1999; Wallace 2005, 2007).

Haplogroup L mtDNAs correspond with the tropics and subtropics; M- and N-derived mtDNAs with the temperate zone; and A, C, D, and X mtDNAs with the Arctic. These region-specific lineages also correspond with differences in functional mtDNA variants (Mishmar et al. 2003; Wallace et al. 2003; Ruiz-Pesini et al. 2004). Therefore, the geographic distribution of haplogroups would appear to reflect adaptation to increasingly cold climates in the temperate zone after 65,000 YBP, when people left Africa. After 10,000 YBP, the climate in the temperate zones warmed and the invention of agriculture provided unlimited calories. These changes put radically different demands on human physiology and the selective factors acting on the various mtDNA haplogroups. The resulting disconnect between mtDNA genotype and environment might then explain the rise in adverse phenotypes such as diabetes, obesity, cardiovascular disease, and so forth.

In the presence of excess calories and with minimal exercise, the ETC of tightly coupled mitochondria would become fully reduced. This would drive the storage of the excess calories as fat leading to obesity and stimulate the mitochondria to produce more ROS. The chronically elevated ROS production would be destructive to the individual's tissues, making the individual more prone to degenerative diseases. By contrast, individuals with partially uncoupled mitochondria would burn the excess calories to generate heat. Consequently, they would tend to stay lean, generate less ROS, and thus be less prone to degenerative diseases. However, should the environment change, such that calories once again become limiting, then individuals with tightly coupled mitochondria would have enough ATP to survive, whereas those with loosely coupled mitochondria would starve.

The relationship between OXPHOS coupling and ROS production may impact on a variety of phenotypes. The apparently more tightly coupled European haplogroup H mtDNAs (Ruiz-Pesini et al. 1998; Montiel-Sosa et al. 2006) are associated with increased resistance to sepsis (Baudouin et al. 2005), presumably because they produce more ROS, which is bactericidal. By contrast, the apparently less coupled haplogroups J, T, and U, which would produce less ROS, are protective of neurodegenerative diseases such as Alzheimer disease (AD) (Carrieri et al. 2001) and Parkinson disease (PD) (van der Walt et al. 2003; Ghezzi et al. 2005; Pyle et al. 2005) and are also associated with increased lifespan (De Benedictis et al. 1999; Ross et al. 2001; Bonafe et al. 2002; Niemi et al. 2003). Certain Asian mtDNA haplogroups are associated with reduced risk of developing type 2 diabetes and metabolic syndrome (Fuku et al. 2007; Tanaka et al. 2007), and elite European long-distance runners have a higher frequency of putatively more coupled mtDNA haplogroups, whereas sprinters have a higher frequency of less coupled haplogroups (Niemi et al. 2005; Castro et al. 2007).

Therefore, variation that is advantageous in one context can be disadvantageous in another. Because human environments can change, often abruptly, either because of planetary fluctuations or human intervention, it follows that mtDNA variation that appears to be maladaptive today might be beneficial when the environment changes. Therefore, to increase the probability of the survival of future generations, we must preserve the genetic diversity of the human gene pool today. By this evolutionary logic, genetic diversity is good and efforts to restrict diversity to enhance individual success in the local environment are bad.

HOW DO WE ASSSESS INTERVENTIONAL GENETICS AND EVOLUTIONARY ETHICS?

Employing this evolutionary logic to evaluate the ethics of interventional human genetics, we can conclude that the application of genetic technology to avoid generating conceptuses that inherit lethal unit-character Mendelian genes is consistent with the evolutionary program and is justifiable. This is particularly true in evolutionary terms, because having severely affected children reduces the parents' potential for having other children that could more successfully pass their genes on to the next generation.

Today, the most common technologies used to avoid offspring with severe Mendelian disorders employ molecular diagnostic analysis of fetal tissues, chorionic villi, or amniocytes. Discovery that the conceptus carries a severe genetic defect might then lead the parents to terminate the pregnancy (Slack et al. 2006; Bellissimo 2007). However, termination of pregnancies is both emotionally stressful and to some morally objectionable. Therefore, more recent genetic technologies have focused on preimplantation genetic testing. In these cases, the prospective parent's eggs and sperm are united in vitro. Then a polar body or blastomere is collected from the various embryos and used to determine the genotype of the embryo. Only those conceptuses free of the genetic defect are implanted into the womb (Fasouliotis and Schenker 1998; Xu et al. 1999; Harper et al. 2002). Although potentially powerful for avoiding single-gene disorders, this procedure is not practical for polygenic traits, as insufficient time would be available to screen for multiple interacting loci before the embryos become nonviable.

Mitochondrial Genetics and Eugenics

Preimplantation genetics is not likely to be a viable approach for manipulating polygenic Mendelian traits, but it could be a viable approach for manipulating complex traits resulting from mtDNA variation. Because the mtDNA is located outside the nucleus and is semiautonomous, a couples' zygote nucleus generated by in vitro fertilization could be removed from the mother's oocyte using a micropipette and injected into an enucleated oocyte from another woman with a more desirable mtDNA genotype. Indeed, because maternal inheritance dictates that all of the children will inherit the mother's mtDNA and thus have a high probability of inheriting the mutant mtDNAs, nuclear transplantation might be the only way that biotechnology can help women who harbor deleterious mtDNA mutations at a high percentage of heteroplasmy to have normal children (Wallace 1987). Simply testing chorionic villi and amniocytes for heteroplasmic mtDNA mutations is unlikely to be helpful for heteroplasmic mtDNA mutations because the random segregation of the mutant mtDNA during the mitotic replications of development could result in different tissues having different levels of the mutant mtDNA. Thus, although the peripheral fetal cells available for testing might have a low level of mutant mtDNA, the tissues that are more prone to adverse symptoms but are not available for testing might harbor a high percentage of mutant mtDNA. Hence, the genetic test would lead to an erroneous prognosis. Therefore, from an evolutionary perspective, the use of nuclear transplantation to avoid transmission of debilitating mtDNA mutations should be acceptable.

Although nuclear transplantation could promise hope for women with pathogenic mtDNA mutations, this same technology could also be used by another couple to increase the probability that their offspring would have a socially desirable phenotype. For example, parents might wish for their children to be better long-distance runners and thus attempt to switch the mother's loosely coupled mtDNAs for those of a woman with more tightly coupled mtDNAs. Alternatively, because loosely coupled mitochondria have been associated with decreased risk for developing late-onset metabolic (obesity and diabetes) and degenerative (AD and PD) diseases, nuclear transplantation to switch a mother's tightly coupled mtDNA for a loosely coupled mtDNA might be argued to be a reasonable approach to increase the probability of assuring her offspring's long-term health.

From the evolutionary perspective, however, the use of nuclear transplantation to manipulate nonlethal mtDNA variation is not appropriate. The late-onset phenotypes associated with ancient mtDNA polymorphisms may well be the result of a mismatch between adaptive mtDNA variants and the current local environment. Therefore, the elimination of these mtDNAs would arbitrarily reduce the species' genetic diversity and thus compromise the potential survival of future generations, should the environment change.

Evolutionary Ethics and Eugenics

Ethics is intertwined with evolutionary principles (Darwin 1882, Chapter V), suggesting that perhaps our genes are defining our ethics. From the perspective of evolution, using biotechnology to avoid creating a child with devastating genetic diseases is simply choosing an alternative to natural selection for eliminating deleterious genes, and thus is "ethical." By contrast, using biotechnology to limit diversity for local benefit is counter to the species survival strategy and is "unethical."

This evolutionary prospective now permits us to reassess eugenics. A 21st century eugenics must divide traditional eugenic concerns into two parts: (1) the endorsement of the application of biotechnology to increase the reproductive potential of a couple, but (2) the restraint of the use of biotechnology for applications that would reduce ancient genetic diversity for local gains.

However, conflict can arise between maximizing the genetic capacity of the individual and optimizing the genetic potential of the population. Inherent in DNA replication and the resulting transmission of biological information to the next generation is the necessity for the individual, the purveyor of the genes, to survive and reproduce. But in our species, this primary directive for individual survival is augmented by a powerful new adaptive strategy, that of group cooperation. The hunting group and later the society have generally proven to be more successful at competing for limiting resources and in protecting the young than the individual has alone (Darwin 1882, Chapter V). Hence, a new genetic program has been added that enhances group association and cooperation. However, this new adaptive strategy requires sublimating maximum individual gain for optimal group success. As a result, humans have evolved two contrasting survival strategies: (1) personal application of effort and resources to maximize individual survival and reproduction and (2) allocation of some individual effort and resources to the group success in hopes that the group will enhance the probability of individual survival and reproduction.

These two survival strategies create a fundamental conflict within the individual, the consequences of which could be envisioned as the root of "good" and "evil" within the societal context. Because the genetic and behavioral traits of individuals vary, individual and societal behavior will also vary. A successful balance between the individual and societal constraints and benefits that increases the overall genetic success of the population is judged as good. Excessive individual selfish behavior or excessive mistreatment of a subgroup by the society is judged as bad. Examples of good individual behavior include paying taxes or adhering to the Chinese one child per family limit. Good societal practices include food redistribution programs, monogamy rules, and the U.S. Bill of Rights. Bad individual behavior includes theft, rape, and murder. Bad societal structures would include racial discrimination and racial cleansing.

In the past, resources within the human niche were generally limiting, and thus intraspecific competition for those resources was required for individual and group survival. Therefore, it was beneficial to the genes and gene pool of one group to restrict the access to the resources of adjacent groups (Darwin 1882, Chapter V). Consequently, the resulting capacity for intergroup conflict over limited resources then became a part of our genetic program.

As populations expanded, cooperative groups became larger, and competition for resources became more intense and sophisticated. This led to warfare and the destruction of individuals from opposing groups. The sanctioning of killing the members of another group while punishing the killing of members of one's own group is then justified by defining the other group as fundamentally different. Commonly exploited differences include "ethnicity," geography, religion, etc. Warfare has now become sufficiently sophisticated that it could easily destroy the entire human race and thus the human gene pool. The question for our species is, can our genetic programs permit defining all people as members of the same group? If so, it may be possible for our species to redirect our efforts from intraspecific conflict to developing a stable human population that can be indefinitely sustained within the limits of Earth's resources? If this proves not to be possible, then the population and the gene pool will eventually be catastrophically diminished by ecological collapse, war, and infectious disease.

The evolutionary advantages of genetic diversity and the large stochastic component of human genetics now provide a compelling justification for social justice and beneficence with respect to genetic diversity and inherited disease. Because the functionally important genetic variants of the species must be distributed throughout the individuals in the population, each individual acquires at conception only a small subset of the array of population variants. The specific combination of variants that the individual inherits plus the environment into which he/she is born will then determine the individual's phenotype. Because the assortment and behavior of the genes is highly stochastic, every new couple has a significant probability (5–10%) of conceiving a child whose hereditary constitution will result in his/her phenotype departing from the current environment's genetic and social norm, such that the child will be classified as having a genetic disease (Gelehrter et al. 1998, pp. 1–3). The corollary to this fact is that all individuals alive inherited his/her strengths and weaknesses by chance. Therefore, neither the parents nor offspring merit either credit or blame for the nature of their genetic repertoire. It is to the species' and societies' advantage to maintain genetic diversity; thus, every member of the society should share the benefits of the able and assist with the costs of those with "special needs." This then becomes society's price tag for ensuring the survival of our progeny and our species.

REFERENCES

Akide-Ndunge O.B., Ayi K., and Arese P. 2003. The Haldane malaria hypothesis: Facts, artifacts, and a prophecy. *Redox Rep.* **8:** 311–316.

Baudouin S.V., Saunders D., Tiangyou W., Elson J.L., Poynter J., et al. 2005. Mitochondrial DNA and survival after sepsis: A prospective study. *Lancet* **366:** 2118–2121.

Bellissimo D.B. 2007. Practice guidelines and proficiency testing for molecular assays. *Transfusion* **47:** 79S–84S.

Bonafe M., Barbi C., Olivieri F., Yashin A., Andreev K.F., et al. 2002. An allele of HRAS1 3′variable number of tandem repeats is a frailty allele: Implication for an evolutionarily-conserved pathway involved in longevity. *Gene* **286:** 121–126.

Cai W.W. 2007. Detection of DNA copy-number alterations in complex genomes using array comparative genomic hybridization. *Methods Mol. Biol.* **381:** 105–120.

Cann R.L., Stoneking M., and Wilson A.C. 1987. Mitochondrial DNA and human evolution. *Nature* **325:** 31–36.

Carrieri G., Bonafe M., De Luca M., Rose G., Varcasia O., et al. 2001. Mitochondrial DNA haplogroups and *APOE4* allele are non-independent variables in sporadic Alzheimer's disease. *Hum. Genet.* **108:** 194–198.

Castro M.G., Terrados N., Reguero J.R., Alvarez V., and Coto E. 2007. Mitochondrial haplogroup T is negatively associated with the status of elite endurance athlete. *Mitochondrion* **7:** 354–357.

Cavalli-Sforza L.L. and Bodmer W.F. 1971. *The genetics of human populations.* W.H. Freeman, San Francisco.

Darwin C. 1871. *The descent of man, and selection in relation to sex.* John Murray, London.

Darwin C. 1871. *The descent of man, and selection in relation to sex.* D. Appleton and Company, New York.

De Benedictis G., Rose G., Carrieri G., De Luca M., Falcone E., et al. 1999. Mitochondrial DNA inherited variants are associated with successful aging and longevity in humans. *FASEB J.* **13:** 1532–1536.

Fan W., Waymire K., Narula N., Li P., Rocher C., et al. 2008. A mouse model of mitochondrial disease reveals germline selection against severe mtDNA mutations. *Science* (in press).

Fasouliotis S.J. and Schenker J.G. 1998. Preimplantation genetic diagnosis principles and ethics. *Hum. Reprod.* **13:** 2238–2245.

Feinberg A.P. 2007. Phenotypic plasticity and the epigenetics of human disease. *Nature* **447:** 433–440.

Frazer K.A., Ballinger D.G., Cox D.R., Hinds D.A., et al.; International HapMap Consortium. 2007. A second generation human haplotype map of over 3.1 million SNPs. *Nature* **449:** 851–861.

Fuku N., Park K.S., Yamada Y., Nishigaki Y., Cho Y.M., et al. 2007. Mitochondrial haplogroup N9a confers resistance against type 2 diabetes in Asians. *Am. J. Hum. Genet.* **80:** 407–415.

Garber K., Smith K.T., Reines D., and Warren S.T. 2006. Transcription, translation and fragile X syndrome. *Curr. Opin. Genet. Dev.* **16:** 270–275.

Gelehrter T.D., Collins F.S., and Ginsburg D. 1998. *Principles of medical genetics*, 2nd ed. Williams and Wilkins, Baltimore.

Ghezzi D., Marelli C., Achilli A., Goldwurm S., Pezzoli G., et al. 2005. Mitochondrial DNA haplogroup K is associated with a lower risk of Parkinson's disease in Italians. *Eur. J. Hum. Genet.* **13:** 748–752.

Giles R.E., Blanc H., Cann H.M., and Wallace D.C. 1980. Maternal inheritance of human mitochondrial DNA. *Proc. Natl. Acad. Sci.* **77:** 6715–6719.

Harper J.C., Wells D., Piyamongkol W., Abou-Sleiman P., Apessos A., et al. 2002. Preimplantation genetic diagnosis for single gene disorders: Experience with five single gene disorders. *Prenat. Diagn.* **22:** 525–533.

Ijssel P. and Ylstra B. 2007. Oligonucleotide array comparative genomic hybridization. *Meth. Mol. Biol.* **396:** 207–222.

Jenuth J.P., Peterson A.C., Fu K., and Shoubridge E.A. 1996. Random genetic drift in the female germline explains the rapid segregation of mammalian mitochondrial DNA. *Nature Genet.* **14:** 146–151.

Jiang Y.H., Bressler J., and Beaudet A.L. 2004. Epigenetics and human disease. *Annu. Rev. Genomics Hum. Genet.* **5:** 479–510.

Johnson M.J., Wallace D.C., Ferris S.D., Rattazzi M.C., and Cavalli-Sforza L.L. 1983. Radiation of human mitochondria DNA types analyzed by restriction endonuclease cleavage patterns. *J. Mol. Evol.* **19:** 255–271.

Lander E.S., Linton L.M., Birren B., Nusbaum C., Zody M.C., et al.; International Human Genome Sequencing Consortium. 2001. Initial sequencing and analysis of the human genome. *Nature* **409:** 860–921.

Merriwether D.A., Clark A.G., Ballinger S.W., Schurr T.G., Soodyall H., et al. 1991. The structure of human mitochondrial DNA variation. *J. Mol. Evol.* **33:** 543–555.

Mishmar D., Ruiz-Pesini E.E., Golik P., Macaulay V., Clark A.G., et al. 2003. Natural selection shaped regional mtDNA variation in humans. *Proc. Natl. Acad. Sci.* **100:** 171–176.

Montiel-Sosa F., Ruiz-Pesini E., Enriquez J.A., Marcuello A., Diez-Sanchez C., et al. 2006. Differences of sperm motility in mitochondrial DNA haplogroup U sublineages. *Gene* **368C:** 21–27.

Niemi A.K. and Majamaa K. 2005. Mitochondrial DNA and ACTN3 genotypes in Finnish elite endurance and sprint athletes. *Eur. J. Hum. Genet.* **13:** 965–969.

Niemi A.K., Hervonen A., Hurme M., Karhunen P.J., Jylha M., and Majamaa K. 2003. Mitochondrial DNA polymorphisms associated with longevity in a Finnish population. *Hum. Genet.* **112:** 29–33.

Orr H.T. and Zoghbi H.Y. 2007. Trinucleotide repeat disorders. *Annu. Rev. Neurosci.* **30:** 575–621.

Pyle A., Foltynie T., Tiangyou W., Lambert C., Keers S.M., et al. 2005. Mitochondrial DNA haplogroup cluster UKJT reduces the risk of PD. *Ann. Neurol.* **57:** 564–567.

Ross O.A., McCormack R., Curran M.D., Duguid R.A., Barnett Y.A., et al. 2001. Mitochondrial DNA polymorphism: Its role in longevity of the Irish population. *Exp. Gerontol.* **36:** 1161–1178.

Ruiz-Pesini E. and Wallace D.C. 2006. Evidence for adaptive selection acting on the tRNA and rRNA genes of the human mitochondrial DNA. *Hum. Mutat.* **27:** 1072–1081.

Ruiz-Pesini E., Mishmar D., Brandon M., Procaccio V., and Wallace D.C. 2004. Effects of purifying and adaptive selection on regional variation in human mtDNA. *Science* **303:** 223–226.

Ruiz-Pesini E., Diez C., Lapena A.C., Perez-Martos A., Montoya J., et al. 1998. Correlation of sperm motility with mitochondrial enzymatic activities. *Clin. Chem.* **44:** 1616–1620.

Ruiz-Pesini E., Lott M.T., Procaccio V., Poole J., Brandon M.C., et al. 2007. An enhanced MITOMAP with a global mtDNA mutational phylogeny. *Nucleic Acids Res.* **35:** D823–D828.

Sabeti P.C., Varilly P., Fry B., Lohmueller J., Hostetter E., et al. 2007. Genome-wide detection and characterization of positive selection in human populations. *Nature* **449:** 913–918.

Saxena R., de Bakker P.I., Singer K., Mootha V., Burtt N., et al. 2006. Comprehensive association testing of common mitochondrial DNA variation in metabolic disease. *Am. J. Hum. Genet.* **79:** 54–61.

Schaefer A.M., Taylor R.W., Turnbull D.M., and Chinnery P.F. 2004. The epidemiology of mitochondrial disorders—Past, present and future. *Biochim. Biophys. Acta* **1659:** 115–120.

Schaefer A.M., McFarland R., Blakely E.L., He L., Whittaker R.G., et al. 2007. Prevalence of mitochondrial DNA disease in adults. *Ann. Neurol.* [Epub ahead of print at http://dx.doi.org/10.1002/ana.21217].

Scott L.J., Mohlke K.L., Bonnycastle L.L., Willer C.J., Li Y., et al. 2007. A genome-wide association study of type 2 diabetes in Finns detects multiple susceptibility variants. *Science* **316:** 1341–1345.

Slack C., Lurix K., Lewis S., and Lichten L. 2006. Prenatal genetics: The evolution and future directions of screening and diagnosis. *J. Perinat. Neonatal Nurs.* **20:** 93–97.

Sladek R., Rocheleau G., Rung J., Dina C., Shen L., et al. 2007. A genome-wide association study identifies novel risk loci for type 2 diabetes. *Nature* **445:** 881–885.

Tanaka M., Fuku N., Nishigaki Y., Matsuo H., Segawa T., et al. 2007. Women with mitochondrial haplogroup N9a are protected against metabolic syndrome. *Diabetes* **56:** 518–521.

van der Maarel S.M., Frants R.R., and Padberg G.W. 2007. Facioscapulohumeral muscular dystrophy. *Biochim. Biophys. Acta* **1772:** 186–194.

van der Walt J.M., Nicodemus K.K., Martin E.R., Scott W.K., Nance M.A., et al. 2003. Mitochondrial polymorphisms significantly reduce the risk of Parkinson disease. *Am. J. Hum. Genet.* **72:** 804–811.

Wallace D.C. 1987. Maternal genes: Mitochondrial diseases. In *Medical and experimental mammalian genetics: A perspective* (ed. V.A. McKusick et al.), pp. 137–190. A.R. Liss, Inc., New York, for the March of Dimes Foundation.

Wallace D.C. 2005. A mitochondrial paradigm of metabolic and degenerative diseases, aging, and cancer: A dawn for evolutionary medicine. *Annu. Rev. Genet.* **39:** 359–407.

Wallace DC. 2007. Why do we have a maternally inherited mitochondrial DNA? Insights from evolutionary medicine. *Annu. Rev. Biochem.* **76:** 781–821.

Wallace D.C., Brown M.D., and Lott M.T. 1999. Mitochondrial DNA variation in human evolution and disease. *Gene* **238:** 211–230.

Wallace D.C., Lott M.T., and Procaccio V. 2007. Mitochondrial genes in degenerative diseases, cancer and aging. In *Emery and Rimoin's principles and practice of medical genetics*, 5th ed. (ed. D.L. Rimoin et al.), pp. 194–298. Churchill Livingstone Elsevier, Philadelphia.

Wallace D.C., Ruiz-Pesini E., and Mishmar D. 2003. mtDNA variation, climatic adaptation, degenerative diseases, and longevity. *Cold Spring Harbor Symp. Quant. Biol.* **68:** 479–486.

Wang E.T., Kodama G., Baldi P., and Moyzis R.K. 2006. Global landscape of recent inferred Darwinian selection for *Homo sapiens*. *Proc. Natl. Acad. Sci.* **103:** 135–140.

Willer C.J., Bonnycastle L.L., Conneely K.N., Duren W.L., Jackson A.U., et al. 2007. Screening of 134 single nucleotide polymorphisms (SNPs) previously associated with type 2 diabetes replicates association with 12 SNPs in nine genes. *Diabetes* **56:** 256–264.

Williams T.N., Mwangi T.W., Wambua S., Peto T.E., Weatherall D.J., et al. 2005. Negative epistasis between the malaria-protective effects of alpha+-thalassemia and the sickle cell trait. *Nature Genet.* **37:** 1253-1257.

Xu K., Shi Z.M., Veeck L.L., Hughes M.R., and Rosenwaks Z. 1999. First unaffected pregnancy using preimplantation genetic diagnosis for sickle cell anemia. *J. Am. Med. Assoc.* **281:** 1701–1706.

Zeggini E., Weedon M.N., Lindgren C.M., Frayling T.M., Elliott K.S., et al. 2007. Replication of genome-wide association signals in U.K. samples reveals risk loci for type 2 diabetes. *Science* **316:** 1336–1341.

Psychiatric Genetics in an Era of Relative Enlightenment

Daniel R. Weinberger

Genes, Cognition and Psychosis Program, National Institute of Mental Health

David Goldman

Laboratory of Neurogenetics, National Institute of Alcohol Abuse and Alcoholism

Editors' Note: Risk is a part of everyday life. In the quote below, Davenport was onto something in recognizing that perhaps some of the risk in our lives is genetic, although his view at the turn of the 20th century was necessarily simplistic. At the turn of the 21st century, we still recognize genetic risk, although the problem has become much more complex. Drs. Weinberger and Goldman set out the precise nature of the complexity of genetic risk and its implications, especially in the realms of human behavior and mental illness. The importance of polymorphisms, or genetic variations, in such key genes as *COMT, ADLH1B* and *ALDH2, SERT, DAOA*, and *BDNF* is set forth, as are the study methods for relating these variations to the risk for such conditions as psychosis, alcoholism, depression, and problems with learning and memory. The authors emphasize that there are no genes for mental illness per se, rather that neurobiological variation related to genetic variability may lead to small increases in the odds of developing a mental illness or behavioral problem. That view is not very different from Davenport's statement that "insanity is a *result* merely and not a specific trait" (p. 24; Davenport's emphasis).

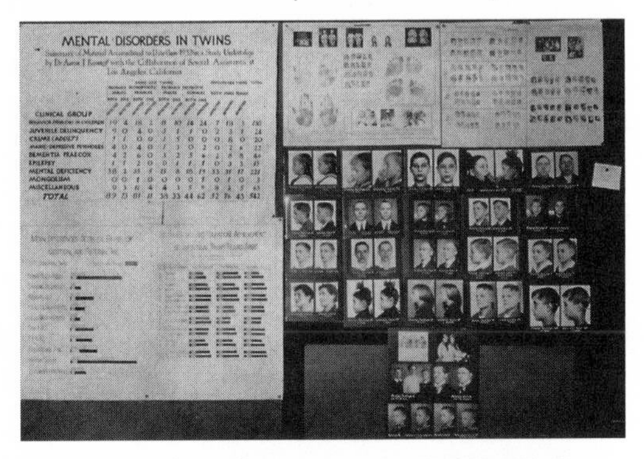

Of course this idea of genetic risk has far-reaching ethical and social implications. What should be done if a person harbors such risk genes—for example, a risk gene like *MAOA* for aggressive behavior and violence? In the case of this gene, links have been made with a low-expressing variant of *MAOA* and differences in the limbic circuitry for emotional regulation. These findings represent the upside of psychiatric genetics because such knowledge opens the door to developing treatments that are based on these genetic profiles. But singling out people with such genetic risk is an ethical minefield, because people with risk gene variants may never develop the condition indicated. And their contribution to human behavioral diversity is important—eliminating this diversity, as the authors point out, would impoverish us.

> *Before considering the inheritance of specific nervous diseases, it may be pointed out that what is inherited is often a general nervous weakness—a neuropathic taint—showing itself now in one form of psychosis and now in another.*
>
> DAVENPORT, P. 93

THE RESEARCH HORIZON OF THE laboratories at Cold Spring Harbor at the turn of the 20th century revealed the first light of a new century of scientific enlightenment, inspired by dramatic insights about the mechanisms of heredity and the role of inherited variation in evolution. Mendel's work had been rediscovered, T.H. Morgan and his students were working on the first genetic linkage maps, the pattern of inheritance of many human diseases was being identified, and it was becoming understood that at least some genetic diseases had a biochemical basis. Francis Galton had founded a science of quantitative genetics for complex traits encompassing major facets of human interindividual variation. However, these scientific breakthroughs also encouraged extreme distortions of what little was known at the time about human behavior and human disease. If plants and animals could be bred to select for desirable traits, why could not humans? This question was at the core of Charles Davenport's treatise on eugenics—

Heredity in Relation to Eugenics. Darwin had explained the origin of species as a result of the relentless winnowing of the genetically unfit. Humans had domesticated and bred animals for better characteristics. Could not humans improve on the natural selection of man? Could not the procreation of individuals with undesirable traits be prevented? Davenport was particularly concerned with what he considered to be the social, spiritual, economic, and biologic ills attributed to human deviancy, especially to behavioral and intellectual disorders. Davenport's principal concerns were feeblemindedness, psychosis, and addiction. His incorrect assessment of the origins of these problems led him and the followers of the eugenics movement to propose a false cure: These characteristics could and should be "bred out" of the human "race."

PSYCHIATRIC GENETICS IN PERSPECTIVE

Such negative eugenics was primarily a creature of the first four decades of the 20th century, eventually discredited by the racist/genocidal program of the Nazis. Curiously, the political demise of eugenics actually preceded its scientific obliteration. Up until World War II, human genetics was primarily statistical and observational, and the science of behavior was also primarily an observational science. The genetics of other species, such as *Drosophila*, was beginning to advance rapidly on the basis of controlled experiments, but genetics was faltering when applied to complex traits in humans, a species that was not available for experimental manipulation. To this day, the study of complex traits remains the most difficult frontier in human genetics.

Studies of the quantitative inheritance of behavioral traits and psychiatric diseases began in the mid-20th century and they continue to the present day in the form of twin, family, and adoption studies that estimate trait variance attributable to inheritance (heritability). These heritability studies improved on the naïve efforts of the prewar period in many ways: sampling, statistical power, phenotype characterization, and mathematical models. However, the possibility of a human behavioral eugenics was not ended by a better appreciation of heritability. Most of the common psychiatric diseases and trait variations in personality and cognition are, in fact, substantially heritable. Instead, in the postwar era the principles of eugenics were buried and psychiatric genetics was reborn because of a more enlightened understanding of the multivalent functionalities of genes in behavior and the widespread population distributions of the DNA variants that are responsible. Only after World War II was it discovered that DNA carried the genetic code; the code was deciphered, the human sequence was determined, and tools became available to manipulate genes. Also, enormous strides were made in elucidating the neurobiological basis of behavior and in accessing the brain at the systems, neuronal, and molecular levels. Brain imaging, cellular electrophysiology, and neurochemistry are all paradigms that emerged during the latter half of the 20th century, and neuropsychology, despite the prior invention of IQ and personality tests, is primarily a discipline of this same era. Integration of these sciences has now led to some understanding of the contribution of individual genes to variation in human behavior and to mental illness. Meanwhile, we have begun to understand the role in behavior and cognition of environmental exposures such as birth trauma, nutritional deficits, and stress.

Although a few of Davenport's ideas about certain inherited brain diseases (e.g., Huntington's disease and some forms of epilepsy) are consistent with current knowledge, almost none of his ideas about the genetics of mental illness and mental disabilities remain scientifically credible. Davenport and his followers confused characteristics that run in families with inherited traits. Language and religion "run in families," but clearly both are learned behaviors. He defined racial groups on the basis of superficial physical or cultural differences, which we now know do not predict relative genetic homogeneity as would be characteristic of a "race." Studies of genetic variation at the level of genetic polymorphism that would consistently reveal that ethnic and interpopulation variance comprises 15%, or less, of human variation came well after Davenport's time. Nor could he know that the extant human species, as a whole, shares a common ancestor "out of Africa" only some 50,000 to 100,000 years ago.

Davenport's concepts of behavioral and mental deviance were charged with moralistic and racist beliefs, which are inconsistent with current scientific evidence about the origins of neuropsychiatric disorders. His views about the inheritance patterns of mental disorders and abilities are not consistent with current scientific evidence about complex genetics. In this essay, we contrast some of the assumptions made by Davenport 100 years ago and our current understanding of the genetics of psychiatric conditions.

MENTAL DEFICIENCIES CAN BE INHERITED

Davenport felt that mental deficiencies, which he lumped into a heterogeneous catalog of conditions including mental retardation, psychosis, addiction, promiscuity, and criminality, were inherited as missing attributes. In his view, "Two mentally defective parents will produce only mentally defective offspring" (p. 66). As we will emphasize, human behavioral disorders frequently arise not from deficit but from variation—and variants at the same genetic locus can confer risk for different behavioral problems and, sometimes under other circumstances, disease-associated variants can lead to behavioral strengths. Neither variant is "good" or "bad." Davenport also believed that the missing qualities followed simple inheritance patterns in families, analogous to the physical traits of Mendel's peas. In his typical examples, affected individuals had two affected parents and two affected parents produced only affected offspring, a pattern of uniform inheritance potentially consistent with a recessive model and complete penetrance of gene effect. Davenport's view was based on anecdotal and biased reports that could not distinguish between familial aggregations of characteristics based on environmental factors from true genetic traits. As will be seen, the study of systematically ascertained twins and families has revealed that such families and such apparently simple patterns of inheritance are a rarity in psychiatry rather than the rule.

Twin Studies

By the middle of the 20th century, research strategies had been developed for disambiguating genetic from environmental factors, and these were being applied to many human characteristics, including behavioral and psychiatric disorders. One strategy involved comparing the rates of concordance for a given characteristic in pairs of twins, some pairs being monozygotic (MZ) ("identical") and other pairs being dizygotic (DZ) ("fraternal"). Because MZ twins are virtually genetic clones and share all of their DNA sequence, whereas DZ twins are like any sibling pair sharing only 50% of their DNA sequence, concordance rates for MZ twins should be greater if the trait is genetically determined. Because MZ and DZ twins tend to share the same environments, they would not show different concordance rates if the characteristics were largely nongenetic. This approach has shown that although most psychiatric disorders are strongly influenced by genetic factors, none of them are determined only this way. For example, in the case of schizophrenia, the term used today to refer to many of the psychotic and "feebleminded" states mentioned by Davenport, genetic concordance in MZ twins is about 50%, meaning that even genetic identity to a co-twin ill with schizophrenia only leads to 50–50 odds of this illness (Kendler 2000). However, this risk in the MZ co-twin of a schizophrenia patient represents a 50-fold increase in risk over the general population. In DZ co-twins of schizophrenia probands (an affected index patient in a family) and in other siblings, there is a four- to fivefold increased risk of schizophrenia. Thus, genes are important but are not by themselves sufficient. Similar data have emerged from twin studies of many other psychiatric diseases, including alcoholism and other addictions, anxiety disorders, bipolar illness, eating disorders, and obsessive–compulsive disorder. Similar data have also emerged for many cognitive and behavioral traits, some of which may more closely reflect the activity of the brain—for example, differences in the electroencephalogram, differences in dimensions of personality, and differences in cognitive performance.

Segregation Analysis

Another strategy for analyzing how tightly genes predict behavioral disturbance in families is to characterize inheritance patterns in systematically ascertained pedigrees and then mathematically model how risk varies depending on the degree of relationship to ill members of the family. The application of so-called segregation analysis to families with psychiatric disorders has demonstrated that none of the clinically defined psychiatric diseases, including schizophrenia, manic–depressive illness (now called bipolar disorder), alcohol abuse, and anxiety disorders, appear to be explained by single-gene inheritance. Thus, there is not likely to be a gene, in any dose, that guarantees any of these disorders. Indeed, such studies have shown that if a child has two parents with the diagnosis of schizophrenia, the maximum risk to the child is no greater than one in five (Kendler 2000). For several psychiatric diseases, including schizophrenia, if an individual is a second-degree relative of an ill family member, the risk falls off much more steeply than would be predicted by simple dominant or recessive gene effects. Thus, psychiatric conditions appear to follow a more complicated pattern of inheritance and to involve many potential genes (i.e., they are multigenic) as well as nongenetic factors. For some diseases (e.g., schizophrenia and autism) family studies show that the disease may not only be multigenic but also polygenic (more than one genetic variant required for illness in an individual).

The evidence from twin and family studies led to a model of multifactorial causality of mental illness, similar to the causality of obesity, high blood pressure, and other common, complex diseases: Genes increase susceptibility, but do not cause disease, per se. Thus, the genetic and nongenetic risk factors are multiple, different risk factors are operative in different families, and these risk factors presumably must coalesce within an individual and in some cases exert their effects at critical times in development to surpass a critical threshold (i.e., for them to be penetrant in genetic terms). The patterns of familial transmission observed, and explained by a complex genetic model, contradict the simple patterns of inheritance illustrated by Davenport.

RISK FACTORS CAN BE INHERITED

Psychiatric disorders are not inherited simply as missing attributes or as mental deficiencies. Rather, the genetics of mental disorders frequently involves inheritance of genetic variants that account for neurobiologic variation and that also lead to small increases in the odds of illness or behavioral disturbance, thus making them risk factors as well. This suggests that there are no genes *for* mental illness per se. Davenport himself appreciated that "insanity is a *result* merely and not a specific trait" (p. 24; Davenport's emphasis). In the past few years, several genes have been identified that increase risk for psychiatric disorders, and in each case such genes affect multiple biologic processes, even in individuals who manifest no behavioral or psychiatric disturbance. These results suggest that if the eugenic goal of eliminating risk alleles were actually carried out, the disastrous consequence would be the elimination of traits that are favorable under many circumstances and important for human behavioral diversity. Also, in some cases both major alleles are actually unfavorable under the wrong circumstances, defying a rational eugenics for such genes.

One gene that illustrates these points is the gene that codes for the enzyme catechol-*O*-methyltransferase—the *COMT* gene. Inherited variations in *COMT* have been associated with increased risk of both psychosis and anxiety-related conditions. COMT is an enzyme that inactivates certain chemical neurotransmitters in the brain, particularly the neurotransmitter dopamine. COMT appears to be especially important for regulating the actions of dopamine in the frontal lobe, where dopamine plays a crucial role in attention and working memory (Weinberger et al. 2001). *COMT* occurs in two common genetic forms, which in Europeans exist in virtually a 50–50 ratio of frequency. In other ethnic groups, the frequency of the different forms ("alleles") varies considerably. One form of the gene, called "Val," is associated with less prefrontal dopamine and poorer working memory and frontal lobe function (Egan et al. 2001). This form of the gene also is weakly associated with increased risk of psychosis. The other form of the gene, called "Met," tends to be associated with better working memory and frontal lobe function but also with increased expression of anxiety-related characteristics (Enoch et al. 2003), especially in the context of resilience to stress—for example, when a person experiences pain (Zubieta et al. 2003). These cognitive and behavioral correlates of genotype are found in normal people with no history of psychiatric disturbance, illustrating that the gene relates not to a psychiatric condition, per se, but to subtle variation in brain function that has cognitive and psychological implications. It is assumed, analogous to the role of genes in other common medical disorders such as high blood pressure, obesity, and cancer, that the biological effects of COMT variation must interact with other factors for illness to occur—*COMT* is a "susceptibility gene." It also should be noted that the high frequency of both alleles across different human ethnic groups suggests that there has been evolutionary pressure to maintain both in the human species.

There are likely two explanations for the persistence of the Val allele, which underscore the wrongness of Davenport's simplistic eugenics for psychiatric disease. First, Val individuals tend to perform on average only about 3% worse on frontal lobe cognitive tests than Met individuals, not much overall. Second, there may be some advantages to having the Val allele. It is conceivable that Val allele individuals may have been more adaptable to certain environments—for example, those related to aggressiveness and social instability. There also may be environmental circumstances in which the Met allele is disadvantageous, for example, when it is critical to be more stress resilient, mentally pliable, and fluid under conditions that can provoke anxiety and a decrement in performance. These last points also illustrate why genetic variations associated with mental disorders do not necessarily translate into missing attributes (i.e., a genetics of deficits): Environmental context may lead to an advantage of alleles that are liabilities in other contexts. The *COMT* story also highlights that the genetic variations associated with risk for psychiatric disorders contribute to human behavioral diversity and represent part of the normal human genetic landscape, not a backwater of genetic aberration.

Whereas family studies had suggested that psychiatric disorders are genetically complex, the discovery of specific genes for such disorders has confirmed their multigenic nature and has shown that many susceptibility genes, by themselves, account for a small increase in risk. Recent studies of the genetics of schizophrenia have identified already upward of ten genes that appear be susceptibility factors, and in no instance so far has the allele of the gene associated with schizophrenia increased the odds of being affected more than twofold. Many of these genes were not even known to scientists until a few years ago. They have names like *dysbindin, neuregulin1, DISC1, GRM3, GAD1, G72,* and *RGS4* (Harrison and Weinberger 2005). In most instances, the association is based on a statistical relationship between the clinical diagnosis of schizophrenia and variation in the DNA sequence of the gene, but it is important to note that, as of yet, how any of these genes are altered functionally is not clear. In the recently reported Wellcome Trust Study (2007), seven common complex diseases including bipolar disorder and other medical illnesses such as diabetes and Crohn's disease were analyzed by the whole-genome association approach, 2000 affected individuals with each disease being compared to 3000 controls. The result of this powerful study was that few common variants can be identified that account for common, complex diseases, and most of these variants have effect sizes of less than 2. In the case of bipolar disorder, only one gene was reliably identified, and it was of modest effect.

THE GENETICS OF MENTAL DISORDERS

Davenport believed that "probably no imbecile is born except of parents who, if not mentally defective themselves, both carry mental defect in their germ plasm" (p. 67). The implication of this statement is that genes related to mental disorders run primarily in those families that have multiple individuals with these conditions. Recent discoveries about the genetic origins of risk for many neuropsychiatric conditions have shown that this assumption is categorically wrong. First, psychiatric disorders are very common in world societies. Schizophrenia exists in approximately 1% of the world population and its geographical distribution is widespread; no culture or population is spared. Indeed it is likely that virtually all families and individuals carry within them genetic variants that are liability factors for psychiatric disease under some circumstances. The lifetime prevalence of major depression is approximately 15% in western cultures. Anxiety disorders affect upward of 20% of the population over the life span. Alcoholism is also very common, especially in certain economic and cultural groups. The occurrence of such high prevalence rates across diverse societies means that it would be virtually impossible to prevent the risk genes for these conditions from diffusing into the general population of humans. Indeed, the evidence about specific susceptibility genes for these conditions indicates that they are common. Rather than being confined to only subgroups of humans, genetic variations that increase risk for what Davenport called feeblemindedness and we now call psychiatric illness exist in relatively high frequency in most world populations. As noted above, both the Val and Met alleles of the *COMT* gene are very common in individuals of European ancestry, most of whom do not manifest psychiatric disorders, even though both alleles affect aspects of brain function to increase susceptibility to different psychiatric diseases.

Genetic Risk for Alcholism

Alcoholism, although the object of social approbation and approximately 50% heritable in societies around the world, is not a social malady suitable for elimination by selective breeding. Across societies and within populations, the prevalence of alcoholism has varied based more on public policy regulating alcohol price and availability and by customs affecting alcohol use than by genes. Simply stated, in societies where individuals typically drink larger quantities of alcohol, alcoholism is more common. Also, childhood trauma (e.g., childhood sexual abuse) plays a specific role in dramatically elevating risk of alcoholism (as well as other psychiatric disorders) (Koss et al. 2003). The vital role of environmental exposure and environmental risk factors in alcoholism has even led some (Merikangas and Risch 2003) to suggest that we should not expend extensive resources toward understanding the genetic and neurobiologic bases of individual differences in alcoholism vulnerability.

Nevertheless, although environmental modification strategies may have a major impact on risk reduction, an understanding of causation and practical medical intervention will be aided by gene discovery, which has had some recent successes. Two genes called alcohol dehydrogenase (*ADH*) and aldehyde dehydrogenase (*ALDH*) are involved in the metabolism of alcohol. In millions of individuals of Asian genetic origin, these two genes are now recognized to have protein-coding genetic variants (*ADH1B* His47Arg and *ALDH2* Glu487Lys) that exert probabilistic effects on risk of alcoholism, each variant conferring a three- to tenfold protective effect. The protection is due to the aversive flushing reaction that results from accumulating acetaldehyde, the intermediate metabolite in the conversion of ethanol to acetate. Interestingly, individuals who carry two copies of the *ALDH2* Lys487 allele ("homozygotes") are nearly completely protected. The protective effect of these variants closely resembles the action of two drugs, disulfiram (Antabuse), used to prevent alcohol consumption, and metronidazole (Flagyl), used to treat protozoal infections but also having the ability to inhibit the aldehyde dehydrogenase enzyme. Alcohol produces an aversive flushing reaction in individuals who take these drugs, or who carry the genetic variants (Goldman and Enoch 1989). Charles Davenport might have been surprised to learn that the genetic variants that help to protect individuals from the flushing reaction to alcohol and the social depravity of alcoholism are rare in most Caucasian groups, but are found in millions of Chinese, Japanese, and Koreans.

Serotonin Transporter and Risk of Depression

Another example of the role of common genes in risk for major psychiatric disorders is the gene for the serotonin transporter (SERT), the protein responsible for removing serotonin from synapses in the brain and the protein targeted by most antidepressant drugs. The *SERT* gene is the first gene identified that appears to increase risk for depression. There is a common sequence variation in a regulatory domain of the gene that does not vary the protein structure (i.e., the amino acid composition) but that is involved in turning the gene on for eventual translation into protein. The *SERT* variation comes in several forms. One particularly common form, called the s allele, has been associated with increased risk for depression and related anxiety traits (Murphy et al. 2004). At least one dose of this form of the gene is found in approximately 50% of Caucasian individuals. Thus, the allele associated with psychiatric morbidity is again almost the rule, not the exception. Interestingly, the s form of the gene is found in even greater frequency in certain populations (e.g., Japanese). The s allele associated with risk of anxiety and depression does not actually exert a very strong effect on risk, at least acting alone. In fact, each copy of the s allele only contributes about one tenth of a standard deviation (<5% of a person's anxiety ranking) to anxiety.

In a landmark study of people in New Zealand, it was shown that the gene increases risk for depression by sensitizing individuals to the cumulative effects of multiple negative environmental experiences (Caspi et al. 2003). In other words, the gene increases risk for the emergence of clinical depression not because it causes depression, but because it impacts on an individual's resiliency to rebound from personal losses (e.g., loss of a loved one, a job, a home). Examined as an individual factor, environmental trauma was actually a more powerful predictor of depression than the genotype, but more information was gained by evaluating the combination of gene and environment than either alone. This study exemplifies an important general concept about the genetics of mental illness: Psychiatric susceptibility genes bias how the environment is experienced. In the case of *SERT*, the mechanism of the effect of the gene appears to involve brain functions related to the processing of negative emotion.

Using neuroimaging as a research tool for studying brain function in living people, it has been shown that the s allele changes the excitability of a brain system focused around the amygdala, a structure deep in the brain that is important throughout mammalian evolution in vetting the potential danger of environmental stimuli (Hariri et al. 2002). This genetic effect on amygdala activity is related to the development of a circuit in the brain that normally quiets the amygdala after it has been activated and allows us to extinguish negative affect. Individuals with s alleles of the *SERT* gene appear to have a less effective amygdala-silencing circuit (Pezawas et al. 2005). In other words, normal humans with s alleles of *SERT* appear to be biologically biased at the level of amygdala excitation to overrespond to potential environmental threat. This effect may explain why individuals with reduction-of-function serotonin transporter alleles appear to be sensitized to negative experiences and at increased risk of depression as such experiences accumulate; however, excessive amygdala responsivity may also have advantageous implications in certain contexts, thus accounting for the persistence of the s allele as a common human genetic variation.

Strongly arguing for the selective advantage of persistence of *SERT* variation is the fact that a nonhuman primate species, the rhesus macaque monkey, has an orthologous functional *SERT* variant that confers similar stress-modified effects on behavior (Barr et al. 2003).

Why, in both the human and the macaque monkey, is it advantageous that the *SERT* gene be functionally polymorphic? We do not fully understand the answer to this question; however, the answer may lie in recent studies (Pezawas et al. 2005) that reveal that the gain-of-function and reduction-of-function *SERT* alleles lead to differences in brain connectivity and overall differences in the modulation of emotions. Furthermore, both rare and common gain-of-function alleles at SERT have now also been associated with a psychiatric disease, namely obsessive–compulsive disorder (Ozaki et al. 2003; Hu et al. 2006), a common, severe, and often intractable psychiatric illness.

Meiosis and Sporadic Mental Deficiency

One of Davenport's tenaciously advocated principles was that mental deficiency was inherited from parents who had mental deficiency genes. He acknowledged that one exception to this dictum was a relatively common form of mental retardation, Down syndrome, which is a sporadic condition based on a defect in maternal meiotic chromosomal segregation. In fact, there are a number of other forms of mental deficiency that also are not the results of genetic mutations passed from parents to their offspring. Two relatively common examples are Williams syndrome and velo-cardio-facial syndrome (VCFS). Both of these disorders involve slight decreases in general intellectual function and peculiar behavioral abnormalities. Both arise because of DNA mutations, but usually occur sporadically rather than being inherited from parents in the typical sense. These syndromes are members of a family of congenital disorders that result from a failure of normal DNA processing during meiosis. Meiosis involves separation of chromosomal pairs into sperm or egg cells. In the process of this uncoupling of chromosome pairs, errors are occasionally made between homologous regions of DNA, and deletions in the DNA sequence arise. VCFS involves a large deletion on chromosome 22, and it occurs in about 1 out of every 4000 births (Lindsay 2001). Williams syndrome involves a large deletion of chromosome 7, occurring in approximately 1 in 20,000 births (Mervis 2003). Individuals with VCFS, in addition to having mild mental retardation, have a high prevalence of psychiatric disorders, especially psychosis (in ~30% of cases). Williams syndrome, which also is associated with mild mental retardation, is characterized by unusual social exuberance and loquaciousness. Because these conditions typically involve spontaneous chromosomal aberrations, most individuals with these conditions do not have parents with any sign of the same disorders. The DNA of the parents of children with these syndromes harbors no signs of the conditions that affect their children.

BEHAVIORAL DISORDERS ARE COMPLEX PHENOTYPES

One of Davenport's observations that has withstood the tests of time and scientific progress is that within families with psychiatric illness the symptoms and the diagnoses are often inconsistent from one affected individual to another. "[W]hat is inherited is often a general nervous weakness—a neuropathic taint—showing itself now in one form of psychosis and now in another" (p. 93).

If one member of a family has either schizophrenia, depression, or alcoholism, other individuals in the family are more likely to be afflicted by other psychiatric conditions. Davenport interpreted this to indicate that a general "weakness" was inherited, which could take various forms. Our current understanding is that behavioral disorders are complex phenotypes that reflect the action of multiple causative factors, both genetic and environmental. They also reflect the complex neurobiological origin of behavior. The brain is not conveniently compartmentalized by psychiatric disease, and neither is the genome. Both genes and brain lead to pleiotropy of gene effects—that is, a tendency to diverse and multifaceted phenotypes from the same risk factor. A gene influencing frontal cortical function (e.g., *COMT*) can influence the risk for various diseases—schizophrenia, alcoholism, anxiety disorders, obsessive–compulsive disorder—for which frontal cortical function plays an important role and can also influence risk for multiple psychiatric diseases (actually, most psychiatric diseases) where resiliency in the face of stress exposure is of key importance. For complex traits, genetic complexity and variation in environmental exposure contribute to pleiotropy within families. First-degree family members share only on average 50% of their genes, and they also vary in developmental experience. For example, risk for schizophrenia is determined as much by obstetrical complications as by any single gene (McGrath and Murray 2003). Variations in psychiatric diagnosis also reflect, to some degree, the imprecision of the diagnostic process, which depends entirely on the subjective experience of a patient and his or her manifest behavior. Subjective experience and manifest behavior are not likely to reflect unitary biological processes.

Despite the apparent variability of psychiatric phenomena in families, the patterns are not random, which probably reflects segregation of specific genetic factors. Thus, psychosis, which is a breakdown in distinguishing reality from inner perception and is often manifest as hallucinations and odd ideas, appears to run in families along with related personality traits. When schizophrenia occurs in a family, other family members who are not affected with this diagnosis may have related symptoms, such as odd behavior, a tendency toward paranoid ideas, social awkwardness, and even transient distortions of perception. These so-called schizotypal or schizoid traits are more frequent in relatives of patients with schizophrenia than in the general population. This phenomenon supports the assumption that unaffected family members share some of the risk genes that translate into subtle changes in brain processes related to the origins of the disorder, but either not enough risk factors or enough protective factors to spare clinical decompensation.

One of the genes that has been associated with a general tendency toward psychosis is called *G72* or *DAOA*. This gene was originally found in individuals who were diagnosed with schizophrenia, but subsequently it has been shown to be associated with affective psychosis also (Harrison and Weinberger 2005). This latter term refers to psychotic behavior that occurs in the context of manic–depressive illness or bipolar disorder. *COMT* may be another gene that increases risk for psychosis of various forms.

Alcoholism and antisocial personality also run together in some families, but alcoholism and other problems—depression and anxiety—run together in others. In the families with two or more of these psychiatric diseases, the diagnoses are not only often co-morbid (showing a tendency to be found in the same individuals), they are cross-transmitted. These different types of patterns of co-morbidity in individuals and cross-transmission in families seem to indicate that there are different genetic and environmental origins for alcoholism. In contrast, schizophrenia and alcoholism do not occur in families more frequently than would be predicted by the chance co-occurrence based on population prevalence, suggesting that risk factors for these disorders are independent. Likewise, and presumably for similar reasons, depression and schizophrenia do not co-occur in families more often than would be predicted by the base rates of these disorders in the general population. Thus, genetic epidemiology, that is to say the patterns of familial transmission and co-morbidity of various psychiatric diseases, has given us a better understanding of the relative role of shared and unshared risk factors, and the heterogeneity within current diagnoses. Although this knowledge informs us that it would be useless to try to round up and sterilize all the "defectives," it is fair to say that the benefits to patients are largely still unrealized.

PSYCHIATRIC DISEASES REMAIN STIGMATIZING

Psychiatric diseases, including alcoholism and other addictions, tend to be lifelong disorders that are relatively resistant to treatment. Often, individuals will not seek treatment if it is not highly effective. Even the well-meaning person may tend to stigmatize individuals who show deviant behavior and cannot be helped, especially if we realize that we share their vulnerability at some level. Psychiatric patients, including the drug addicted, remain among the most stigmatized in our culture. In this important regard, perhaps we have not advanced so much since the time of Davenport. One task of biological psychiatry, of which psychiatric genetics is a key part, is to subclassify diseases based on etiology in order to develop new treatments and to better target treatments to patients likely to respond. An important result of better understanding and of better treatment is destigmatization.

Lessons From the *BDNF* Gene and Risk for Psychiatric Illness

The search for common genetic variants exerting effects on behaviors found in all of us, including the psychiatrically ill, recently revealed the interesting example of the gene for brain-derived neurotrophic factor or BDNF. This gene has been associated with increased risk for schizophrenia and for bipolar psychosis. BDNF is a small protein secreted by nerve cells that serves a critical role in brain development, in the continuously changing pattern of connections that nerve cells make with each other during life and in learning and memory. The role of BDNF in development and learning is related to the process of neuronal plasticity. These dynamic alterations that neurons make in the complex architecture of their dendrites and in their connections with other cells in response to changes in the stimuli that they receive involve complex molecular adaptations to changes in the environment. *BDNF*, like most genes, carries many sequence variations, which have accumulated in humans over evolution; but two alleles, in particular, have been shown to have a major impact on the function of BDNF in neuronal plasticity and in learning and memory.

The most common form of *BDNF* is the "Val" allele form, which has a frequency of about 80% in individuals of European ancestry. Another form, the "Met" allele, is less common (~20%), but because people carry two copies of *BDNF*, it is found in almost one-third of Caucasians. The Met allele is associated with abnormal cellular function of BDNF. The Met allele appears to misinform the cell about how to handle the BDNF protein after it is made and leads to a failure to put BDNF in the right molecular package and deliver it to the right intracellular address. Thus, Met BDNF is not secreted normally in response to signals from one cell to another and it alters the role of BDNF in neuroplasticity. Consistent with this molecular variation, individuals with *BDNF* Met alleles, even normal individuals, show subtle changes in brain morphology and do not remember information as effectively as individuals with Val alleles, although the difference in memory performance is slight (Egan et al. 2003).

Surprisingly, the association to psychiatric illness is with the Val allele (i.e., the common allele that shows normal BDNF function). In other words, individuals with the less frequent Met allele, which causes altered BDNF function, are less susceptible to manifest psychosis, as if the Met allele protects against it.

Davenport was from an era uninformed by any of this knowledge about the identity and mechanisms of functional genetic variation, but from his perspective it would have been expected that the Met allele (i.e., the "weak" allele), would be the risk form of BDNF. However, from the perspective of modern molecular science, the association with the Val allele is consistent with the complex interplay of genes and environment that account for psychiatric disorders. BDNF is an effector protein involved in mediating a final common pathway responsible for neuronal plasticity, important in normal development but also in stress, the effects of drugs, and brain injury.

It has been suggested that the Met allele, just as it may be disadvantageous in not responding appropriately to normal molecular signals involved in learning and memory, also does not "hear" signals that initiate neuroplasticity related to negative experience. Thus, Met BDNF may protect against modifications in brain circuitry that develop because of negative factors, including other psychiatric risk genes and adverse environments. The *SERT* gene, which modifies the sensitivity of an individual to negative environmental experience, may interact with BDNF. Serotonin is a neurotransmitter signal from one cell to another related in part to emotionality, especially fear and anxiety. Stress early in development may sensitize the brain circuitry related to negative emotion by altering serotonin signaling (Gross and Hen 2004). This process involves BDNF, which responds to the serotonin neurotransmitter signal. The *BDNF* Met allele may protect against the effect of SERT signaling associated with stress early in development because it does not respond to it, and it may also protect against the additional negative effect of the s *SERT* alleles (and other genes that similarly prime the negative mood circuit). The net effect of this interaction would be to reduce the chances of a psychiatric illness emerging later in life. Because Val *BDNF* alleles are much more common, most individuals with especially stressful developmental environments or with s *SERT* alleles will not be protected in this way.

The interaction of *BDNF* and *SERT* illustrates the complex interplay of multiple genes and the environment in varying risk for psychiatric disorders. The more we learn about the genetic and molecular origins of human behavioral variation and of psychiatric illness, the more it is clear that mental disorders are not reducible to the stigma of biologic weakness, and that mental illness frequently emerges from the actions of alleles that are not disease alleles per se, but are alleles crucial to human behavioral variation that makes us more adaptable as a species, more individual, and perhaps more interesting.

Psychiatric genetics is poised on a great frontier of scientific discovery. Although there is still a huge void in knowledge and many missing pieces in understanding even the few genes that have been identified thus far, the discovery of objective causative factors for several psychiatric illnesses holds enormous promise. It is clear that our understanding of psychiatric disorders and our approach to further research has been changed profoundly and permanently. The scourge of stigma and prejudice that prevailed for much of the 20th century has been challenged by scientific data and by facts. We now know that the seeds for understanding the genetics of psychiatric illness were found in Mendel's studies, but the patterns of causation are usually not amenable to simple Mendelian analysis, being far more elaborate and complex than the inheritance patterns of the seven physical characteristics Mendel studied. Furthermore, human behavior may not be readily amenable to modification by controlled breeding. But most importantly the multivalent nature of genetic variation altering human behavior makes the goal of behavioral improvement by eliminating deleterious alleles an illusory goal. Through selection of so-called "good" alleles, we might well be able to impoverish human behavioral diversity, but it is unlikely that we would thereby improve the mental health of populations.

REFERENCES

Barr C.S., Newman T.K., Becker M.L., Champoux M., Lesch K.P., Suomi S.J., Goldman D., and Higley J.D. 2003. Serotonin transporter gene variation is associated with alcohol sensitivity in rhesus macaques exposed to early-life stress. *Alcohol Clin. Exp. Res.* **27:** 812–817.

Caspi A., Sugden K., Moffitt T.E., Taylor A., Craig I.W., Harrington H., McClay J., Mill J., Martin J., Braithwaite A., and Poulton R. 2003. Influence of life stress on depression: Moderation by a polymorphism in the 5-HTT gene. *Science* **301:** 386–389.

Egan M.F., Goldberg T.E., Kolachana B.S., Callicott J.H., Mazzanti C.M., Straub R.E., Goldman D., and Weinberger D.R. 2001. Effect of *COMT* Val108/158 Met genotype on frontal lobe function and risk for schizophrenia. *Proc. Natl. Acad. Sci.* **98:** 6917–6922.

Egan M.F, Kojima M., Callicott J.H., Goldberg T.E., Kolachana B.S., Bertolino A., Zaitsev E., Gold B., Goldman D., Dean M., Lu B., and Weinberger D.R. 2003. The *BDNF* val66met polymorphism affects activity-dependent secretion of BDNF and human and human memory and hippocampal function. *Cell* **112:** 257–269.

Enoch M.A., Xu K., Ferro E., Harris C.R., and Goldman D. 2003. Genetic origins of anxiety in women: A role for a functional catechol-*O*-methyltransferase polymorphism. *Psychiatr. Genet.* **13:** 33–41.

Goldman D. and Enoch M. 1989. Genetic epidemiology of the ethanol metabolic enzymes: A role for selection. In *Genetic variation and nutrition* (ed. A.P. Simopoulos and B. Childs), pp. 143–160. Karger, Basel.

Gross C. and Hen R. 2004. The developmental origins of anxiety. *Nat. Rev. Neurosci.* **5:** 545–552.

Hariri A.R., Mattay V.S., Tessitore A., Kolachana B., Fera F., Goldman D., Egan M.F., and Weinberger D.R. 2002. Serotonin transporter genetic variation and the response of the human amygdala. *Science* **297:** 400–403.

Harrison P.E. and Weinberger D.R. 2005. Schizophrenia genes, gene expression and neuropathology: On the matter of their convergence. *Mol. Psychiatry* **10:** 40–68.

Hu X.Z., Lipsky R.H., Zhu G., Akhtar L.A., Taubman J., Greenberg B.D., Xu K., Arnold P.D., Richter M.A., Kennedy J.L., Murphy D.L., and Goldman D. 2006. Serotonin transporter promoter gain-of-function genotypes are linked to obsessive–compulsive disorder. *Am. J. Hum. Genet.* **78:** 815–826.

Kendler K.S. 2000. Schizophrenia: Genetics. In *Comprehensive textbook of psychiatry*, 7th ed. (ed. B.J. Sadock and V.A. Sadock), pp. 1147–1159. Lippincott, Williams, & Wilkins, New York.

Koss M.P., Yuan N.P., Dightman D., Prince R.J., Polacca M., Sanderson B., and Goldman D. 2003. Adverse childhood exposures and alcohol dependence among seven Native American tribes. *Am. J. Prev. Med.* **25:** 238–244.

Lindsay E.A. 2001. Chromosomal microdeletions: Dissecting del22q11 syndrome. *Nat. Rev. Genet.* **2:** 858–868.

McGrath J.J. and Murray R.M. 2003. Risk factors for schizophrenia: From conception to birth. In *Schizophrenia II* (ed. S.R. Hirsch and D.R. Weinberger), pp. 232–250. Blackwell, London.

Merikangas K.R. and Risch N. 2003. Genomic priorities and public health. *Science* **302:** 599–601.

Mervis C.B. 2003. Williams syndrome: 15 years of psychological research. *Dev. Neuropsychol.* **23:** 1–12.

Murphy D.L., Lerner A., Rudnick G., and Lesch K.P. 2004. Serotonin transporter: Gene, genetic disorders, and pharmacogenetics. *Mol. Interv.* **4:** 109–123.

Ozaki N., Goldman D., Kaye W.H., Plotnicov K., Greenberg B.D., Lappalainen J., Rudnick G., and Murphy D.L. 2003. Serotonin transporter missense mutation associated with a complex neuropsychiatric phenotype. *Mol. Psychiatry* **8:** 933–936.

Pezawas L., Meyer-Lindenberg A., Drabant E.M., Verchinski B.A., Munoz K., Kolachana B.S., Egan M.F., Mattay V.S., and Weinberger D.R. 2005. 5-HTTLPR polymorphism impacts human cingulated-amygdala interactions: A genetic susceptibility mechanism for depression. *Nat. Neurosci.* **8:** 828–834.

Wellcome Trust Case Control Consortium. 2007. Genome-wide association study of 14,000 cases of seven common diseases and 3000 shared controls. *Nature* **447:** 661–678.

Weinberger D.R., Egan M,F., Bertolino A., Callicott J.H., Mattay V.S., Lipska B.K., Berman K.F., and Goldberg T.E. 2001. Prefrontal neurons and the genetics of schizophrenia. *Biol. Psychiatry* **50:** 825–844.

Zubieta J.K., Heitzeg M.M., Smith Y.R., Bueller J.A., Xu K., Xu Y., Koeppe R.A., Stohler C.S., and Goldman D. 2003. *COMT* val158met genotype affects μ-opioid neurotransmitter responses to a pain stressor. *Science* **299:** 1240–1243.

GENES IN MIND?

Lindsey Kent

Bute Medical School, St. Andrews University

Simon Baron-Cohen

Departments of Psychiatry, Cambridge University

Editors' Note: Few topics engender debate as heated as the question of genetic versus environmental influence. Davenport's hero Francis Galton was an early adapter of the idea of using twins to disentangle nature (genetics) and nurture (environment), especially in the study of the heritability of intelligence. In his 1875 article "The History of Twins, as a Criterion of the Relative Powers of Nature and Nurture," Galton emphasized the importance of twin studies in investigating mental heredity and concluded that "nature prevails enormously over nurture." Twin studies have been fraught with controversy, particularly those studies that were motivated by eugenics, such as those in 1920–1940 Germany. Other social issues have also been caught up in the nature/nurture question, such as the politics of race in Nazi Germany and, according to some, race and intelligence in Herrnstein and Murray's controversial 1994 book *The Bell Curve*. The nature/nurture issue has even entered popular culture, as in the 1983 Hollywood movie *Trading Places*. And there is the steady announcement in news reports of "intelligence genes," such as *IGF-2R*, *DTNBP1*, and *CHRM2*.

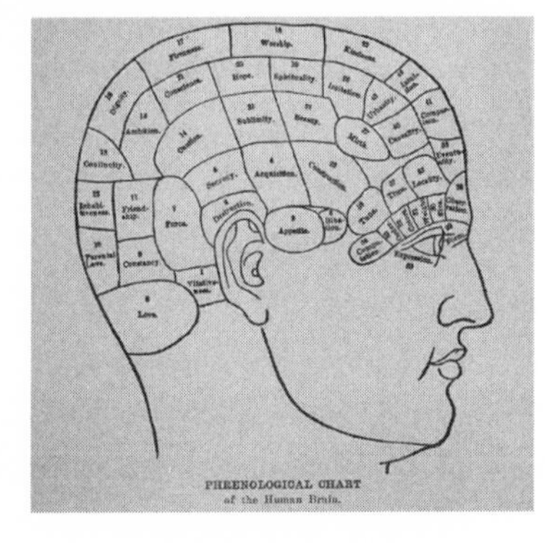

PHRENOLOGICAL CHART of the Human Brain.

In this essay, Drs. Baron-Cohen and Kent disentangle the nature/nurture controversy and present the current science exploring the nature of the human mind. They explain the issues of single-gene versus complex traits and the current use of twin studies. They also present molecular genetic studies, using the particular examples of the roles of *COMT* and *BDNF* genes in relation to mental ability. The effects of variations associated with differences in intelligence, heterosis, and quantitative trait loci (QTLs) are also explored.

Are we making progress in teasing out the effects of genes on cognitive ability versus those of the environment? Some investigators say that there are probably about 100 or so genes that influence intelligence. Do investigators even agree on the definition of intelligence and methods for its measurement? Is intelligence a function of the number of genetic variants affecting intelligence combined with environmental influences? Hopefully, the combination of molecular genetics and behavioral genetics described here will bring us closer to the answers.

> *Though a man pride himself on the freedom of his will his every action is determined by his protoplasmic makeup, plus the modification it has received through experience, plus the relative vigor and quality of the stimulus he receives.*
>
> DAVENPORT, PP. 264–265

DO GENES INFLUENCE OUR MINDS? And if so, how important is genetic influence compared to the shaping role of the environment? These are old questions that are no less important today. Charles Davenport's book in 1911 from Cold Spring Harbor, Long Island, New York, provides a fascinating look into views about genetics from one century ago, views that strike the modern reader as both familiar and alien: familiar, because Davenport's important idea that genes play an important role in creating human characteristics (both physical and mental) is now widely accepted; alien, because Davenport uses the term "eugenics" in a way that is incompatible with much modern thinking surrounding the ethical use of genetics. Later in this essay, we update Davenport's important idea that genes influence the mind. We do this by examining the results of modern twin and molecular genetic studies. Twin studies—a method Davenport does not seem to have used—allow nature and nurture to be disentangled. And the new molecular studies are the jewels in nature's crown. To anticipate, these studies suggest that genes play a modest role in the mind. In contrast, a raft of studies of the effects of environmental deprivation show how nurture plays the dominant role. But before we turn to such evidence, we tackle head-on the alien ideas of eugenics, and reflect on how incompatible these beliefs are in the modern context.

EXPERIMENTS IN BREEDING WERE PROPOSED IN THE EARLY 20TH CENTURY

Eugenics is, Davenport reminds us, all about "better breeding" (p. 1)—not for its own sake but to improve the race. Proponents of the eugenics movement took seriously that with the control that breeding experiments had exercised over crop and animal quality in agriculture, it was perfectly logical to apply the same approach to improving human races. That most eugenicists were white Americans or Europeans may be significant, given the history of European and American imperialism. Progress, according to the eugenicists, was only possible by securing "the best blood" (p. 1). According to this brave new science, what was true for how one could improve a horse or a crop holds true for humans in every detail.

The well-intentioned doctors and scientists who joined the ranks of the eugenics movement were taken in by the force of this logic. As thinking, intelligent policy-makers, with a knowledge of genetics, they argued that it was our responsibility to help reduce the burden of (presumed) genetic weaknesses in society. Why, Davenport asked, should the country "have to support about half a million insane, feeble-minded, epileptic, blind and deaf, 80,000 prisoners, and 100,000 paupers at a cost of over 100 million dollars per year?" (p. 4). This, then, was a health-economics argument, which, according to its advocates, had a clear solution. One needed to improve the race, and this could be achieved through two steps: First, one had to induce "young people to make a more reasonable selection of marriage mates; to fall in love intelligently" (p. 4). And second, one had to introduce controls by the state to prevent "the propagation of the mentally incompetent" (p. 4).

For the eugenicists, marriage and mating were out of control, representing a near-random process that inevitably would give rise to physically and psychologically defective and socially undesirable traits in the offspring. "Matings... seem to be made at haphazard" (p. 7). The logical alternative, Davenport and others urged almost 100 years ago, in a voice that seems reasonable but today ethically very worrying, was "contemplated union," which would result in healthy offspring.

For eugenicists, marriage is "*an experiment* in breeding" (p. 7; Davenport's emphasis), with children the result of the experiment. What was recommended was to move marriage from being a mere experiment, where the results are unpredictable, to being a science, where the results followed highly predictable laws: "As we now know how to make almost any desired combination of the characters of guinea-pigs, chickens, wheats, and cottons so may we hope to do with man" (p. 7).

Today, such "designer babies" recommendations trigger important ethical caution as a reaction, because the eugenics movement led not only to the compulsory sterilization of thousands of adults with below-average IQ in the United States but also to the policies of the Nazi regime in Germany and occupied Europe during the 1930s and early 1940s: Jews, Gypsies, black people, and gay people, as well as patients with psychiatric conditions and learning disabilities, were hunted down, rounded up, transported to death camps, and exterminated or used as experimental subjects, such as the infamous Josef Mengele twin experiments at Auschwitz. The eminently reasonable-sounding eugenics proposal turned out to be the thin end of the wedge, because once one buys into the notion of genetic cleansing at all, it is but a small step to keep extending such policies to exclude any unwanted group in society.

It is instructive to look at the groups that Davenport and the early eugenicists wished to exclude. The "feeble-minded" were at the top of the list, and it taxed these intellectuals as to how to define "feeble-minded." Being incapable of independence? Being more than 3 years below one's own age on the Binet test of mental abilities? Being a "moral imbecile" (p. 9)? Being deficient in some socially important trait? These alternatives were all considered and found insufficient as criteria. Indeed, the targets that such criteria were intended to cover included "the sexually immoral, the criminalistic, those who cannot control their use of narcotics, those who habitually tell lies by preference, and those who run away from school or home" (p. 9). In contrast, the traits that were to be encouraged included the converse of these, as well as abilities such as mathematics, music, science, art, and literature.

So this is the alien voice of Davenport. Eugenics is now an electric fence, triggering public debate about the risks of genetic interventions, and the introduction of checks and balances. No longer could we expect in a modern debate to hear the following (Abstract on the Report of the Immigration Commission on Recent Immigration in Agriculture, p. 41, cited on p. 216):

> The Hebrew on the land is peaceable and law abiding, but he does not tamely submit to what he believes to be oppression and he has a highly developed sense of personal rights, civil and economic.

Nor would we expect a scientist like Davenport to go on to say "the mass of recent Hebrew immigrants occupy a position intermediate between the slovenly Serbians and Greeks and the tidy Swedes, Germans and Bohemians" (p. 216). Such generalizations about supposed genetic characteristics of ethnic groups are rightly seen as racist and, with the hindsight of the Second World War, dangerously so. But that does not mean that, because Davenport and the eugenics movement were wrong in their methods or their policies, modern genetics has not benefited from the basic idea that genes might influence the mind. In the next sections, we summarize how the field of cognitive and behavior genetics has moved beyond stereotyping ethnic groups, and beyond collecting pedigrees, to the study of twins to determine the extent of genetic influences in the mind, and to use molecular genetic techniques to isolate such influences.

ARE THERE SINGLE VERSUS COMPLEX INFLUENCES IN THE MIND?

Following Mendel, Davenport assumed disorders of the mind (e.g., epilepsy, bipolar affective disorder, alcoholism, and low intelligence) arose from *single*-gene defects. Some of these, such as Huntington's chorea, are indeed single-gene disorders, but the vast bulk of psychological traits are *complex* traits, that is, inherited in a quantitative way. The differences between individuals (known as the phenotypic variance) are shaped by multiple genetic and environmental influences and the interactions between the two. For example, bright children are more likely to seek out challenging environments, which subsequently influence their development. Similarly, individuals with antisocial personality traits are more likely to trigger negative reactions in others, which subsequently impacts on their antisocial behavior. Therefore, unlike single-gene disorders that are *deterministic* (if you possess the genetic defect, you develop the disorder), quantitative complex traits are *probabilistic* (possessing susceptibility genetic variants merely increases the probability of developing that disorder). Each susceptibility genetic variant may be neither necessary nor sufficient for developing the disorder, and environmental risk factors typically play an important part in susceptibility.

The quantitative trait loci (QTLs) concept proposes that complex traits (such as personality and cognitive abilities) are *polygenic* (more than one gene). Inheritance of personality and cognitive abilities appear not to be compatible with Mendelian genetics, but each of the genetic variants are indeed inherited by Mendelian principles. If there are multiple genes that affect complex heritable traits, it is likely that the traits are distributed quantitatively as dimensions (Plomin and Spinath 2002). If this is correct, then the QTL that contributes to cognitive ability in the normal range should be the same QTL that contributes to the entire dimension of that cognitive ability. A QTL for reading *ability* should also be associated with reading *disability*. A QTL for personality disorder should represent the extreme of the distribution of that personality trait in the population. QTLs studies are a current major focus in psychological genetics. The extremes of some personality and behavioral traits are associated genetically with some psychiatric disorders—for example, hyperactivity symptoms and attention-deficit hyperactivity disorder (Levy et al. 1997), neuroticism and anxiety and depression (Eaves et al. 1989), and schizotypal personality disorder and schizophrenia (Farmer et al. 1987). Nevertheless, the genetic links between normal variation of some traits and possible related disabilities are not yet well known. For example, we do not yet know the genetic basis of the link between normal sociability on the one hand and autistic spectrum social impairments on the other.

TWIN STUDIES ARE A VALUABLE APPROACH

Twin studies can estimate the degree to which genes play a role in continuously distributed traits in the mind. Monozygotic (MZ) twins are genetically identical and therefore share 100% of their genes. Dizygotic (DZ) twins share on average only 50% of their genes—no more than is shared by full siblings. Assuming that both types of twin share environmental effects to a similar extent (an assumption not without its critics), the more similar MZ twins are for a given mental characteristic, the more one can infer this similarity is due to genetic influences. This allows researchers to determine the *heritability* of a trait. Heritability refers to the genetic contribution to individual differences for that trait.

So what do twin studies reveal? In Table 1, we list heritability estimates for some different aspects of cognition and behavior. (Where there are conflicting results from different studies, we quote the highest estimate of genetic influence.) These data are summarized in Robert Plomin's seminal text (Plomin and Nesselroade 1990). What is immediately apparent is that such twin studies show that very few personality or cognitive traits and few psychological disorders are highly heritable (>80% heritable, where the environment plays a minor role if at all). The vast number of psychological traits or disorders have low-to-modest heritability, indicating that the environment is playing a substantial role. It is important to note that heritability estimates for most cognitive traits increase with age.

THE ENVIRONMENT PLAYS A ROLE IN SHAPING THE MIND

Some of the results in Table 1 may not surprise you; for example, that autism and bipolar affective disorder are strongly heritable. Other results may seem counterintuitive. Why should your propensity to di-

Table 1. *Heritability estimates for a range of psychological traits or disorders*

Characteristic	Heritability (%)
Bipolar affective disorder	85
Autism	80
Schizophrenia	80
Antisocial personality disorder	60
Hyperactivity	60
IQ	50
Divorce	50
Tobacco smoking	50
Male alcoholism	50
Spatial ability	45
Mental speed	40
Neuroticism	40
Juvenile delinquency	35
Memory	30
Verbal fluency	30
Anxiety	30
Female alcoholism	30
Creativity	25
Sociability	25
Heterosexual mate preference	0
Reaction time	0

vorce or to smoke be moderately heritable? Can there really be "genes for divorce" or "genes for smoking"? This is unlikely, but there may be genes that contribute to addictive behavior or instability in relationships, and the challenge for modern genetics is to identify what the ultimate mechanisms are that might produce such results. Clearly, divorce and smoking take place only in environments where divorce is legal or tobacco is available, illustrating the obvious point that such genes must interact with environments. But what of the mental characteristics at the bottom of the list, which have almost no heritability? How could the environmental influence be so strong?

We now know that your early childhood is astonishingly important in determining your emotional state in later life. For example, if you were maltreated or abused, your adult cortisol levels (your stress hormone) will be abnormally high or low (Bremne and Vermetten 2001), you are more likely to experience a later trauma (such as rape) (Nishith et al. 2000), you have a significantly increased risk of adult depression (Bifulco et al. 1998), you are more likely to become an abuser yourself (Widom 1989), you are more likely to become a convicted violent offender (Widom 1989), and you have a raised risk of developing a personality disorder (Johnson and Al 1997).

Or take a different environmental factor: 17% of the general population lost a parent in childhood, but among adult criminals, 32% lost a parent in childhood. This strongly suggests loss of a parent in childhood doubles your risk of developing criminality. Similarly, 30% of those who develop juvenile delinquency lost a parent in childhood, as did 27% of those with adult depression (Albert 1983). Interestingly, loss of a parent in childhood has also been shown to be a strong environmental influence for success in life, for some people. Thus, 35% of U.K. Prime Ministers, 34% of U.S. Presidents, and 30% of U.K entrepreneurs lost a parent in childhood (Albert 1983). Early loss can put you onto a path of instability, insecurity, and mental health problems, or it can drive you to achieve the highest levels of success—perhaps as a defense against such insecurity. Such examples underscore the major role the environment can play in the mind.

MOLECULAR GENETIC STUDIES OF PERSONALITY AND COGNITIVE PROCESSES ARE THE NEXT STEP

With the knowledge that genes play *some* role in the mind, researchers are now focusing on identifying the actual genetic variants involved. Genetic investigation of personality traits has studied candidate genes within the serotonergic and dopaminergic (neurotransmitter) systems. Genetic variation within the serotonin transporter gene has been linked to neuroticism, and variation in the dopamine receptor gene *DRD4* has been linked to novelty-seeking behavior (Paterson et al. 1999). Many studies have, however, failed to replicate these findings, partly because many genes are likely to underlie complex traits, each with a relatively small effect, and many studies do not yet have sufficiently large enough sample sizes to detect such effects.

Some cognitive disabilities arise from single-gene disorders or chromosomal abnormalities, such as Down syndrome and Fragile X syndrome. Traits such as memory, IQ, calculating ability, and reading ability are clearly complex polygenic traits. There are some rare single-gene mutations that on first sight might be thought to be complex polygenic traits. These are often severe disorders, but they clearly demonstrate a Mendelian pattern of inheritance. For example, a mutation in the *FOXP2* gene is responsible for a severe form of speech and language impairment in a family known as KE (Lai et al. 2001). Interestingly, this gene does not seem to have a role in language impairment more generally.

Catecholamines (such as dopamine and noradrenaline) are neurotransmitters involved in attention and information processing. Studies examining genetic variation within the gene encoding catechol-*O*-methyltransferase (COMT) and brain function were among the first to be examined. COMT is the major mammalian enzyme involved in the breakdown of released dopamine in the brain's frontal cortex. The *COMT* gene contains a well-studied functional genetic variant, such that one form of the produced enzyme has reduced activity. Two independent studies (Egan et al. 2001; Malhotra et al. 2002) demonstrated that the low-activity version of this enzyme is associated with better performance on tests of frontal brain function. Subsequent functional neuroimaging and neurophysiology studies also support the association between the *COMT* genotype and dopamine-mediated frontal function (Gallinat et al. 2003; Mattay et al. 2003).

A further avenue for interesting results has been the investigation of brain-derived neurotrophic factor (BDNF). BDNF is involved in the survival and differentiation of dopaminergic neurones in the developing brain (Hyman et al. 1991) and is known to be involved in learning and memory processes and neuronal plasticity. It is of interest that in a recent study (Egan et al. 2003), a particular genetic variant of BDNF was associated with poor episodic memory. Clearly, single genetic variants do not function alone in isolation, however, and much further work is needed to understand how both genetic and environmental influences interact.

Reading ability has a heritability of around 50%, which has prompted the search for "dyslexia susceptibility genes" (and by the same token, reading ability loci). Linkage studies have reported possible susceptibility genes for dyslexia on chromosomes 2p, 6p, 15q, and 18p. Interestingly, around one-third of children with dyslexia also have attention-deficit hyperactivity disorder (ADHD). There is some evidence suggesting that a QTL on chromosome 6p may account for some of this overlap (Willcutt et al. 2002). The implication here is that the same genetic variant is involved in increasing susceptibility to more than one disorder (pleiotropy).

One particular area that has attracted considerable attention is the genetics of autistic spectrum conditions. Autism impairs social development and communication and is accompanied by narrow interests or "obsessions" and repetitive activity. The modern view is that there is an autism "spectrum," comprised of several subgroups, the largest of these being classic autism and Asperger syndrome. The most strongly implicated chromosomal regions harboring susceptibility genes for autism are 7q, 2q, and 15q. In addition there is an increasingly large literature focusing on candidate gene analysis in autism. A number of genetic variants have been studied that have demonstrated association with autism including, among others, polymorphisms in the serotonin transporter gene (5HTT) (for review, see Veenstra-Vanderweele and Cook 2004). However, the majority of findings have not been widely replicated and there is as yet no clear evidence implicating a particular gene in autism susceptibility.

HETEROSIS EXPLAINS THE PERSISTENCE OF SOME DISEASE GENES

Why should gene variants for disorders persist in the gene pool? The idea of heterosis is that although it may be maladaptive to have the entire set of genes needed for a disorder, possessing some of the genes may confer some selective advantage to the individual in certain environments. An intriguing pattern that is emerging from the field of autism is that the first-degree relatives of people with autism, who have some autism genes but not all, may excel in the fields of engineering, science, technology, and other "systemizing" fields (Baron-Cohen et al. 1997, 2003). The genes in such first-degree relatives may allow for a talent in concentrating on a narrow topic in great detail and allow the development of expertise in understanding systems. When an individual possesses all the susceptibility genes for a given disorder, such talents may appear as highly restrictive obsessions or major impairment in sociability and as a diagnosable medical condition (autism). Equally, in the field of bipolar affective disorder, first-degree relatives are often overrepresented in the areas of creativity, especially in poetry and the humanities (Andreasen and Glick 1988). It is an intriguing possibility that here, too, a mild combination of the genotype may confer advantages, but a major combination may be deleterious. Such a model may account for why classic autism or bipolar affective disorder genes do not become less common over time.

NATURE VERSUS NURTURE RECONSIDERED

Let us take stock. In the century since Davenport's book was published, modern geneticists have reached surprising conclusions: "Most behavioural variability among individuals is environmental in origin" (Plomin and Daniels 1987). "The wonderful diversity of the human species is not hard-wired in our genetic code. Our environments are critical" (Craig Venter, lead investigator of one of the human genome maps, *The Observer*, 11.2.01, p. 1). These genetics researchers are reminding us that, although genes do influence our behavior and personality, their influence is only modest for many traits. Notable exceptions exist, such as the strongly genetic conditions of autism and bipolar affective disorder.

So, why should science be spending so much money and time trying to find the genes that influence the mind? We conclude that this is money well spent, because such genetics research has the potential to teach us how we—and our brains—are made. Genes code for specific biological functions, and the pursuit of such knowledge is worthwhile in its own right. Genetics may lead to some important medical breakthroughs in treatment. Modern genetics must learn the lessons of its shameful eugenics past, and we reject any use of genetics to control who can breed with whom, or in the eradication of "types" of individuals from the population. We concur with Plomin and Venter that, if we have learned anything, it is that money is equally needed to support environmental interventions to reduce the risks of mental health problems in society. Quality of education, family support, early parental care, and economic stability are of major importance in shaping the mind.

ACKNOWLEDGMENTS

The authors were supported by the Medical Research Council, the National Alliance of Autism Research (NAAR), Cure Autism Now (CAN), and the Nancy Lurie Marks Family Foundation, during the period of this work. We are grateful to Martin Yuille, Frank Dudbridge, Nigel Goldenfeld, and Sally Wheelwright for discussions of these ideas.

REFERENCES

Albert R.S., ed. 1983. Family positions and the attainment of eminence. In *Genius and eminence: The social psychology of creativity and exceptional achievement.* Pergamon, Oxford.

Andreasen N.C. and Glick I.D. 1988. Bipolar affective disorder and creativity: Implications and clinical management. *Compr. Psychiatry* **29:** 207–217.

Baron-Cohen S., Wheelwright S., Stott C., Bolton P., and Goodyer I. 1997. Is there a link between engineering and autism? *Autism* **1:** 153–163.

Baron-Cohen S., Richler J., Bisarya D., Gurunathan N., and Wheelwright S. 2003. The Systemising Quotient (SQ): An investigation of adults with Asperger Syndrome or High Functioning Autism and normal sex differences. *Philos. Trans. R. Soc. (Lond.) B; Special issue on "Autism: Mind and Brain"* **358:** 361–374.

Bifulco A., Brown G.W., Moran P., Ball C., and Campbell C. 1998. Predicting depression in women: The role of past and present vulnerability. *Psychol. Med.* **28:** 39–50.

Bremne J.D. and Vermetten E. 2001. Stress and development: Behavioral and biological consequences. *Dev. Psychopathol.* **13:** 473–489.

Eaves L.J., Eysenck H., and Martin N.G. 1989. *Genes, culture, and personality: An empirical approach.* Academic Press, London.

Egan M.E., Kojima M., Callicott J.H., Goldberg T.E., Kolachana B.S., Bertolino A., Zaitsev E., Gold B., Dean M., Lu B., and Weinberger D.R. 2003. The BDNF val66met polymorphism affects activity-dependent secretion of BDNF and human memory and hippocamal function. *Cell* **112:** 257–269.

Egan M.F., Goldberg T.E., Kolachana B.S., et al. 2001. Effect of COMT $Val^{108/158}Met$ genotype on frontal lobe function and risk for schizophrenia. *Proc. Natl. Acad. Sci.* **98:** 6917–6922.

Farmer A.E., McGuffin P., and Gottesman I.I. 1987. Twin concordance for DSM-III schizophrenia. Scrutinizing the validity of the definition. *Arch. Gen. Psychiatry* **44:** 634–641.

Gallinat J., Bajbouj M., Sander T., Schlattmann P., Xu K., Ferro E.F., Goldman D., and Winterer G. 2003. Association of the G1947A COMT ($Val^{108/158}Met$) gene polymorphism with prefrontal P300 during information processing. *Biol. Psychiatry* **54:** 40–48.

Galton F. 1875. The history of twins, as a criterion of the relative powers of nature and nurture. *Fraser's Magazine* **12**: 566–576.

Herrnstein R.J. and Murray C. 1994. *The bell curve: Intelligence and class structure in American life.* Free Press, New York.

Hyman C., Hoffer M., Barde Y.-A., Juhasz M., Yancopoulos G.D., Squinto S.P., and Lindsay R.M. 1991. BDNF is a neurotrophic factor for dopaminergic neurons of the substantia nigra. *Nature* **350:** 230–232.

Johnson J.G. and Al E. 1997. Childhood maltreatment increases risk for personality disorders during early adulthood. *Arch. Gen. Psychiatry* **56:** 600–606.

Lai C., Fisher S.E., Hurst J.A., Varga-Khadem F., and Monaco A.P. 2001. A novel forkhead-domain gene is mutated in a severe speech and language disorder. *Nature* **413:** 519–523.

Levy F., Hay D., McStephen M.C.W., and Waldman I. 1997. Attention deficit hyperactivity disorder: A category or a continuum? Genetic analysis of a large scale twin study. *J. Am. Acad. Child Adolesc. Psychiatry* **36**: 737–743.

Malhotra A.K., Kestler I.J., Mazzanti C., and Al E. 2002. A functional polymorphism in the *COMT* gene and performance on a test of prefrontal cognition. *Am. J. Pyschiatry* **159:** 652–654.

Mattay V.S., Goldberg T.E., Fera F., Hariri A.R., Tessitore A., Egan M.F., Kolachana B., Callicott J.H., and Weinberger D.R. 2003. Catechol *O*-methyltransferase val^{158}*-met* genotype and individual variation in the brain response to amphetamine. *Proc. Natl. Acad. Sci.* **100:** 6186–6191.

Nishith P., Mechanic M.B., and Resick P.A. 2000. Prior interpersonal trauma: The contribution to current PTSD symptoms in female rape victims. *J. Abnorm. Psychol.* **109:** 20–25.

Paterson A.D., Sunohara G.A., and Kennedy J.L. 1999. Dopamine D4 receptor gene: Novelty or nonsense. *Neuropsychopharmacology* **21:** 3–16.

Plomin R. and Daniels D. 1987. Why are children from the same family so different from each other? *Behav. Brain Sci.* **10:** 1–16.

Plomin R. and Nesselroade J.R. 1990. Behavioral genetics and personality change. *J. Pers.* **58:** 191-220.

Plomin R. and Spinath F.M. 2002. Genetics and general cognitive ability. *Trends Cogn. Sci.* **6:** 169-176.

Veenstra-Vanderweele J. and Cook E.H., Jr. 2004. Molecular genetics of autism spectrum disorder. *Mol. Psychiatry* **9:** 819–832.

Widom C.S. 1989. The intergenerational transmission of violence. In *Pathways to criminal violence* (ed. N.A. Weiner and M. Wolfgang). Sage, California.

Willcutt E.G., Pennington B.F., Smith S.D., Cardon L.R., Gayan J., Knopik V.S., Olson R.K., and DeFries J.C. 2002. Quantitative trait locus for reading disability on chromosome 6p is pleiotropic for attention-deficit/hyperactivity disorder. *Am. J. Med. Genet.* **114:** 260–268.

Davenport and Heredity Counseling

Philip R. Reilly
Interleukin Genetics, Inc.

Editors' Note: Davenport collected hundreds of pedigrees to support his analysis of scores of genetic conditions, both real and spurious, in *Heredity in Relation to Eugenics.* As it turns out, his analyses of pedigrees for Huntington's chorea, hereditary ataxia, and albinism, to name a few, were correct, and his advice in these cases represents one of the earliest instances of genetic counseling. Of course, not all of his analyses were correct, and his advice in many cases was anything but palatable.

Pedigrees remain an important tool for genetic counselors, but their additional resources, such as amniocentesis, chorionic villus sampling, testing for genetic markers, and, soon, genome sequencing, offer additional reliable data for providing information to those people who need to make a crucial decision about their futures and their children. These resources have also given rise to new issues regarding the uses of such genetic information. Dr. Reilly presents the complexities of heredity counseling today against the backdrop of Davenport's foray into this area and offers some thoughts on revising the role of today's genetic counselors.

After we get precise knowledge of the methods of inheritance of the commoner important traits we shall be in a position to advise, at least in respect to these traits.

DAVENPORT, P. 26

ALTHOUGH HE WAS NOT TRAINED AS A physician, Charles Benedict Davenport's book, *Heredity in Relation to Eugenics*, has a solid claim to be remembered as the first handbook on genetic counseling. A curious amalgam of projects in genetic research, clinical medicine, training, and public policy, the Eugenics Record Office that Davenport (with the support of some of the nation's wealthiest families) founded in 1910 has an even stronger claim as the first center to provide advice on recurrence risk based on family history of disease. Twenty-five years would elapse before the University of Michigan opened the first "heredity clinic" in a medical school in 1940–1941. With support from the Dight Institute, Sheldon Reed, a non-physician geneticist, opened the third such clinic in 1947 at the University of Minnesota. It was Reed who coined the term "genetic counseling" to replace the earlier term "genetic hygiene," by then tainted with the horrors committed in Nazi Germany in the name of "racial hygiene." (As a side note, Reed had an interesting connection to the ERO in that he arranged for the transfer of the ERO records after the Office's closure to the Dight Institute, material that included approximately 40,000 pedigrees and a cross-index of about 2 million cards.)

According to F. Clarke Fraser, a leading human geneticist of the last third of the 20th century, as recently as 1955 there were only 15 genetic counseling centers in the United States, of which only four were directed by physicians (Fraser 1979). Today, there are hundreds of such clinics as well as countless patient support groups, most of which have close ties to medical and counseling experts. Genetic counseling has become a routine aspect in the general practice of obstetrics (prenatal screening and diagnosis) and pediatrics (newborn screening for inborn errors of metabolism). In fact, over the last quarter century, genetic counseling has become a profession in its own right. In the United States there are about 1500 board-certified genetic counselors (most of them women with a master's degree who work at major medical centers), but only a tiny number engage in private practice.

DAVENPORT'S BOOK WAS REVOLUTIONARY

From the perspective of a century, Davenport's decision to write *Heredity in Relation to Eugenics* can be viewed from two contrasting perspectives. Some might conclude that it was extraordinarily innovative, perhaps revolutionary; others might decide it was premature and based on inadequate research. I take the former perspective.

Although Davenport will be forever linked to the eugenics movement, as an impatient, exuberant exemplar of the Progressive Era, he held no illusions about the magnitude of the scientific work that had to be completed before genetic information could be routinely used in clinical practice to assist couples seeking the healthiest possible children. Consider this from the first page of the Preface to *Heredity in Relation to Eugenics* (1911): "A vast amount of investigation into the laws of inheritance of human traits will be required before it will be possible to give definite instruction as to fit marriage matings." Despite the caveat, Davenport believed that enough was known about the role of genes in a sufficient number of human conditions that it would be valuable to assemble that information and, where the facts permitted, offer broad advice to at-risk families about how to behave.

In its organization the book clearly reflects the era in which it was written. The short first chapter defines eugenics and states its goal: "to improve the race by inducing young people to make a more reasonable selection of marriage mates" (p. 4). Restating his goal another way, Davenport (coining a fine oxymoron) writes of the need "to fall in love intelligently" (p. 4). The brief second chapter sets forth the scientific principles on which eugenics is based. It is a panegyric to Mendelism, written with the naïve faith that single genes drive complex disorders.

More than half of the text (155 pages) is devoted to a single chapter (III) in which Davenport considers 41 human traits or disease categories. The topical organization mirrors the era. The first six sections, devoted to eye color, hair color, hair shape, skin color, height, and weight, address the key morphological metrics that dominated the academic study of race in the 19th century. In 1911, there was not quite yet universal agreement among anthropologists that there was only one human species (let alone the number of races encompassed by *Homo sapiens*), and some of those who acceded to the idea of a single species held that racial differences were pronounced. Physical anthropology was preoccupied with arguments about racial typologies until the mid-20th century.

Sections 7–13 of Chapter III discuss highly valued abilities (e.g., music, art, mathematics) that have long been recognized to run in families. Davenport strove to find golden Mendelian ratios in families where both parents evidenced prowess in one of the arts. For example, he suggested that when both parents had strong literary abilities "all (or nearly all)" (p. 54) of the offspring would be so endowed. He suspected the gene(s) driving artistic traits were dominants. In a mating between two dominantly affected persons, one would predict that the trait would manifest on average in at least three-quarters of the offspring.

Sections 19–40 survey human diseases, largely by organ system. Herein, Davenport reviews the state of genetic knowledge of about 75 disorders. Not surprisingly, disorders of the eyes, ears, and skin command disproportionate attention. Directly visible to the examiner's eye, clinicians had more success in categorizing them than they had with disorders of the internal organs. Davenport covers 15 eye disorders, but only two diseases of the heart.

Davenport begins the specific disease part of Chapter III by focusing on nervous disorders. From the perspective of almost 100 years, his five disease entries—epilepsy, insanity, pauperism, narcotism, and criminality—seem bizarre. But his entries are congruent with contemporary thinking in neurology in the last quarter of the 19th century. Most academic neurologists embraced the view that aberrant social behaviors derived from a "neuropathic taint"—a primary disorder of the brain that impaired the rational powers and/or the moral sensibility. With the conceptualization of the gene as a unit character that carried information with high fidelity through the germ line (keyed to the rediscovery of Mendel's observations around 1900), the divide between neurology and genetics was quickly bridged. Indeed, more than a few researchers thought that these five disorders of neuropathic taint all derived from the same genetically malformed circuitry.

In 1911, the publication year of *Heredity in Relation to Eugenics,* the eugenics movement in the United States was drifting from the enthusiasm for positive eugenics (how to beget better babies) that the English polymath, Francis Galton, had popularized, to a focus on negative eugenics (how to reduce the risk of having bad babies). The chapter reflects the social anxieties of the day. Despite Davenport's desire to help individual families, he (and many other writers in the eugenic movement) was worried about the social costs of behaviors that he suspected were genetically driven. Only after discussing the five neuropathic disorders listed above did Davenport address obviously clinical topics such as stroke, cerebral palsy, multiple sclerosis, and ataxia. These he covered under the rubric, "Other Nervous Diseases."

In reviewing the evidence of the role of heredity (the word "gene" was not yet widely used) in disease, Davenport had few options to offer his readers. One, he suggested repeatedly, is to marry unrelated persons (pp. 186–188, 260). One hundred years ago, consanguineous marriages in the United States were considerably more common than today. If one assumes that recessive genes cause some fraction of a disease, then advice to first cousins to avoid marriage is rational.

Unfortunately, Davenport's assumptions about the role played by genes in disease were often based on flimsy evidence. For example, he counsels persons in families in which a member has cerebral palsy to avoid cousin marriages. To this day, we are still uncertain about the manifold causes of this disorder, but we do know that it is not commonly associated with single genes. Another example of a claim based on inadequate facts is Davenport's assertion (p. 111) that the common eye condition, senile cataract, is a dominantly inherited disorder. Although there are a few, rare forms of cataract that follow that pattern, the vast majority of cases arise without a discernible genetic influence. Under the influence of extremely limited evidence (apparently a single pedigree), Davenport, who acknowledged that the disease is not usually considered hereditary, also incorrectly proclaimed that multiple sclerosis is sometimes expressed as a sex-linked disorder, and that unaffected sisters of affected men "should not have children" (p. 99).

One troubling aspect of the reproductive advice scattered through *Heredity in Relation to Eugenics* is that Davenport does not seem to appreciate the impact of his most frequently given advice—to refrain from having children—on the readers that would rely on his expertise. Consider glaucoma, a usually late-onset form of vision impairment that over many years can cause complete blindness. After outlining his claim that glaucoma is primarily a genetic disorder, Davenport asserts that, "certainly affected persons should avoid having children, while non-affected may marry if the disease first appeared in the grandparents at 50 or after" (pp. 114–115). Because most cases of glaucoma manifest after age 50 and run a slow course, the advice does not seem to match the disorder. It certainly did not anticipate that over the next decades there would emerge new tests and new treatments for this disorder.

Davenport offered better reproductive advice about disorders for which clinical knowledge was more advanced. For example, in the 1880s, large epidemiological studies by Alexander Graham Bell provided the foundation for counseling that marriage between a deaf person and a hearing person held little risk of having deaf children. Today, we know that Bell was correct. Much hereditary deafness is caused by individually uncommon autosomal recessive disorders. Today, the typical advice to a couple planning marriage in which one person is congenitally deaf (but has hearing parents) and the other person has normal hearing is that the risk of having a deaf child is very low (generally <1%). Even in 1911 (a time when geneticists had at least rudimentary familiarity with recessive disorders), it should have been possible to make this point.

PARADIGMS IN GENETIC COUNSELING

What is the biggest difference between Davenport's approach to genetic counseling and that practice today? Since about 1970, the paradigm in clinical genetics has been "nondirective counseling." Genetic counselors perceive their role as one of providing information and support, not of guiding decisions. If Davenport were alive today, I think it likely that he would find this position to be irrational and possibly even unethical. Consider one single-gene disorder that he knew well—Huntington's disease, a late-onset, incurable, dominantly inherited, neurodegenerative disorder. Davenport's advice was unequivocal: "persons with this dire disease *should not have children*" (p. 102; Davenport's emphasis). Here he erred. Once the disease manifests, few affected persons become parents. Because he knew the condition was a late-onset dominant disorder that often did not present until after age 40, what Davenport probably intended to say was that persons who are at one in two risk of being affected should not have children. Because only a tiny fraction of cases arise from new mutations, the incidence of this disorder would drop dramatically in one generation if at-risk persons refrained from becoming biological parents. Few clinicians in the modern era have made this argument. One of the few such papers, written by former President of the American Society of Human Genetics, Dr. Margery Shaw, more than 20 years ago, garnered no support (Shaw 1987).

Although Davenport's directive counseling would be roundly booed today, it merits an unbiased reevaluation. As the number of disorders for which a genetic influence is important mushrooms, as scientific understanding of the impact of genetic variation on disease risk grows, as the health-care delivery system continues to operate under serious time constraints, as the threat of malpractice lawsuits proliferates, as the ability of laypersons to grasp the subtlety of the counseling message remains limited, a cautious reconsideration of the genetic counseling process may be of value. Perhaps there is a rational moral argument to permit clinical geneticists and genetic counselors to reply cautiously to the question, "What would you do if you were in my situation?" I think that at the least Davenport would argue that there is nothing wrong with a professional providing advice to a layperson when it is requested.

REFERENCES

Fraser F.C. 1979. Introduction: The development of genetic counseling. In *Genetic counseling: Facts, values, and norms* (ed. A.M. Capron et al.), pp. 5–16. Alan R. Liss, Inc., New York.

Shaw M.W. 1987. Testing for the Huntington gene: A right to know, a right not to know, or a duty to know. *Am. J. Med. Genet.* **26:** 243–246.

Genetics and Equality

Ronald Dworkin

New York University School of Law

Editors' Note: For Charles Davenport, the central issue was "What can be done to reduce the frequency of the undesirable mental and bodily traits which are so large a burden to our population?" (pp. 255–256). He regarded this, in the first place, as a question for individuals who would be encouraged to exercise common sense and good judgment in their selection of mates; to, as Davenport put it, "fall in love intelligently" (p. 4). Marriage was to him "an experiment in breeding" (p. 7). But if people were incapable of exercising good judgment, or were known to be defective in mind or body, Davenport held that the state should intervene. Indeed, he held that society had an obligation to act: "Society has not only the right, but upon it devolves the profound duty, to know the nature of the germ plasm upon which, in last analysis, the life and progress of the state depend" (p. 267). His disciple Harry Laughlin was more forthright: "Society must look upon germ-plasm as belonging to society and not solely to the individual who carries it" (Laughlin 1914, p. 16). Such views strike at the very heart of what it means to be a free individual. Ronald Dworkin offers a very different view.

WANTED:

Better Babies:

PRESIDENT HOOVER says progress is the margin by which the next generation excels this.

All photos by Underwood and Underwood

ELLSWORTH HUNTINGTON

Co-author, *The Builders of America*

Any scheme for obtaining a more favorably balanced birth rate by economic methods must be judged by at least two main criteria. First, how far does it exercise the right kind of selection and thereby satisfy the requirements of eugenics? Second, how fully does it satisfy the economic requirement of insuring the selected families against the decline in the standard of living which is often the penalty of having children? For our present purpose, and in the present nebulous state of knowledge, the right kind of selection means one that increases the number of children in families where both parents rise well above the average in intelligence, strength of character, and general value as members of society. Insurance against a decline in the standard of living means more than relief of the sudden financial strain which often accompanies the birth of a child. It means also that as the number of children increases up to reasonable limits, the family is not obliged to economize to a degree that is painful or humiliating, but can live essentially as before. Some sacrifice on the part of parents for the sake of children is doubtless desirable, but it is obviously too much to ask ordinary human beings to step down to a lower economic level and build a new set of social relationships because they have three or four children.

It has been suggested that some kind of insurance might solve the economic phase of the problem of the dangerously low birth rate among the finest of our middle classes. Such insurance might provide for the payment of specific sums whenever a child is born, or for the education of the child after it leaves the public schools. There may be many advantages in such a system, but it does not satisfy the two criteria mentioned above. It is easy to imagine a form of compulsory insurance in which all persons engaged in gainful occupations pay something each year toward a fund for

(Concluded on page 48)

EUGENE ROBISON

Insurance Counselor

The possibility of family insurance which would provide benefits for the birth of children is one of the topics for discussion. In order to apply the principle of insurance we must reduce the uncertain to the certain through the law of averages. When the uncertainties can become controlled events to any appreciable degree insurance merely becomes a savings account against a more or less certain date in the future.

It seems to me, therefore, that family insurance which would provide benefits for the birth of children is not an insurable subject and falls more into the field of savings account. The idea has been advanced that married couples should take insurance at the time of marriage, the premium for which would insure them a certain sum at the birth of each child. To an actuary this becomes nothing more than a savings account, and has the further drawback of being a tax upon marriage.

However, one does not want to beg the question of equalizing the financial burden at the time of the birth of children, and consequently I am offering for discussion the following plan:

Suppose it were possible to grant from the state funds a sum of $100 for each Ph.D. who is the parent of a child. Thus the child of a man and woman who each held such a degree would entitle them to the sum of $200 and then upon some careful scale a college graduate would receive perhaps $90, a high school graduate $75, a grammar school graduate $50 and a moron nothing. Any combination of these groups would result in a corresponding average benefit to be provided by the state at the time of birth. For a father who wished,

This essay was excerpted and modified with permission of the author from Dworkin R. 2000. Playing God: Genes, clones, and luck. In *Sovereign virtue: The theory and practice of equality*, pp. 427–452. Harvard University Press, Cambridge, Massachusetts.

In Davenport's time, what society could do to reduce such "undesirable traits" was limited by lack of knowledge of the genetic basis of disorders and the limited strategies available for correcting matters. Today, however, tests can identify genetic predictors of disease or of predisposition to disease. It may become possible to clone human beings or radically to alter the chromosomes of an early fetus to make the resulting child taller or more intelligent or less aggressive. These new technologies will present a great variety of moral, social, and political problems in the coming years; how far and when should genetic tests be allowed, or required, or forbidden, to be used? In what, if any, circumstances should genetic manipulation of fetuses be permitted? In Dr. Dworkin's view, such questions threaten our moral compass and make us "fear losing our grip on what is wrong." His discussion of chance and choice in this essay delineates the boundaries by which we structure our values and provides a starting point for approaching these genetic issues.

> *The commonwealth is greater than any individual in it. Hence the rights of society over the life, the reproduction, the behavior and the traits of the individuals that compose it are, in all matters that concern the life and proper progress of society, limitless, and society may take life, may sterilize, may segregate so as to prevent marriage, may restrict liberty in a hundred ways.*
>
> DAVENPORT, P. 267

NO OTHER CHAPTER OF OUR SCIENCE, including cosmology, has been more exciting in recent decades than genetics, and none has been remotely as portentous for the character of the lives our descendants will lead. We must improve our understanding of these rapid changes in the basic science of genetics and also of the developing techniques for applying that basic science in medical diagnosis, prognosis, and therapy. We also must better appreciate how government and commerce—interacting in ways that range from funding grants to patent policy to statutory prohibitions and regulation—fuel, restrain, and shape these developments.

And we must aim, above all, to try to identify and assess the great variety of moral, social, and political problems that the new technology will present to the new century. To some degree, those problems are evident and pressing now. Tests can identify genetic predictors of disease, or of predisposition to disease, and new tests of that kind are coming on-line with increasing speed. So we already face extremely difficult issues about how far and when these tests should be allowed, or required, or forbidden, to be used, and how far, if at all, employers and insurance companies should be allowed to demand or ask for their results. Some problems are more speculative, because we shall face them only if science develops in a particular way. If it becomes possible to clone human beings, for example, or radically to alter the chromosomes of an early fetus to make the later child more intelligent or less aggressive, then people will have to decide whether, in some or all circumstances, these interventions are undesirable, and, if so, whether they should be forbidden by law.

I shall concentrate on certain of the moral and political problems, both evident and speculative, that the new genetics may generate in the 21st century. I shall not discuss all such problems that I believe will arise in the discussion: In particular I will not say much here about the feasibility, propriety, or character of government regulation of research and commerce. Instead I shall discuss the issues that seem to me fundamental and pervasive, and therefore likely to cut across the different topics and days into which our discussions have been organized.

WE NEED TO DISTINGUISH DERIVATIVE AND DETACHED VALUES

I will make use of an important distinction which has not been much canvassed in the moral or philosophical literature, at least in the terms in which I shall draw it, and which I should therefore introduce here. This is a distinction between two kinds of values to which we might appeal in evaluating the implications of new technology. The first set of values, which I shall call derivative values, are parasitic on the interests of particular people. We must ask, in considering whether any new technique should be permitted or regulated or forbidden, the likely impact of any such decision on individual interests. Who will be better off and who worse off in virtue of any such decision? Then we must evaluate implications in that dimension: We must ask whether any particular decision or practice is "cost–benefit efficient"—do the gains to some outweigh, on some scale of interpersonal comparison, the losses to others? We must also ask, in that dimension, whether the outcome is "fair" or "just"—is it right that some should lose and others gain in that way?

The second set of values that will figure in our argument constitute what I have elsewhere called "detached" value: These are values that do not derive from the interests of particular people but are rather intrinsic to objects or events in some other way. Many people think that great works of art have intrinsic value, that their value does not depend on whether they actually give pleasure or enlightenment, and many think that it is intrinsically wrong when a distinctive animal species becomes extinct, that this is bad quite apart from the impact on the interests of actual people. These are examples of what many people, at least, take to be detached values.

The controversy over abortion brings out the importance of the difference. If, as I have elsewhere argued, an early fetus can have no interests of its own, then the argument that abortion is wrong because it is against someone's interests is indefensible. But it nevertheless makes sense to believe, as a great many people do, that abortion is always morally problematic, and at least in some cases morally wrong, because it offends an intrinsic or detached value—the "sanctity" of human life in any form. Advances in genetic science raise many problems, as we shall see, about derivative interests: These are problems of efficiency and justice. But I shall argue that the sharp negative reactions that both people and governments have displayed to some of the more speculative genetic techniques, particularly cloning and radical genetic engineering, are not best understood as appealing to derivative values of that kind, although they are often, rather lamely, presented in that dress. They are much better understood as deep and instructive appeals to intrinsic, detached, value.

HUMAN CLONING AND GENETIC ENGINEERING CREATE FEAR AND LOATHING

Why Not?

The most arresting of the possibilities geneticists are now exploring would give scientists and doctors the power to choose which human beings there will be. People gained that power long ago, in a broad and clumsy way, when they came to understand that allowing certain people rather than others to mate would have consequences for the kind of children they produced. Eugenics, which was supported by George Bernard Shaw and Oliver Wendell Holmes as well as Adolph Hitler, was modeled on that simple insight. But genetic science now holds out the possibility, at least as comprehensible fantasy, of creating particular human beings who have been designed, one by one, according to a detailed blueprint, or of changing existing human beings, either as fetuses or later, to create people with chosen genetic properties.

Even the fantasy of this, when the technology was first described, was greeted with shock and indignation, and that shock crystallized when scientists in Scotland cloned an adult sheep, and other scientists and publicists speculated that the technique could be used to clone human beings. Committees hurriedly appointed by governments and international bodies immediately denounced the very idea. President Clinton ruled that federal funds could not be used to finance research into human cloning, and the United States Senate is on the verge of forbidding, through preposterously overbroad and panicky legislation, any and all such research. The possibility of comprehensive genetic engineering—altering a zygote's genetic inheritance to produce a battery of desired physical, mental, and emotional propensities in the way contemplated by the 1997 film, *Gattaca*—has also aroused great fear and revulsion, and any success in engineering mammals, comparable to the creation of Dolly, would undoubtedly provoke a similar official response. (In this discussion I shall often use the phrase "engineering" to include both comprehensive genetic alteration and human cloning, the latter being treated as a special case of the former. Of course, engineering and cloning are very different techniques, but many of the social and moral issues they raise are the same.)

The rhetoric of the European Parliament is not untypical of the reaction that prospects of genetic engineering have produced. In its October 28, 1993 "resolution on the cloning of the human embryo," that body declared its "firm conviction that the cloning of human beings, whether on an experimental basis, in the context of fertility treatments, preimplantation diagnosis, for tissue transplantation, or for any other purpose whatsoever, is unethical, morally repugnant, contrary to respect for the person, and a grave violation of fundamental human rights which can not under any circumstances be justified or accepted." How might we justify, or even explain, this blunderbuss reaction? We might explore three grounds of objection that are frequently mentioned. First, great caution might be thought appropriate in the face of possible danger. We do not know that, if human cloning or other comprehensive genetic engineering is possible at all, research into it or attempts at it will not pose great dangers to public health and safety. We do not know whether, for example, attempted engineering would result in an unacceptable number of miscarriages, or in the birth of an unacceptable number of deformed children. Second, resistance to engineering might be thought grounded in worries about social justice. Cloning, if available, is bound to be hideously expensive for a long time, and hence would be available only to rich people who would want, out of vanity, to clone themselves, increasing the unfair advantages of wealth. (So those horrified by the prospect of cloning have cited the specter of thousands of Rupert Murdochs or, perhaps even worse, Donald Trumps.) Third, we might hope to explain the reaction as generated, in some part, by a detached and reasonably familiar aesthetic value. Engineering, if available, might well be used to perpetuate now desired traits of height, intelligence, color, and personality, and the world would be robbed of the variety that seems essential to novelty, originality, and fascination. We must discuss each of these supposed justifications for a ban on research and development, but in my own view, at least as now informed, they do not separately or together justify the dogmatic strength of the reaction I described.

Security

It is unclear how far the Dolly precedent should be relied on in predicting the likely results of experimentation into human cloning. On the one hand, technical skill will presumably improve; on the other, human cloning may prove more difficult than cloning sheep. Several hundred attempts were necessary to produce one sheep, but, as I understand it, the rest were lost through early miscarriage, and no deformed but viable sheep was produced. Nor is there much reason to think that either cloning or engineering would produce germ-line damage threatening generations of deformity, or deformity that would not appear for generations. In any case, however, these dangers are not enough, on their own, to justify forbidding the further research that could refine our appreciation of them, and perhaps our ability to forestall or reduce whichever threats are in fact genuine. True, the sudden appearance in 1998 of Dr. Richard Seed, in the headlines and on the screen, promising to clone anyone for a high price, was enough to terrify anyone. But regulation can reign him in, along with the thousands of other cloning cowboys who would be bound to appear, without closing down research altogether. If we are assessing the risks of damage that experimentation or testing might produce, moreover, we must also take into account the hope that advancing and refining the techniques of genetic engineering will vastly decrease the number of defects and deformities with which people are now born or into which they inexorably grow. The balance of risk may well be thought to tilt in favor of experimentation.

Justice

We can easily imagine genetic engineering becoming a perquisite of the rich, and therefore as exaggerating, possibly even exponentially, the already savage injustice of both prosperous and impoverished societies. But these techniques have uses beyond vanity, and these may justify research and trials, even if we decide that vanity is an inappropriate and forbidden motive. We noticed, earlier, the important medical gains that have already been achieved through selective engineering, and more comprehensive engineering can confidently be expected enormously to expand these. Cloning may prove to have particularly dramatic medical benefits. Parents of a desperately sick child might want another child, whom they would love as much as any other, but whose blood or marrow might save the life of the sick child from which it was cloned. Cloning individual human cells, rather than an entire organism, might have even more evident benefits. A reengineered and then heavily cloned cell, for example, taken from a cancer patient, might prove to be a cure for that cancer when the clones were reintroduced. We must also count, moreover, benefits beyond the narrowly medical.

Childless couples, for example, or single women, or single men, might wish to procreate through cloning, which they might think better than the alternatives available. Or they may have no alternative at all.

Perhaps, as I just suggested, we could regulate engineering to screen out all but approved motives. If this is possible, does justice demand it, even if we assume that there are no other objections to it? I do not believe so. We should not, as I have already said, seek to improve equality by leveling down, and, as in the case of more orthodox genetic medicine, techniques available for a time only to the very rich often produce discoveries of much more general value for everyone. The remedy for injustice is redistribution, not denying benefits to some with no corresponding gain to others.

Aesthetics

We already have clones —genetically identical multiple births (which have increased as a result of infertility treatment) produce clones—and twins and other genetically identical children show that identical genes do not produce identical phenotypes. We may have underestimated nature in years past, but nurture remains important too, and the reaction to the prospect of engineering has underestimated its importance in turn. Nevertheless, people do fear that if we replace the genetic "lottery" with engineered reproduction, the welcome diversity of human types will be progressively replaced with uniformity dictated by vogue. To some degree, of course, greater uniformity is unambiguously desirable: There is no value, aesthetic or otherwise, in the fact that some people are doomed to a disfigured and short life. But it is widely believed better that, within limits, people look and act different in ways that might well be the consequence of different alleles. This thought appeals to a derivative value: That it is better for everyone to live in a world of differences. But it might also be seen as appealing to a detached value: Many people think that diversity is a value in itself, so that it would remain valuable even if, for some reason, people came to prefer uniformity.

What is not plain, however, is how far engineering, even if it were freely and inexpensively available, would actually threaten desirable diversity. Presumably all parents, if given a choice, would wish their children to have the level of intelligence and other skills that we now regard as normal, or even that we now believe superior. But we cannot regard that as undesirable: It is, after all, the object of education, ordinary as well as remedial, to improve intelligence and skill levels across the boards. Do we have good reason to fear that if parents had the choice between sexual reproduction, which produces a child of both, and cloning one of them—or a third person altogether—they might often make the latter choice? Or make it for reasons other than to exclude damaging alleles, or because they were incapable of sexual reproduction? That seems unlikely. Do we have reason to fear (as many people are concerned) that parents will engineer a reproductive zygote in order to make it a male rather than a female child, for example? It is true that in certain communities—in Northern India, for example—male children are preferred to female ones. But that preference seems so sensitive to economic circumstances, as well as to shifting cultural prejudices, that it offers no ground for supposing that we will suddenly be swamped, around the world, with a generation dominated by males. Selective abortion for sex has been available, as a result of amniocentesis and liberal abortion laws, for some time now, and no such general trend seems to have been established. In any case, we would not be justified in stopping experimentation on the basis of such thin speculation.

The fear, however, goes beyond a fear of sexual asymmetry: It is a fear that one phenotype—say, blond, conventionally good-looking, nonaggressive, tall, musically talented, and witty—will come to dominate a culture in which that phenotype is particularly valued. We should pause to notice the scientific assumptions embedded in that fear: It supposes not only that comprehensive genetic design is possible, but that the various properties of the preferred phenotype can be assembled in the same person through that design, as if each property were the product of a single allele whose possession made the property at least very highly likely, and that could be specified, and would have that consequence, independently of the specification of or phenotypic expression of other alleles. Each of those assumptions seems improbable, and their combination highly so. It seems much more likely that even parents with state of the art engineering at their disposal would have choices to make, and risks about the impact of nurture and experience to run, that they would make these choices differently in response to the very differences among them that we now celebrate, and that the later impact of differing personal choices by offspring, perhaps in search of individuality, would enlarge on those differences.

The basic motivational assumptions behind the fear seem equally as dubious as the scientific assumptions, moreover. Most people delight in the mysteries of reproduction—that value is, after all, at the root of the very objection we are considering—and many, and perhaps the great bulk, of people would forgo engineering, beyond trying to eliminate obvious defects and handicaps, as distasteful. If all this is right, the aesthetic objection is overblown and, at best, premature. We would need much more information, of a kind that could only be produced through research and experimentation, before we could even judge the assumptions on which the objection is founded, and it would therefore seem irrational to rely on that objection to prevent that research.

ARE WE PLAYING GOD?

The arguments and objections we have so far been canvassing do not provide what T.S. Eliot called an "objective correlative" for the immediate and largely sustained repulsion that I described. People feel some deeper, less articulate, ground for that revulsion, even if they have not or perhaps cannot fully express that ground but can only express it in heated and logically inappropriate language, like the bizarre reference to "fundamental human rights" in the European Parliament resolution I quoted earlier. We will not adequately appreciate the real power of the political and social resistance to further research into genetic engineering, or the genuine moral and ethical issues that that research presents, until we have succeeded in better understanding that deeper ground, and we might begin with another piece of rhetoric. It is wrong, people say, particularly after more familiar objections have been found wanting, to play God.

This objection appeals to what I called a detached rather than a derivative value. Playing God is thought wrong in itself, quite apart from any bad consequences it will or may have for any identifiable human being. Nevertheless it is deeply unclear what the injunction really means—unclear what playing God is, and what, exactly, is wrong with it. It cannot mean that it is always wrong for human beings to attempt to resist natural catastrophes or to improve upon the hand that nature has dealt them. People do that—always have done that—all the time. What is the difference, after all, between inventing penicillin and using engineered and cloned genes to cure even more terrifying diseases than penicillin cures? What is the difference between setting your child strenuous exercises to reduce his weight or increase his strength and altering his genes, while an embryo, with the same end in view?

These are not rhetorical questions. We must try to answer them, but we must begin at some distance from them, in the overall structure of our moral and ethical experience. For that structure depends, crucially, on a fundamental distinction between what we are responsible for doing or deciding, individually or collectively, and what is given to us, as a background against which we act or decide, but which we are powerless to change. For the Greeks, this was a distinction between themselves and their fate or destiny, which was in the hands or the laps of the Gods. For people, even today, who are religious in a conventional way, it is a distinction between how God designed the world, including our natural condition in it, and the scope of the free will he also created. More sophisticated people use the language of science to the same effect: For them the fundamental distinction falls between what nature, including evolution, has created, by way of particles and energy and genes, and what we do in that world and with those genes. For everyone, the distinction, however they describe it, draws a line between who and what we are, for which either a divine will or nothing but a blind process is responsible, and what we do with that inheritance, for which we are indeed, separately or together, responsible.

That crucial boundary between chance and choice is the spine of our ethics and our morality, and any serious shift in that boundary is seriously dislocating. Our sense of a life well-lived, for example, is fundamentally shaped by supposed givens about the upper limits of human life span. If people could suddenly be expected to live ten times as long as we now do, we would have to recreate the whole range of our opinions about what an attractive kind of life would be, and also our opinions about what activities that carry some risk of accidental death for others, like driving, are morally permissible. History already offers, in our own time, less dramatic but nevertheless profound examples of how scientific change radically dislocates values. People's settled convictions about the responsibilities of leaders to protect their own soldiers in war, at any cost, changed when scientists split the atom and vastly increased the carnage that those convictions could justify. People's settled convictions about euthanasia and suicide changed when deathbed medicine dramatically increased a doctor's power to extend life beyond the point at which that life had any meaning for the patient. In each case a period of moral stability was replaced by moral insecurity, and it is revealing that in both episodes people reached for the phrase "playing God," in one case to accuse the scientist who had dramatically increased our powers over nature by cracking what had been thought fundamental in God's design, and in the other by taking upon themselves a decision that the past limits of medicine had made it easy to treat as God's alone.

My hypothesis is that genetic science has suddenly made us aware of the possibility of a similar, although far greater, pending moral dislocation. We dread the prospect of people designing other people because that possibility in itself shifts—much more dramatically than in these other examples—the chance/choice boundary that structures our values as a whole, and such a shift threatens, not to offend any of our present values, derivative or detached, but, on the contrary, to make a great part of these suddenly obsolete. Our physical being—the brain and body that furnishes for each of us his material substrate—has long been the absolute paradigm of what is both devastatingly important to us and, in its initial condition, beyond our power to alter and therefore beyond the scope of our responsibility, either individual or collective. The popularity of the phrase "genetic lottery" itself shows the centrality of our conviction that what we most basically are is a matter of chance not choice. I must be careful not to be misunderstood. I do not mean that genetic continuity provides the key to the technical philosophical problem of personal identity, although some philosophers have indeed thought this. I mean to make a psychological point: People think that the very essence of the distinction between what God or nature provides and what they are responsible for making of or with that provision is to be defined physically, in terms of what is in "the genes" or, in a metaphor reflecting an older science, "the blood."

If we were to take seriously the possibility we are now exploring—that scientists really have gained the capacity to create a human being having any phenotype that they or their prospective parents choose—then we could chart the destruction of settled moral and ethical attitudes starting at almost any point. We use the chance/choice distinction not simply in our assignments of responsibility for situations or events, for example, but in our assessments of pride, including pride in what nature has given us. It is a striking phenomenon, now, that people take pride in physical attributes or skills they did not choose or create, like physical appearance or strength, but not when these can be seen to be the results of the efforts of others in which they played no part. A woman who puts herself in the hands of a cosmetic surgeon may rejoice in the result, but can take no pride in it; certainly not the pride she would have taken if she had been born into the same beauty. What would happen to pride in our physical attributes, or even what we made of them, if these were the inexorable results not of a nature in whose pride we are allowed, as it were, to share, but of the decision of our parents and their hired geneticists?

But the most dramatic use of the fundamental chance/choice distinction is in the assignment of personal and collective responsibility, and it is here that the danger of moral insecurity seems greatest. We now accept the condition in which we were born as a parameter of our responsibility—we must make the best of it that we can—but not as itself a potential arena of blame, except in those special cases, themselves of relatively recent discovery, in which someone's behavior altered our embryonic development, through smoking, for example, or drugs. Otherwise, though we may curse fate for how we are, as Richard Crookback (better known as Richard III) did, we may blame no one else. The same distinction holds, at least for most people, and for many reflective moral philosophers, for social responsibility as well. We feel a greater responsibility to compensate victims of industrial accidents and of racial prejudice, as in both cases they are victims, although in different ways, of society generally, than we feel to compensate those born with genetic defects or those injured by lightning, or in those other ways that lawyers and insurance companies call, in an illuminating phrase, acts of God.

How would all this change if everyone was as he is through the decisions of others, including the decision of some parents I described earlier to let nature take its course? Change it must. But how and why? Once again, these questions are not rhetorical. I do not know the answers and can hardly guess at them. But that is the point. The terror many of us feel at the thought of genetic engineering is not a fear of what is wrong; it is rather a fear of losing our grip on what is wrong. We are not entitled—it would be a serious confusion—to think that even the most dramatic shifts in the chance/choice boundary somehow challenge morality itself—that there will one day be no more wrong or right. But we are entitled to worry that our settled convictions will, in large numbers, be undermined, that we will be in a kind of moral free fall, that we will have to think again against a new background and with uncertain results. Playing God is, in that way, playing with fire.

Suppose that this hypothesis, at least as it might be corrected and improved, makes sense, and accounts for the powerful surge in people's emotional reaction to genetic engineering that is not accounted for by the more discrete grounds we first examined. Have we then discovered not only an explanation but a justification for the objection, a reading of "don't play God" that shows why, at least in this instance, we should not? I think not. We would have discovered a challenge that we must take up rather than a reason for turning back. For our hypothesis implicates no value—derivative or detached—at all. It reveals only reasons why our contemporary values, of both kinds, may be wrong or at least ill-considered. If we are to be morally and ethically responsible, there can be no turning back once we find, as we have found, that some of the most basic presuppositions of these values are mistaken. Playing God is indeed playing with fire. But that is what we mortals have done, since Prometheus, the patron saint of dangerous discovery. We play with fire, and take the consequences, because the alternative is an irresponsible cowardice in the face of the unknown.

REFERENCE

Laughlin H.H. 1914. Report of the Committee to Study and Report on Best Practical Means of Cutting of the Defective Germplasm in the American Population. I. The scope of the Committee's work. *ERO Bull.* **10A:** 16.

Genetics and Human Nature

Lewis Wolpert
University College, London

Editors' Note: Causality as a philosophical concept has occupied human discourse since the time of Aristotle. The 18th century philosopher David Hume's ideas about causality still influence us today. For him, "causation is the basis of all our reasoning concerning matters of fact" (Hume 1975a, pp. 26–27) and "by means of it I paint the universe in my imagination" (Hume 1975b, pp. 67–68). As a physical concept, causality is a basic tenet of the scientific exploration of the physical world.

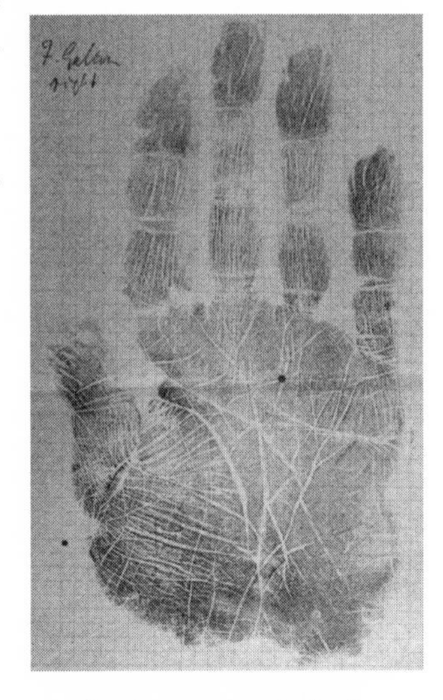

In *Heredity in Relation to Genetics*, Charles Davenport attributed a long list of human traits to heredity, some genuine and some spurious. In this essay, Dr. Wolpert makes the case for the genetic basis of the causal beliefs—the relationship between cause and effect—humans have about the physical world. He proposes that causal belief is the attribute that defines human nature and separates us from the other primates, because once humans could understand the nature of cause and effect, they were able to manipulate the natural world. That our brains are hardwired for causality, according to Dr. Wolpert, is evidenced by the fact that even infants display such beliefs.

Dr. Wolpert also discusses the influences of environmental effects versus heredity, a topic to which Charles Davenport gave very little space in *Heredity in Relation to Genetics*. Davenport was a strong believer in heredity as the basis of our human traits, but he did recognize that environmental issues might play some role, as evidenced by the opening quote for this essay. Dr. Wolpert argues that important as environmental influences may be, they still operate on a substrate developed under the control of genes. And this brings up the question of whether we can genetically alter our human nature? Dr. Wolpert's discussion of this issue is wide-ranging and emphasizes the need for public education about the consequences of genetic techniques as well as the importance of distinguishing between science and technology. As Dr. Wolpert points out, "Dangers and ethical issues arise only when science is applied to technology."

> *So long as we regard heredity and environment as opposed so long will we experience endless contradictions in interpreting any trait, behavior or disease. The truth seems to be that for human phenomena there is not only the external or environmental cause but also an internal or personal cause. The result is, in most cases, the reaction of a specific sort of protoplasm to a specific stimulus.*
>
> DAVENPORT, P. 252

OUR INCREASING KNOWLEDGE OF GENETICS should be changing our beliefs about human nature, the basic attributes of human life that distinguishes it from other animal life—but has it? To what extent has knowledge of how genes can determine so many of human characteristics affected everyday thinking about human nature? Darwin's theory of evolution is now supported by discoveries in genetics and other scientific investigations, but does the general public really accept that we humans have evolved by a process based on random events—namely, mutations in genes?

The study of genetics, together with cell and developmental biology, has transformed our understanding of the biological world, not only of how it evolved but also of how genes can affect so many human characteristics. These include physical characteristics such as our height and the shape of our noses as well as many neural and psychological ones that include intelligence and mental disorders. Yet there is a resistance to such understanding. This is illustrated by the 1998 report of the Nuffield Council on Bioethics on genetic screening (www.nuffieldbioethics.org), which claimed that one must view a human as a whole and cannot identify the effect of genes as compared to environmental influences. The scientific evidence is completely against this, but the report reflects the hesitation in assigning causes to genes. Perhaps it reflects a widespread distaste for the unromantic notion that the genetic molecule DNA has so much influence on our nature and yet is itself just a long double helix.

But genes have determined the nature of human nature—not just physical characters but mental too. We are descended from apelike ancestors and genes played a fundamental role. But what is it about human nature that distinguishes us from other primates like chimpanzees? Do not believe the widely quoted claim that human genes differ very little, about 1.5%, from that of chimpanzees, as recent studies suggest that there may well be thousands of genes that either encode an altered protein or are expressed differently in humans and chimpanzees. But what is it that makes humans fundamentally different from other primates? I suggest that the main feature that characterizes humans is a belief in cause and effect that came with toolmaking—technology has driven human evolution and that is a result of genes changing our brains.

BELIEF IN CAUSALITY DISTINGUISHES HUMANS FROM OTHER ANIMALS

In 1739, David Hume put forward his doctrine about causality. Our idea of causality, he claimed, is that there is a necessary connection among things, particularly actions. However, this connection cannot be directly observed and can only be inferred from observing one event always following another. Thus, he argued that a causal relationship inferred from such observations cannot in fact be rationally inferred. Although this may be philosophically true, it is a problem for philosophers that need not concern us, as it is obvious what the cause is if, for example, I cut my hand with a knife.

David Premack, the primate language investigator, has pointed out there are two classes of causal beliefs. One, as Hume suggested, is based on one event being linked to another and can be called weak or "arbitrary," for there need not be any obvious connection between them, like switching on a light. Animals can learn connections by the pairing of events by the process of associative learning. The other, which is uniquely human, is strong or "natural" causality and is programmed into our brains so that we have evolved the ability to have a concept of forces acting on objects. Such strong causal beliefs are already present in human infants. A key question is how this type of belief evolved. Causal beliefs are a fundamental characteristic of humans; animals, by contrast, have very few causal beliefs.

Anthropologist Kenneth Oakley made clear that humans may be distinguished as the toolmaking primate. This is a key to understanding the origins of causal belief and tool use to change the environment, defend, and attack. No animal other than a human being has ever properly realized this, and although there is evidence that some primates do use simple tools, they have no clear understanding of cause and effect, nor of the differences between tools and how they may be used in a variety of ways. No animal uses a container to carry, for example, food or water; pots and bags are totally human and were invaluable for our ancestors when traveling.

Only humans effectively cause one object to interact with another or the environment in a multitude of different ways and invented the technology that drove human evolution. This is due to the evolution of the ability of humans to have causal beliefs, together with the ability to imitate and so have cultural evolution. One cannot make complex tools without a concept of cause and effect. Belief in cause and effect is a developmental primitive in children. A key component of this ability has been suggested to be a causal operator in the brain, which may involve connections between the left frontal lobe and left orientation area.

It should thus not be surprising that causal beliefs are present in human infants and that their acquisition must have a genetic basis. Young children perceive that certain objects have causal properties with a renewable source of energy or force, and this is a sophisticated idea unique to humans. These special objects, or agents, can act in pursuit of goals. They also have a concept of "force," which is a primitive mechanical notion—not the same as the scientific concept of force. The basic idea is that when bodies move they possess a force and this can, on impact, be transmitted to other objects which can receive or resist. By 3 months, infants expect a stationary object to be displaced when hit by a moving object and by 6 months can reliably estimate how far it should move.

While nonhuman primates have an understanding of all kinds of quite complex physical and social concepts and can distinguish the animate from the inanimate, they do not view the world in terms of the kinds of intermediate and often hidden "forces"—underlying causes, reasons, intentions, and explanations—that are so important for human thinking, particularly technology. But the beginning of causal beliefs was there in ancestral apes, and it was from this that our causal and technological skills evolved. It is even possible that language evolved in relation to causal beliefs as so many verbs are causal. Selecting humans to have causal beliefs is one fundamental way in which genes have molded human nature. It is relevant that most mental illnesses involve false causal beliefs.

HOW GENES CONTROL AND INFLUENCE OUR HUMAN NATURE

There is quite a lot of resistance to this type of thinking in which so many aspects of human nature are due to genes. There is hostility to genes influencing human characters as diverse as creativity, intelligence, mental illness, and criminality.

The evidence about the understanding of genetics by the public is disturbing when one notes that a European survey found that about half of those questioned thought that tomatoes did not have genes unless they were genetically modified. Again, many members of the public know that one does not take an antibiotic for a viral illness but do not understand the difference between a virus and a bacterium—that a bacterium is a cell, a living organism, whereas a virus contains the genetic code for making another virus when it is in an appropriate cell.

A central idea in relation to genetics is that cells are the "miracles" of evolution, and it has been clear for some time that we are no more than a complex society of cells, in which genes play a key role in controlling the cells' behavior. It is rarely made clear to the public that genes are rather passive, and it is the proteins they code for that are the workhorses of the cells. This unromantic image of life, which invokes no magical life forces, is not for many people an attractive one and goes against religious beliefs. One might even say it goes against human nature.

The main process by which genes control and influence our nature is by controlling cell behavior during embryonic development from a single cell, the fertilized egg. Genes in the embryo do not provide a blueprint for the adult but a set of instructions as to how the cells must behave, and the interactions involved are complex. Like all science, genetics goes against common sense, and there is nothing in an individual's experience of the world that relates to the way genes influence cells and how we develop from the egg. For many, genes are as mystical as witches and possibly as potentially evil as they can cause so many diseases.

The relationship between genes and human nature is made particularly difficult because of the very complex relationship between gene action and the results of this action. There are very few examples where the relationship between action of a gene and its effect on human shape or behavior are easy to understand. One of the clearest examples is sickle-cell anemia, where a mutation in the gene coding for hemoglobin causes it to aggregate and deform the red blood cell. But such a clear link is very rare, and in most cases many genes are involved and the pathway is very complex. Consider, for example, depression or schizophrenia. Both have a strong genetic basis; in the case of schizophrenia about 80% of the cause is genetic. Many genes are involved and many of these act to determine the behavior of the numerous neurons in the developing nervous system in the embryo. But then the interaction between the myriad nerve cells that result in schizophrenia needs to be understood. There is no simple route from the genes to the illness, and environmental influences are involved, and so this process has an almost mystical character quite outside our normal experience of causal events.

It is all too easy to be misled as to what genes actually do for us. There is no gene, for example, for the eye; many hundreds, if not thousands, are involved, but a fault in just one can lead to major abnormalities. The language in which many of the effects of genes is described leads to confusion. No sensible person would say that the brakes of a car are for causing accidents. Yet, using a convenient way of speaking, there are numerous references to, for example, the gene for homosexuality or the gene for criminality. When the brakes of the car that are there for safe driving fail, then there is an accident. Similarly, if criminality has some genetic basis, then it is not because there is a gene for criminality, but because of a fault in the genetic component that has resulted in this particular undesirable effect. A genetic error could have affected how the brain developed—genes control development of every bit of our bodies. There is no aspect of human nature unaffected by gene action during our embryonic development. Of course later environmental experience will have major effects on the individual, and nature/nurture debates are all too common.

One clear case where genes are widely accepted as being closely related to human nature is in relation to sexual differences. It is the Y chromosome that makes the human embryo develop as a male and not a female. This is a powerful example of the relationship between human nature and genetics, even if the mechanism whereby the Y chromosome exerts its influence is complex. It is also clear that these genetic influences determine the sexual behavior that is so key a feature of human nature, and this raises issues about the nature of homosexuality. The relationship between genetic influence and the environment in determining the nature of a particular individual is complex and quite controversial. A case can be made for mysticism and religion having a strong genetic basis (Wolpert 2006). Even time is embedded in genes that control our circadian clock.

DANGEROUS SCIENCE—FEAR OF GENETICS IS MISPLACED

Has knowledge of genes had an impact on the arts—novels, for example? In general, no, science fiction possibly being an exception. Indeed the whole of Western literature has not been kind to science or scientists and is filled with images of scientists meddling with nature with disastrous results—particularly those meddling in genetics. Just consider Mary Shelley's *Frankenstein, or the Modern Prometheus*, and Aldous Huxley's *Brave New World*, which have had such a powerful negative effect on the image of genetics. The idea that scientific knowledge is dangerous is deeply embedded in our culture. Adam and Eve were forbidden to eat from the Tree of Knowledge. In Milton's *Paradise*

Lost the serpent addresses the Tree as the "Mother of Science"; moreover the archangel Raphael advises Adam to be lowly wise when he tries to question him about the nature of the universe.

There is a fear and distrust of science, particularly genetic engineering and the supposed ethical issues it raises. These include the fears of the effect of this science in diminishing our spiritual values (even though many scientists are themselves religious), the impact of industry using genetically modified organisms in despoiling the environment, and modification and choice of designer babies. Geneticists are repeatedly referred to as "playing God." Many of these criticisms coexist with the hope, particularly in medicine, that their science will provide cures to all major illnesses, like cancer, heart disease, and genetic disabilities like cystic fibrosis. But is a science like genetics dangerous and what are the special social responsibilities of geneticists other than making clear the implications of their work? It is not for them to decide how it should be applied; that decision is for society as a whole.

Another serious problem in thinking about genetics and its applications is the conflation of science and technology. The distinction between science and technology, between knowledge and understanding on the one hand and the application of that knowledge to making something or using it in some practical way, is fundamental. Science produces ideas about how the world works, whereas the ideas in technology result in usable objects. But it is technology that generates ethical issues, from cars to cloning a human. In contrast to technology, reliable scientific knowledge is value-free and has no moral or ethical value. Science tells us how the world is. That we are not at the center of the universe is neither good nor bad, nor is the possibility that genes can influence our intelligence or our behavior. Dangers and ethical issues arise only when science is applied to technology.

Could genetic knowledge be dangerous by, for example, identifying that one group of people are inferior in certain skills? Philosopher Philip Kitcher's main example of the implications of a dangerous scientific finding is related to genetics providing evidence for racial differences that show that one group of people is inferior in certain skills. Such a finding, he is sure, would further disadvantage them, and he may be right. But he does not even consider a society that would positively compensate them as is often done for those with genetic illnesses. Contrary to his view, I regard reliable scientific knowledge as value-free.

Mary Shelley could be both proud and shocked. Her creation of a scientist creating and meddling with human life has become the most potent negative symbol of modern genetics. But shocked because her brilliant fantasy has become so distorted that even those who are normally quite sensible lose all sense of proportion when the idea of cloning humans appears before them. The image of Frankenstein has been turned by the media into genetic pornography whose real aim is to titillate, excite, and frighten. The biological moralists are triumphant. The possibility of cloning of a human so that it has the same genes as the donor of the nucleus inserted into the egg has created hysteria. For many it goes against human nature. The moral masturbators have been out in force telling us of the horrors of cloning. Jeremy Rifkin, an activist regarding biotechnology issues, demanded a worldwide ban and suggested that cloning should carry a penalty "on a par with rape, child abuse and murder." Many others, national leaders included, have joined in that chorus of horror. But what horrors? What ethical issues? In all of the righteous indignation, I have not found a single new relevant ethical issue spelled out. The really important issue is how the child will be cared for.

What fantasy is it about cloning that so upsets people? Is it that it is unnatural? Although genes are very important, so is the environment. Identical twins are not identical as there are differences that arise from differences in position in the womb, and also there is some variation in development even with the same genes. Environmental influences after birth could make the child a very different person from the genetic donor. A cloned child might have some problems about its nature, and this would have to be taken into account. However, this is an issue common to several other types of assisted reproduction, such as surrogate mothers and anonymous sperm donors. I am totally against human cloning at present because it carries a high risk of abnormalities, as numerous scientific studies on other animals show. In fact, it is quite amusing to observe the swing from moralists who denied that genes have an important effect on intelligence to saying that a cloned individual's behavior will be entirely determined by the individual's genetic makeup.

HOW GENES AFFECT OUR HUMAN IMAGE

How has knowledge about genes altered the way people think about themselves and others? Does, for example, knowing the origin of one's illness is genetic make one feel better or worse? Does genetic screening for short stature, for example, diminish the self-image of those who have it? It is far from clear, but there is a tendency to avoid genetic explanations unless the evidence is very strong, as in cystic fibrosis and sickle-cell anemia. But with conditions more complex in origin like depression and schizophrenia there is a strong tendency to look for causes in the environment. It is not clear as to whether people with mental illness, for example, prefer a genetic explanation; perhaps it leaves them with no chance of escape. And there is the problem as to what extent genes can cause mental abnormalities so the afflicted person is no longer responsible for their actions.

In cases where there is a genetic basis, screening of the embryo to avoid a child being born with a genetic disability is possible. Prenatal diagnosis serves a variety of purposes. In some cases, prenatal genetic testing can lead to timely medical intervention to mitigate or eliminate disease. In other cases, prenatal diagnosis allows families and medical professionals to prepare for the care of a child with special needs. Some of those with such a disability feel that such screening diminishes their human nature. A counterexample is Stephen Hawking, the physicist who has amyotrophic lateral sclerosis (Lou Gehrig's disease). Professor Hawking writes, "I am quite often asked: How do you feel about having ALS? The answer is, not a lot. I try to lead as normal a life as possible, and not think about my condition, or regret the things it prevents me from doing, which are not that many" (www.hawking.org.uk/text/disable/disable.html).

CAN GENETICS ALTER HUMAN NATURE?

Could genetics be used to alter human nature? In relation to sport there is the real possibility of introducing genes into muscles to make them stronger. More serious is the possibility that designer babies will be made by both selecting what genes they have and introducing new genes. Anxieties about making designer babies by introducing genes into the embryo are at present premature as it is far too risky, and too little is known about which genes could actually alter human nature. We may have, in the first instance, to accept what Ronald Dworkin has called procreative autonomy, a couple's right "to control their own role in procreation unless the state has a compelling reason for denying them that control" (Dworkin 1993, p. 148). One must wonder why the biological moralists do not devote their attention to other technical advances like that convenient form of transport, the car, which claims more than 50,000 killed or seriously injured each year. Applications of embryology and genetics, in striking contrast, have caused minimal harm. But human nature could be altered by genes in the future. Such genetic alterations should be permitted if safe, and only banned if they can be shown to have definite bad effects on either individuals or society.

Genes have constructed the basic features of human nature. Nurture and environmental influences can have powerful effects, but they can only work through the structures that developed under the control of genes. Separating out the mutual influences is not easy but it is essential not to ignore either contribution. Only in this way may we be able to modify and deal with antisocial behavior and many debilitating conditions.

REFERENCES

Dworkin R. 1993. *Life's dominion: An argument about abortion, euthanasia, and individual freedom*. Harper Collins, London.

Hume D. 1975a. An enquiry concerning human understanding. In *Enquiries concerning human understanding and concerning the principles of morals* (ed. L.A. Selby-Bigge), 3rd ed. Clarendon, Oxford.

Hume D. 1975b. *A treatise of human nature* (ed. L.A. Selby-Bigge), 2nd ed. Clarendon, Oxford.

Wolpert L. 2006. *Six impossible things before breakfast—The evolutionary origin of belief*. Faber, London.

What follows is a facsimile of the 1911 edition of Charles Benedict Davenport's book *Heredity in Relation to Eugenics*, originally published by Henry Holt and Company. The facsimile is identical to the original edition except that five Henry Holt and Company advertising pages have been removed from the end of the book and Plate II now appears on facing pages after page 213 rather than as an insert between pages 218 and 219. Some of the labeling in both Plate I (frontispiece) and Plate II has been reset to improve readability.

PLATE I

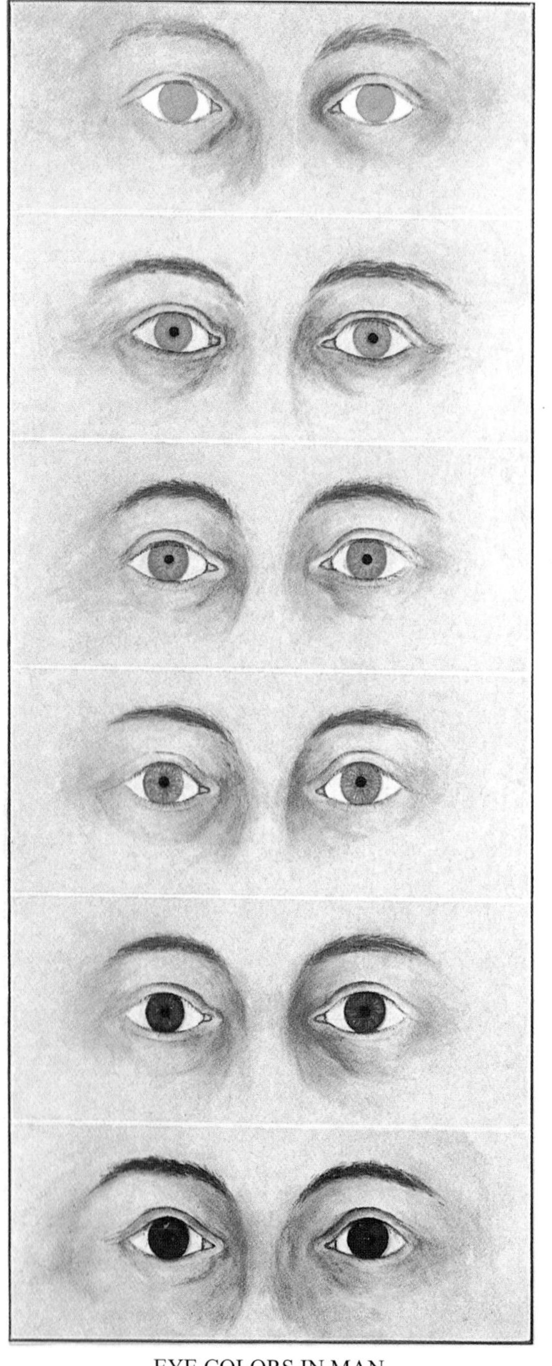

A. PIGMENT OF CHOROID COAT AND PIGMENT OF IRIS ABSENT. 1. The Albino eye. Red from unobscured blood vessels.

B. PIGMENT OF CHOROID PRESENT.

α. IRIS WITHOUT TRUE PIGMENT. 2. BLUE. Due to a purple layer on back of eye.

β. IRIS WITH TRUE PIGMENTS.

a. *Lipochrome or yellow pigment.* 3. GREEN or cat eye. Yellow pigment on blue background.

b. *Melanic or black pigment.* 4. HAZEL or gray eye. Dilute brown pigment around pupil only.

5. BROWN eye. Melanic pigment; various shades from various dilutions.

5. BLACK eye. An abundance of melanic pigment.

EYE COLORS IN MAN

HEREDITY IN RELATION TO EUGENICS

BY

CHARLES BENEDICT DAVENPORT

CARNEGIE INSTITUTION OF WASHINGTON
DIRECTOR, DEPARTMENT OF EXPERIMENTAL EVOLUTION
COLD SPRING HARBOR, LONG ISLAND, N. Y.
SECRETARY OF THE EUGENICS SECTION
AMERICAN BREEDERS' ASSOCIATION

NEW YORK
HENRY HOLT AND COMPANY
1911

PRESS OF T. MOREY & SON
GREENFIELD, MASS., U. S. A.

TO

MRS. E. H. HARRIMAN

IN RECOGNITION OF THE GENEROUS ASSISTANCE
SHE HAS GIVEN TO RESEARCH IN EUGENICS
THIS BOOK IS
GRATEFULLY DEDICATED

PREFACE

RECENT great advances in our knowledge of heredity have revolutionized the methods of agriculturalists in improving domesticated plants and animals. It was early recognized that this new knowledge would have a far-reaching influence upon certain problems of human society —the problems of the unsocial classes, of immigration, of population, of effectiveness, of health and vigor. Now, great as are the potentialities of the new science of heredity in its application to man it must be confessed that they are not yet realized. A vast amount of investigation into the laws of the inheritance of human traits will be required before it will be possible to give definite instruction as to fit marriage matings. Our social problems still remain problems. For a long time yet our watchword must be *investigation.* The advance that has been made so far is chiefly in getting a better method of study.

In this book I have sought to explain this new method. An application of this method to some specific problems, especially to the transmission of various human traits and susceptibilities to disease, has been attempted. The suggestions made are by no means final but are made to illustrate the general method and give the most probable conclusions. Only with much more accurate data can the laws of inheritance of family peculiarities be definitely determined.

Some general consequences of the new point of view for the American population have been set forth in Chapters IV to VI. Their essential truth will, I trust, be generally recognized. In any case it will not be amiss to point out the fundamental difference between the modern eugenical and the contrasted or "euthenical" standpoints. As a matter of fact the eugenic teachings that we think of as new are very old. Modern medicine is responsible for the loss of appreciation of the power of heredity. It has had its attention too exclusively focussed on germs and conditions of life. It has neglected the personal element that helps determine the course of every disease. It has begotten a wholly impersonal hygiene whose teachings are false in so far as they are laid down as universally applicable. It has forgotten the fundamental fact that all men are created *bound* by their protoplasmic makeup and *unequal* in their powers and responsibilities.

As indicated, it is the aim of this book to incite to further investigation. Some space is devoted to the eugenics movement—a movement which it is hoped will, in this country, for the present, take mainly the form of investigation. To this movement the Eugenics Record Office (a branch of the work of the American Breeders' Association) is dedicated. The Eugenics Record Office wishes to get in touch with all persons interested in the eugenics movement. It invites every person who is willing to do so to record his heritage and place the record on file at the Record Office. "Drop a postal card" at once to the Eugenics Record Office, Cold Spring Harbor, New York, and ask for the blank schedule they furnish. It is understood that all data deposited in this way will be held as confidential and be used only for scientific purposes. The data received are carefully preserved in a fireproof vault and indexed so as to be available to the student. Specifically, the Record Office seeks pedigrees of families in which one or more of the following traits appear:—short stature, tallness, corpulency, special talents in music, art, literature, mechanics, invention and mathematics, rheumatism, multiple sclerosis, hereditary ataxy, Ménière's disease, chorea of all forms, eye defects of all forms, otosclerosis, peculiarities of hair, skin and nails (especially red hair), albinism, harelip and cleft palate, peculiarities of the teeth, cancer, Thomsen's disease, hemophilia, exophthalmic goiter, diabetes, alkaptonuria, gout, peculiarities of the hands and feet and of other parts of the skeleton. We do not appeal primarily to physicians for this information but to the thousands of intelligent Americans who love the truth and want to see its interests advanced. At the same time, physicians can aid in the work by inducing persons with bodily or mental peculiarities that run through their families to send to the Record Office for blank schedules on which to record the method of inheritance of the trait in question. Thus every one can share in the eugenics movement.

The Eugenics Record Office will be glad to assist in the establishment of local eugenics societies which shall become centers for the study of local blood-lines and for local instruction. The Office seeks to assist state officials in the study of the classes which are supported and protected by the State, and to assist the States to locate the centers in which their defectives and delinquents are being bred. It is believed that a little money spent in studying the sources of reproduction of persons who are destined to become state wards will prove a highly profitable investment, since it may lead to steps that will diminish such reproduction.

In the preparation of the present volume the author has been aided by many hands. Professor James A. Field, of the University of Chicago, has kindly read the proof and made valuable suggestions. The bibliography and the pedigree charts were largely prepared by Miss Amey B. Eaton, of the Eugenics Record Office. Professor E. B. Wilson has generously granted me the use of Figures 1 to 6 from his invaluable book, "The Cell in Development and Inheritance." Hundreds of persons have voluntarily contributed the data upon which the conclusions that have been drawn are based. My friend and colleague, Mr. H. H. Laughlin, Superintendent of the Eugenics Record Office, has assisted in many points and has contributed the frontispiece. My wife has, as usual, revised the manuscript and prepared it for the printer. The Trustees of the Carnegie Institution have granted me exceptional opportunities for the prosecution of the work. Last, but by no means least, this work and the collection of data out of which it has grown have been made possible by the financial assistance and by the personal stimulus and advice given by the lady to whom, in insufficient recognition, this book is, with her permission, dedicated. To all those who have so kindly assisted me I return thanks. I trust the book will be useful to humanity, so as to justify them for the pains they have taken to bring it to pass.

C. B. D.

Carnegie Institution of Washington
Station for Experimental Evolution
Cold Spring Harbor, N. Y.

CONTENTS

CHAPTER I

EUGENICS: ITS NATURE, IMPORTANCE AND AIMS

CHAPTER II

THE METHOD OF EUGENICS

CHAPTER III

THE INHERITANCE OF FAMILY TRAITS

CHAPTER IV

THE GEOGRAPHIC DISTRIBUTION OF INHERITABLE TRAITS

CHAPTER V

MIGRATIONS AND THEIR EUGENIC SIGNIFICANCE

CHAPTER VI

THE INFLUENCE OF THE INDIVIDUAL ON THE RACE

CHAPTER VII

THE STUDY OF AMERICAN FAMILIES

CHAPTER VIII

EUGENICS AND EUTHENICS

CHAPTER IX

THE ORGANIZATION OF APPLIED EUGENICS

PLATES

HEREDITY IN RELATION TO EUGENICS

CHAPTER I

EUGENICS: ITS NATURE, IMPORTANCE AND AIMS

1. What Eugenics Is

Eugenics is the science of the improvement of the human race by better breeding or, as the late Sir Francis Galton expressed it:—"The science which deals with all influences that improve the inborn qualities of a race." The eugenical standpoint is that of the agriculturalist who, while recognizing the value of culture, believes that permanent advance is to be made only by securing the best "blood." Man is an organism—an animal; and the laws of improvement of corn and of race horses hold true for him also. Unless people accept this simple truth and let it influence marriage selection human progress will cease.

Eugenics has reference to offspring. The success of a marriage from the standpoint of eugenics is measured by the number of disease-resistant, cultivable offspring that come from it. Happiness or unhappiness of the parents, the principal theme of many novels and the proceedings of divorce courts, has little eugenic significance; for eugenics has to do with traits that are in the blood, the protoplasm. The superstition of prenatal influence and the real effects of venereal disease, dire as they are, lie outside the pale of eugenics in its strictest sense. But no lover of his race can view with complaisance the ravages of these diseases nor fail to raise his voice in warning against them. The parasite that induces syphilis is not only hard to kill but it frequently works extensive damage to heart, arteries and brain, and may be conveyed from the infected parent to the unborn child. Gonorrhea, like syphilis, is a parasitic disease that is commonly contracted during illicit sexual intercourse. Conveyed by an infected man to his wife it frequently causes her to become sterile. Venereal diseases are disgenic agents of the first magnitude and of growing importance. The danger of acquiring them should be known to all young men. Society might well demand that before a marriage license is issued the man should present a certificate, from a reputable physician, of freedom from them. Fortunately, nature protects most of her best blood from these diseases; for the acts that lead to them are repugnant to strictly normal persons; and the sober-minded young women who have had a fair opportunity to make a selection of a consort are not attracted by the kind of men who are most prone to sex-immorality.

2. The Need of Eugenics

The human babies born each year constitute the world's most valuable crop. Taking the population of the globe to be one and one-half billion, probably about 50 million children are born each year. In the continental United States with over 90 million souls probably 2½ million children are annually born. When we think of the influence of a single man in this country, of a Harriman, of an Edison, of a William James, the potentiality of these 2½ million annually can be dimly conceived as beyond computation. But for better or worse this potentiality is far from being realized. Nearly half a million of these infants die before they attain the age of one year, and half of all are dead before they reach their 23rd year—before they have had much chance to affect the world one way or another. However, were only one and a quarter million of the children born each year in the United States destined to play an important part for the nation and humanity we could look with equanimity on the result. But alas! only a small part of this army will be fully effective in rendering productive our three million square miles of territory, in otherwise utilizing the unparalleled natural resources of the country, and in forming a united, altruistic, God-serving, law-abiding, effective and productive nation, leading the remaining 93 per cent of the globe's population to higher ideals. On the contrary, of the 1200 thousand who reach full maturity each year 40 thousand will be ineffective through temporary sickness, 4 to 5 thousand will be segregated in the care of institutions, unknown thousands will be kept in poverty through mental deficiency, other thousands will be the cause of social disorder and still other thousands will be required to tend and control the weak and unruly. We may estimate at not far from 100 thousand, or 8 per cent, the number of the non-productive or only slightly productive, and probably this proportion would hold for the 600 thousand males considered by themselves. The great mass of the yearly increment, say 550 thousand males, constitute a body of solid, intelligent workers of one sort and another, engaged in occupations that require, in the different cases, various degrees of intelligence but are none the less valuable in the progress of humanity, Of course, in these gainful occupations the men are assisted by a large number of their sisters, but four-fifths of the women are still engaged in the no less useful work of home-making. The ineffectiveness of 6 to 8 per cent of the males and the probable slow tendency of this proportion to increase is deserving of serious attention.

It is a reproach to our intelligence that we as a people, proud in other respects of our control of nature, should have to support about half a million insane, feeble-minded, epileptic, blind and deaf, 80,000 prisoners and 100,000 paupers at a cost of over 100 million dollars per year. A new plague that rendered four per cent of our population, chiefly at the most productive age, not merely incompetent but a burden costing 100 million dollars yearly to support, would instantly attract universal attention. But we have become so used to crime, disease and degeneracy that we take them as necessary evils. That they were so in the world's ignorance is granted; that they must remain so is denied.

3. The General Procedure in Applied Eugenics

The general program of the eugenist is clear—it is to improve the race by inducing young people to make a more reasonable selection of marriage mates; to fall in love intelligently. It also includes the control by the state of the propagation of the mentally incompetent. It does not imply destruction of the unfit either before or after birth. It certainly has only disgust for the free love propaganda that some ill-balanced persons have sought to attach to the name. Rather it trusts to that good sense with which the majority of people are possessed and believes that in the life of such there comes a time when they realize that they are drifting toward marriage and stop to consider if the contemplated union will result in healthful, mentally well-endowed offspring. At present there are few facts so generally known that they will help such persons in their inquiry. It is the province of the new science of eugenics to study the laws of inheritance of human traits and, as these laws are ascertained, to make them known. There is no doubt that when such laws are clearly formulated many certainly unfit matings will be avoided and other fit matings that have been shunned through false scruples will be happily contracted.

CHAPTER II

THE METHOD OF EUGENICS

1. Unit Characters and their Combination

When we look among our acquaintances we are struck by their diversity in physical, mental, and moral traits. Some of them have black hair, others brown, yellow, flaxen, or red. The eyes may be either blue, green, or brown; the hair straight or curly; noses long, short, narrow, broad, straight, aquiline, or pug. They may be liable to colds or resistant; with weak digestion or strong. The hearing may be quick or dull, sight keen or poor, mathematical ability great or small. The disposition may be cheerful or melancholic; they may be selfish or altruistic, conscientious or liable to shirk. It is just the fact of diversity of characteristics of people that gives the basis for the belief in the practicability of improving the qualities of the "human harvest." For these characteristics are inheritable, they are independent of each other, and they may be combined in any desirable mosaic.

The method of inheritance of these characteristics is not always so simple as might be anticipated. Extensive studies of heredity have, of late years, led to a more precise knowledge of the facts. The element of inheritance is not the individual as a whole nor even, in many cases, the traits as they are commonly recognized but, on the contrary, certain unit characters. What are, indeed, units in inheritance and what are complexes it is not always easy to determine and it can be determined only by the results of breeding. To get at the facts it is necessary to study the progeny of human marriages. Now marriage can be and is looked at from many points of view. In novels, as the climax of human courtship; in law, largely as a union of two lines of property-descent; in society, as fixing a certain status; but in eugenics, which considers its biological aspect, marriage is an experiment in breeding; and the children, in their varied combinations of characters, give the result of the experiment. That marriage should still be only an *experiment* in breeding, while the breeding of many animals and plants has been reduced to a science, is ground for reproach. Surely the human product is superior to that of poultry; and as we may now predict with precision the characters of the offspring of a particular pair of pedigreed poultry so may it sometime be with man. As we now know how to make almost any desired combination of the characters of guinea-pigs, chickens, wheats, and cottons so may we hope to do with man.

At present, matings, even among cultured people, seem to be made at haphazard. Nevertheless there is some evidence of a crude selection in peoples of all stations. Even savages have a strong sense of personal beauty and a selection of marriage mates is influenced by this fact, as Darwin has shown. It is, indeed, for the purpose of adding to their personal attractiveness that savage women or men tattoo the skin, bind up various parts of the body including the feet, and insert ornaments into lips, nose and ears. Among civilized peoples personal beauty still plays a part in selective mating. If, as is sometimes alleged, large hips in the female are an attraction, then such a preference has the eugenic result that it tends to make easy the birth of large, well-developed babies, since there is probably a correlation between the spread of the iliac bones of the pelvis and the size of the space between the pelvic bones through which the child must pass. Even a selection on the ground of social position and wealth has a rough eugenic value since success means the presence of certain effective traits in the stock. The general idea of marrying health, wealth, and wisdom is a rough eugenic ideal. A curious antipathy is that of red haired persons of opposite sex for each other. Among thousands of matings that I have considered I have found only two cases where both husband and wife are red headed, and I am assured by red haired persons that the antipathy exists. If, as is sometimes alleged, red hair is frequently associated with a condition of nervous irritability this is an eugenic antipathy.

In so far as young men and women are left free to select their own marriage mates the widest possible acquaintance with different sorts of people, to increase the amplitude of selection, is evidently desirable. This is the great argument for coeducation of the sexes both at school and college, that they may increase the range of their experience with people and gain more discrimination in selection. The custom that prevails in America and England of free selection of mates makes the more necessary the proper instruction of young people in the principles of eugenical matings.

The theory of independent unit characters has an important bearing upon our classifications of human beings and shows how essentially vague and even false in conception these classifications are. A large part of the time and expense of maintaining the courts is due to this antiquated classification with its tacit assumption that each class stands as a type of men. Note the extended discussions in courts as to whether A belongs to the white race or to the black race, or whether B is feeble-minded or not. Usually they avoid, as if by intention, the fundamental question of definition, and if experts be called in to give a definition the situation is rendered only worse. Thus one expert will define a feeble-minded person as one incapable of protecting his life against the ordinary hazards of civilization, but this is very vague and the test is constantly changing. For a person may be quick-witted enough to avoid being run over by a horse and carriage but not quick enough to escape an automobile. A second expert will define a feeble-minded person as one who cannot meet all (save two) of the Binet test for three years below his own; if he fail in one only he is no longer feeble-minded. But this definition seems to me socially insufficient just because there are moral imbeciles who can answer all but the moral question for their proper age. Every attempt to classify persons into a limited number of mental categories ends unsatisfactorily.

The facts seem to be rather that no person possesses all of the thousands of unit traits that are possible and that are known in the species. Some of these traits we are better off without but the lack of others is a serious handicap. If we place in the feeble-minded class every person who lacks any known mental trait we extend it to include practically all persons. If we place there only those who lack some trait desirable in social life, again our class is too inclusive. Perhaps the best definition would be: "deficient in some socially important trait" and then the class would include (as perhaps it should) also the sexually immoral, the criminalistic, those who cannot control their use of narcotics, those who habitually tell lies by preference, and those who run away from school or home. If from the term "feeble-minded" we exclude the sexually immoral, the criminalistic, and the narcotics such a restriction carried out into practice would greatly reduce the population of institutions for that class. Thus one sees that a full and free recognition of the theory of unit characters in its application to man opens up large social, legal and administrative questions and leads us in the interests of truth, to avoid classifying *persons* and to consider rather their *traits*.

2. The Mechanism of the Inheritance of Characteristics

That traits are inherited has been known since man became a sentient being. That children are dissimilar combinations of characteristics has long been recognized. That characteristics have a development in the child is equally obvious; but the mechanism by which they are transmitted in the germ plasm has become known only in recent years.

We know that the development of the child is started by the union of two small portions of the germ plasm—the egg from the mother's side of the house and the sperm from the father's. We know that the fertilized egg does not contain the organs of the adult and yet it is definitely destined to produce them as though they were there in miniature. The different unit characters, though absent, must be represented in some way; not necessarily each organ by a particle but, in general, the resulting characteristics are determined by chemical substances in the fertilized egg. It is because of certain chemical and physical differences in two fertilized eggs that one develops into an ox and the other into a man. The differences may be called *determiners*.

Determiners are located, then, in the germ cells, and recent studies indicate a considerable probability that they are to be more precisely located in the nucleus and even in the chromatic material of the nucleus. To make this clear a series of diagrams will be necessary.

Figure 1 is a diagram of a cell showing the central nucleus in which runs a deeply staining network—the chromatin. In the division of a cell into two similar daughter cells the most striking fact is the exact division of the chromatin (Fig. 2). We know enough to say that the nucleus is the center of the cell's activity and for reasons that we shall see immediately it is probable that the chromatin is the most active portion of the nucleus.

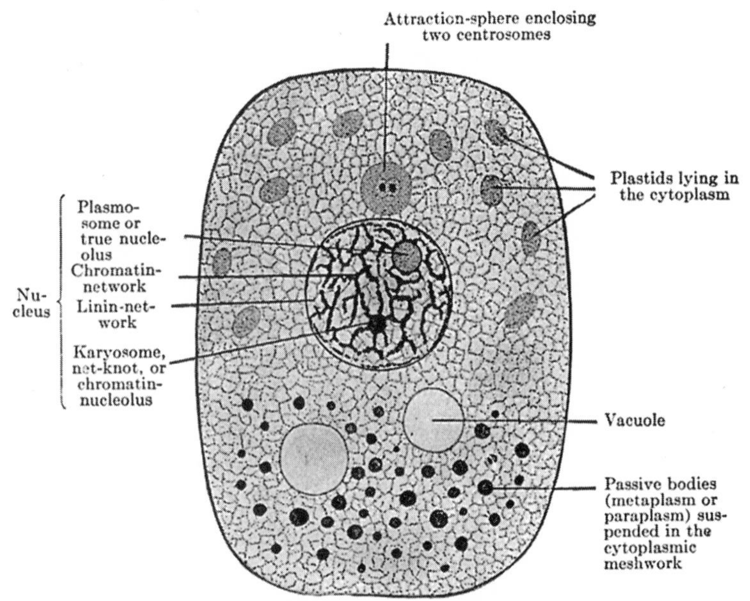

Fig. 1.—Diagram of a cell. Its basis consists of a meshwork containing numerous minute granules (*microsomes*) and traversing a transparent ground substance. From E. B. Wilson: "The Cell in Development and Inheritance."

The fertilization of the egg (Fig. 3) brings together determiners from two germ plasms and we know that, on the whole, the two germ cells play an equal rôle in carrying determiners. Now the germ cells are of very different size in the female (egg) and the male (sperm). Even the nuclei are different; but the amount of chromatic substance is the same. Hence it seems probable that the chromatic substance is the carrier of the equal determiners.

But if determiners from the male are added to those from the female in fertilization it would seem necessary

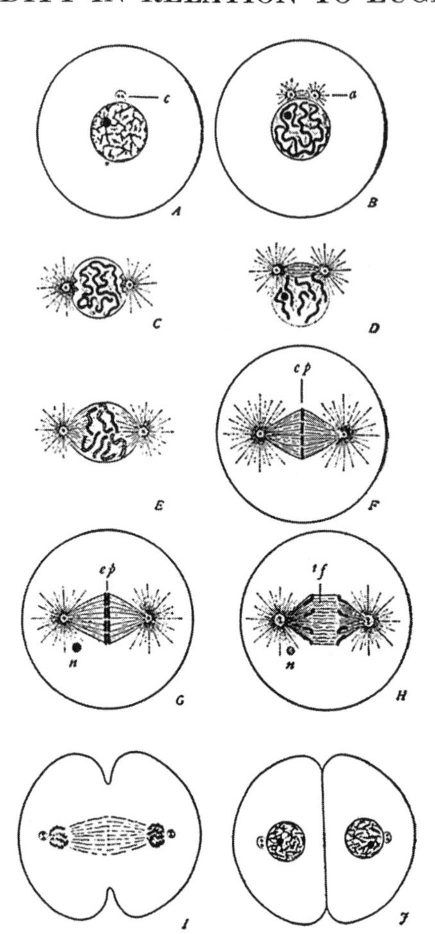

FIG. 2.—Diagrams showing a series of stages in the process of division of the chromosomes during cell division. *A*. Resting cell in which the chromatic material lies (apparently) scattered through the nucleus: at *c* is a pair of recently divided central bodies (*centrosomes*) which come to be the centers of the forces that separate the chromosomes. *B*. The chromatin has fallen into the form of a thick ribbon or sausage-like body, outside of which lies a dark body which is called the "nucleolus." The centrosomes are moving apart. *C*. The centrosomes now lie far apart and the thin membrane around the nucleus is beginning to disappear—a process completed in *D*, where a "spindle" is seen lying between the two centrosomes. The chromosomes are beginning to move under the influence of the new forces centered at the centrosomes. *E*. A later phase in which changes of two sorts are taking place in the chromosomes; first, they are moving to an equatorial position between the two poles, and, secondly, they show their double nature by virtue of which the subsequent

that the number of these determiners should double in every succeeding generation. There must be some special mechanism to prevent this result. An appropriate mechanism is, indeed, ready and had been seen and studied long before its significance was understood; this is the elimina-

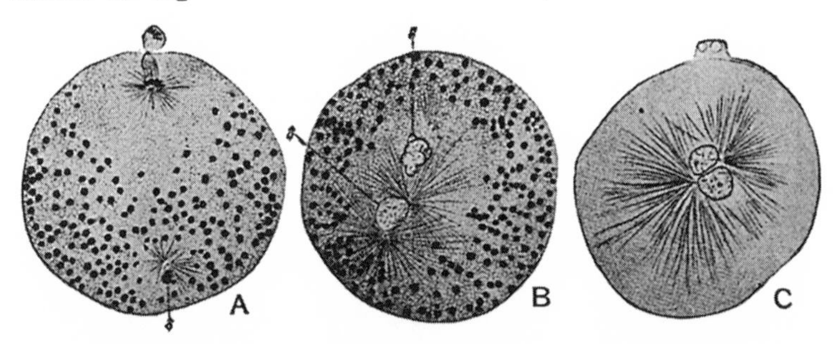

Fig. 3.—Three stages in the fertilization of the egg of a marine ringed worm (*Thalassema*). As seen in thin dyed sections. *A*. At the top of the egg there is occurring a division of the chromosomes that constitutes the ripening or "maturation" of the egg, illustrated in greater detail in Fig. 4. At the bottom a sperm cell (♂) has entered the egg and is penetrating through it toward its center. *B*. The nucleus of the egg is now returning toward the center to meet that of the sperm. *C*. The egg and sperm nuclei are now in contact; henceforth they work in unison; fertilization is completed. After Griffin from E. B. Wilson: "The Cell in Development and Inheritance."

tion from both the immature egg and the immature sperm of half of the chromatic material (Fig. 4). Thus if the immature sex-cell contains four chromatic bodies (chromosomes) each mature sex-cell will contain only two chromosomes. Moreover, each of the chromosomes in the immature sex-cell is double; one half having originated long before in its maternal germ plasm and the other half in its paternal germ plasm. The mechanism for maturation is

process of splitting takes place. *F*. The processes just preceding chromosome division are now completed; the activity of the centers is at its height; the chromosomes now constitute an "equatorial plate," *e. p.* *G*. The chromosomes at the equatorial plate are now beginning to move apart. *H*. The separation of the chromosomes is continuing and in *I* is completed; meanwhile the activity at the centers has declined and division of the body of the cell is beginning. *J*. Division of the cell completed; the nuclei and centrosomes at the condition with which we started at *A*. From E. B. Wilson: "The Cell in Development and Inheritance."

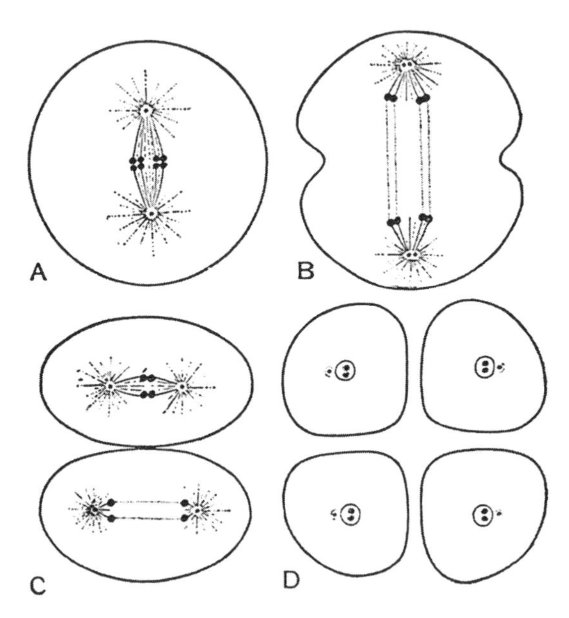

FIG. 4.—Diagrams illustrating the process of *reduction of the chromosomes* by which half of the chromatic material is eliminated from the sex-cell. *A*. The germ cell is beginning its penultimate division—there are four chromosomes but each of them has already begun to divide to go to their respective poles, as seen at *B*. *C*. The last division is taking place, but the four chromosomes do not lie side by side in the equatorial plate as in *A*, but they unite in two pairs and, in the division, the elements of these pairs are sundered again. Thus out of the original cell four ripe sperm-cells (*D*) each with only two chromosomes arise. From E. B. WILSON: "The Cell in Development and Inheritance."

such that either the paternal or maternal component of any chromosome is eliminated in the process, but not both. (Fig. 5). Beyond the condition that one half of each kind of chromosome must go to each daughter cell it seems to be a matter of chance whether the portion that goes to a particular cell be of paternal or of maternal origin. It is even conceivable that one germ cell should have all of its chromosomes of maternal origin while the other cell has all of a paternal origin.

The important point is that the number of chromosomes in the ripe germ cell has become reduced to half and so it is ready to receive an equal half number from the germ cell with which it unites in fertilization.

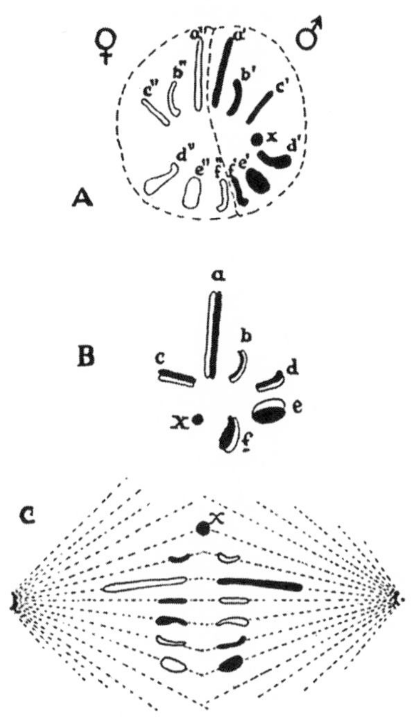

Fig. 5.—Diagram illustrating the mechanism in the chromatic bodies that secures the segregation of determiners. The determiners are assumed to be packed away in the chromosomes. There are equivalent chromosomes (a' and a'', b' and b'', etc.) in the nuclei of the male (♂) and female (♀) germ cells that unite in the fertilized egg (Fig. 3) and these two sets of chromosomes pass into all the embryonic cells—whether of the soma or germ gland—that develop in the young individual. In the division of ordinary body-cells, as illustrated in Fig. 2, each rod a', a'', b', b'', etc., splits lengthwise and half of each goes to each daughter cell. But in a division just before the germ cells become ripe, as in Fig. 4*C*, the like chromosomes unite in pairs as at *B*. Thus a' unites with a'' to form a; b' unites with b'' to form b; etc. Consequently, the number of chromosomes is reduced to half the typical number. When cell-division thereupon occurs (*C*) and the chromosomes split, either the chromosomal element that was derived from the father (black) or that derived from the mother (white) goes, indifferently, to either daughter cell. Consequently, each germ cell contains some chromosomes of maternal and some of paternal origin but not two chromosomes of the same kind. Since, by hypothesis, each chromosome contains particular kinds of determiners it follows that the same germ cell does not contain the (sometimes contrasting) characters of both parents, but some have the paternal character and others the *corresponding* maternal character.

3. The Laws of Heredity

We are now in a position to understand the modern laws of heredity. First of all it will be recognized that nothing is inherited except the determiners in the germ cells; the characters themselves, on the contrary, are not directly inherited. A clear grasp of this fact gives the answer to many questions. Thus the possibility of the transmission of somatic mutilations is seen to depend upon the capacity of such mutilations to modify the determiners in the germ plasm, and such capacity has never been proved. On the other hand, the germ cells receive nutritive and other particles from the blood and they may receive also poisons from it. Hence arises the possibility of depauperization of the germ plasm and of "race poisons;" but these are exceptional and little known phenomena.

To understand the way heredity acts, let us take the case where both germ cells that unite to produce the fertilized egg carry the determiner for a unit character, A. Then in the child that develops out of that fertilized egg there is a *double* stimulus to the development of the unit character A. We say the character is of *duplex* origin. If, on the other hand, only one germ cell, say the egg, has the determiner of a character while the other, the sperm, lacks it, then in the fertilized egg the determiner is *simplex* and the resulting character is of simplex origin. Such a character is often less perfectly developed than the corresponding character of duplex origin (Fig. 6). Finally, if neither germ cell carries the determiner of the character A, it will be absent in the embryo and the developed child. A person who shows a character in his body (soma) may or may not have the determiner for that character in all of the ripe germ cells he carries, but a person who lacks a given unit character ordinarily lacks the corresponding determiner

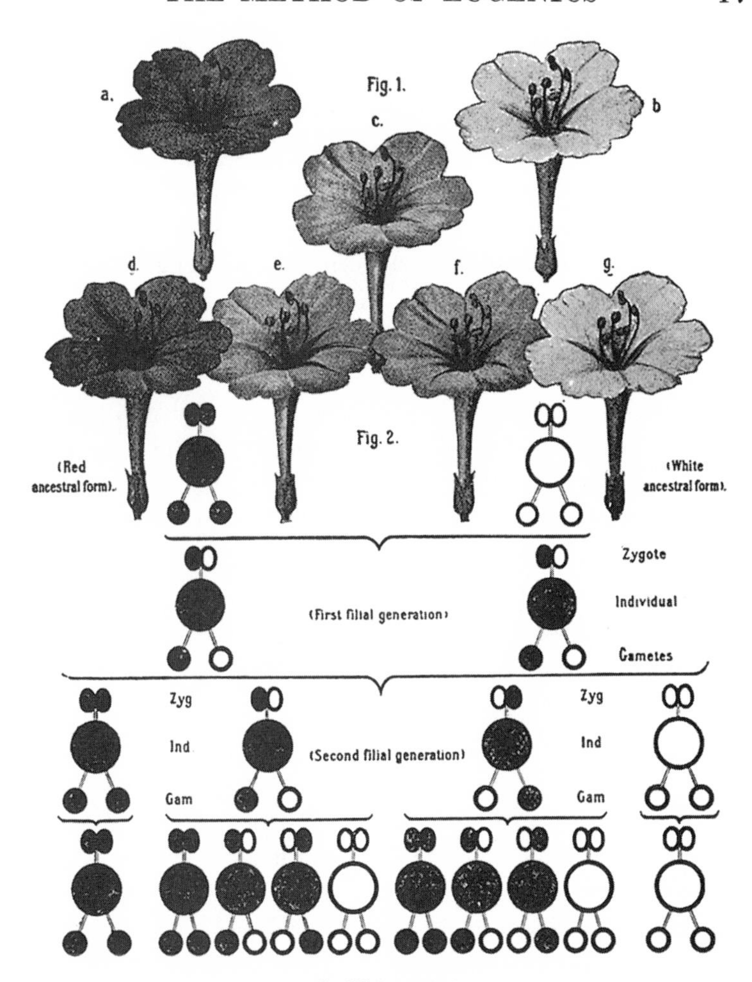

FIG. 6.—Illustration of laws of inheritance drawn from the crossing of red (*a*) and white (*b*) flowered four-o'clocks (*Mirabilis jalappa*). The offspring of this cross, having the determiner for red from one side only, produced pink flowers only (*c*). But when these pink-flowered plants were bred together they produced plants of which one in four had red flowers (duplex, *d*), two in four had pink flowers (simplex, *e. f.*), while one in four had no red pigment (nulliplex, *g*). In the lower part of the chart is a diagram showing for each generation the sort of germ cells involved in the union (zygote), the color of the adult, and the nature of the germ-cells he produces; all carried out to the third generation of descendants. From V. HAECKER: "Wandtafeln zur allgemeinen Biologie" (Nageli: Leipzig).

in all of his germ cells; for, were the determiner present anywhere in his organization (including his germ cells) the corresponding character would ordinarily show in his soma.

In connection with the so-called Mendelian analysis of heredity a nomenclature has grown up which is somewhat different from that here employed. Thus the absent character is often called *recessive*, the present character *dominant* and the condition in the offspring resulting from a crossing of the two is called *heterozygous*, which is the equivalent of simplex. It is to be kept in mind that in this work "absence" does not always imply absolute but only relative absence. Thus the pigmentation of light brown hair is "absent" to "black," and "tow" is absent to light brown; but pigment is present in all these grades of hair. To avoid the confusion between relative and absolute absence the terms recessive and dominant are often used to advantage, wherever a series of grades of a character is under consideration.

These general principles may be rendered clearer by means of a Table of the different sorts of matings of germ cells. And, to focus attention, let us have in mind a concrete example; that of pigment of the iris of the human eye. In the following table P stands for the determiner of brown pigment and p for its absence. Six sorts of unions are possible. See also Plate I, frontispiece.

TABLE I

LAWS OF INHERITANCE OF CHARACTERS BASED ON CONDITIONS OF THE DETERMINERS IN THE PARENTAL GERM PLASMS

	DETERMINERS			
Case	*One parent*	*Other parent*	*Offspring*	*Characteristics of offspring*
1	PP	PP	PP, PP	All with pigmented iris (brown-eyed)
2	PP	Pp	PP, Pp	All pigmented, but half simplex

DETERMINERS—*continued*

Case	*One parent*	*Other parent*	*Offspring*	*Characteristics of offspring*
3	PP	pp	Pp, Pp	All pigmented and all simplex
4	Pp	Pp	PP, Pp, pP, pp	¼ duplex pigmented; ½ simplex; ¼ unpigmented (blue-eyed)
5	Pp	pp	Pp, pp	½ simplex; ½ unpigmented (blue-eyed)
6	pp	pp	pp, pp	All unpigmented (blue-eyed)

In the case of an individual who has received the determiner for one of his unit characters from one side of the house only (say from mother), not only is the character simplex, but when the germ cells mature in that person they are of two types, namely, with the determiner and without the determiner; and these two types are equally numerous (Fig. 5). This is the phenomenon known as segregation of presence and absence in the germ cells. If both parents are simplex in a character, so that they produce an equal number of germ cells with and without the character then in a large number of offspring, 1 in 4 will have the character duplex; 2 in 4 simplex, and 1 in 4 will not have the character at all (nulliplex). This gives in the offspring of such a pair the famous 3 to 1 ratio, sometimes called the Mendelian ratio.

TABLE II

LAW OF CONDITION OF EYE-CHARACTERS IN CHILDREN BASED ON THE CHARACTERS OF THEIR PARENTS

One parent	*Other parent*	*Cases*	*Offspring*
brown	brown	1, 2, 4	Either all of the children have brown eyes, or one fourth have blue eyes
brown	blue	3, 5	Either all children brown-eyed (though simplex) or half blue-eyed
blue	blue	6	All blue-eyed

Now the foregoing rules, which we have illustrated by the case of eye-color, hold generally for any positive determiner or its unit character.

4. Inheritance of Multiple Characters

In the foregoing section we considered the simplest case, namely that in which a single character is taken at a time—*i. e.*, one parent has some character that the other lacks. We have now to consider the cases which are still commoner in nature where the parents differ in respect to two independent characters. Let, for example, the two characters be eye-pigment and hair curliness. Then each one of the six matings given in Table I for eye-color may occur combined with any one of the six matings for hair form; so that there would be a total of 6 times 6 or 36 possible combinations of matings. Similarly Table II would be replaced by one of 9 entries as follows.

Table III

LAW OF COMBINED INHERITANCE OF EYE-COLOR AND HAIR FORM

One parent	*Other parent*	*Offspring*
Brown eye, curly hair	Brown eye, curly hair	Either all brown-eyed and curly-haired; or one-fourth blue-eyed and also one-fourth of all straight-haired (with or without blue eyes)
Brown eye, curly hair	Brown eye, straight hair	All (or all but one-fourth) brown-eyed, and either all or one-half straight-haired
Brown eye, straight hair	Brown eye, straight hair	All (or all but one-fourth) brown-eyed; all straight-haired
Brown eye, curly hair	Blue eye, curly hair	All (or one-half) brown-eyed; all (or three-fourths) curly-haired
Brown eye, curly hair	Blue eye, straight hair	All (or one-half) brown-eyed; all (or one-half) curly-haired
Brown eye, straight hair	Blue eye, straight hair	All (or one-half) brown-eyed; all straight-haired
Blue eye, curly hair	Blue eye, curly hair	All blue-eyed; all (or three-fourths) curly-haired
Blue eye, curly hair	Blue eye, straight hair	All blue-eyed; all (or one-half) curly-haired
Blue eye, straight hair	Blue eye, straight hair	All blue-eyed; all straight-haired

The lessons that this enforces are: first, that characters are often and, indeed, usually, inherited independently and, secondly, that the outcome of a particular mating may be predicted with some precision; indeed, in many matings with certainty.

This study might be extended to cases of three or more independent characters but the tables in such cases become more complex and little would be gained by making them as the principle has been learned by the cases already given. In view of the great diversity of parents in respect to their visible characters the variability of children is readily accounted for.

5. Heredity of Sex and of "Sex-limited" Characters

In most species, as in man, there are two sexes, and they are equally numerous. For a long time this equality has been a mystery; but of late years, through the studies of McClung, Wilson, Stevens and Morgan, the mystery has been cleared up. For there has been discovered in the germ plasm a mechanism adequate for bringing about the observed results. We now know that sex is probably determined strictly by the laws of chance, like the turn of a penny. The cytological theory of the facts is as follows. One sex, usually (and herein taken as) the female, has all cells, even those of the young ovary, with a pair of each kind of chromosome, of which one pair is usually smaller than the others and more centrally placed. The chromosomes of this pair are called the X chromosomes. In the male, on the other hand, the forerunners of the sperm cells have one less chromosome, making the number odd. This odd chromosome [exceptionally paired] is usually of small size and is also known as an X chromosome. In the cell-division that leads to the formation of the mature spermatozoon, this odd chromosome goes *in toto* to one of the two daughter cells (Fig. 5). The X chromosomes are commonly regarded as the "sex-chromosomes." With them are associated various characters that are either secondary sex characters or "sex-limited" characters. Consequently in respect to each and every such character the primordial egg cells are duplex and all the ripe eggs have one sex determiner and its associated characters. The primordial male cells are simplex and consequently, after segregation has occurred, the spermatozoa are of two equally numerous kinds—with and without the sex-determiner. The fertilization of a number of eggs by a number of sperm will result in two equally common conditions—namely a fertilized egg, called *zygote*, that contains *two* sex determiners—such develops into a female; and a zygote that contains only one sex determiner—such develops into a male. The nature of the germ cells in the germ gland of the future child and of the associated secondary sex-characters thus depend on which of the two sorts of sperm cells go into the make-up of the zygote.

Whenever the male parent is characterized by the absence of some character of which the determiner is typically lodged in the sex chromosome a remarkable sort of inheritance is to be expected. This is called sex-limited inheritance. The striking feature of this sort of heredity is that the trait appears only in males of the family, is not transmitted by them, but is transmitted through normal females of the family. Striking examples of this sort of heredity are considered later in the cases of multiple sclerosis (Fig. 64); atrophy of optic nerve (Fig. 77); color blindness (Fig. 88); myopia (Figs. 90, 91); ichthyosis (Figs. 106, 108); muscular atrophy (Fig. 125); and haemophilia (Fig. 134).

The explanation is the same in all cases. The abnormal condition is due to the absence of a determiner from the male X chromosome. Its inheritance can be followed from Figure 7, adapted from Wilson, 1911.

If the trait be a positive sex-limited one, originating either on the father's or the mother's side, its inheritance

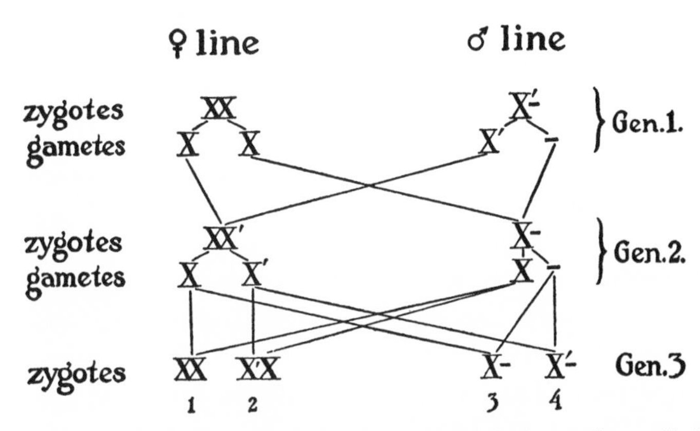

Fig. 7.—Diagram illustrating the method of inheritance in sex limited heredity. *X*, the sex chromosome, double in the female individual, single in the male. When ripe germ cells are formed in the female, each contains the sex determiner, but in the male half of the germ cells have and half lack the determiner (represented by the dash—). Let *X′* represent the sex chromosome of the original male that showed the defect (absence of some unit character). Let such a male be mated with a female of an unaffected strain. Then all children will have the determiner for the positive condition (Gen. 2, zygotes, *i. e.*, fertilized eggs and the individuals that develop from them). In the third generation four kinds of zygotes will appear: 1, the normal female who is not capable of transmitting the defect; 2, the normal female who is capable of transmitting the defect; 3, the normal male who is incapable of transmitting the defect; 4, the defective male. Based on E. B. Wilson, 1911.

will be more irregular; but it can be worked out by the aid of Figure 7.

6. The Application of the Laws of Heredity to Eugenics

If one is provided with a knowledge of the methods of inheritance of unit characters it might seem to be an easy matter to state how each human trait is inherited and to show how any undesirable condition might be eliminated from the offspring and any wished for charácter introduced.

Unfortunately, such a consummation cannot for some time be achieved. The reason for the delay is twofold. First, we do not yet know all of the unit characters in man; second, we can hardly know in advance which of them are due to positive determiners and which to the absence of such.

Unit characters can rarely be recognized by inspection. For example the white coat color of a horse is apparently a simple character, but experimental breeding shows that it is really due to several independently inheritable factors. The popular classification of traits is often crude, lagging far behind scientific knowledge. Thus insanity is frequently referred to a single trait. It is clear, however, that insanity is a *result* merely and not a specific trait. Some cases of insanity indicate an innate weakness of the nervous system such as leads it to break down under the incidence of heavy stress; other cases of insanity are due to a destruction of a part of the brain by a wound as, for instance, of a bullet. In some cases, through infection a wide-spread deterioration of the brain occurs; in other cases a clot in a cerebral blood vessel may occlude it, cut off nutrition from a single locality of the brain and interfere with movements that have their centres at the affected point. Now these four results cannot be said to be due to the same unit defect; and they can hardly be compared in the study of heredity.

On the other hand, the original expectation that progress must wait on a complete analysis of unit characters proves not to be correct. There are a number of forms of insanity that are sharply separable symptomatically and structurally which have a common basis in that they are due to a nervous weakness; and "nervous weakness" may behave in heredity with relation to "nervous strength" like a lower grade, or the absence, of a highly developed character. Even without a complete analysis of a trait into its units we may still make practically important studies by using the principle that when both parents have low grades of a trait-complex the children will have low grades of that complex.

The matter of dependence of a character on a determiner or its absence is of great importance and is not easy to anticipate. For instance, long hair as in angora cats, sheep or guinea pigs is apparently not due to a factor added to short hair but rather to the absence of the determiner that stops growth in short-haired animals. One can only conclude whether a character is due to a determiner or to its absence by noting the effect of breeding likes in respect to the given trait. If all offspring are like the parents in respect to a trait, the trait (if simple) is probably a negative one. But if the offspring are very diverse, the trait (if simple) is probably due to a positive determiner and the germ cells of the parents are of two kinds; some with and some without the determiner.

The determination of unit characters is complicated by the fact that a character due to a simplex determiner often differs from one due to a duplex determiner. In the former case the character is slow in developing and frequently fails of reaching a stage of development found in the latter case. The offspring of red and black-eyed birds may have at first a light iris which gradually darkens. This fact is spoken of as the imperfection of dominance in the simplex condition.

Despite the difficulties in analysis of units of heredity and despite the complications in characters it is possible to see clearly the method of inheritance of a great number of human traits and to predict that many more will become analyzed in the near future.

CHAPTER III

THE INHERITANCE OF FAMILY TRAITS

Before any advice can be given to young persons about the marriage that would secure to them the healthiest, strongest children it will be necessary to know not only the peculiarities of their germ plasms but also the way in which various characters are inherited. The work of the student of eugenics is, consequently, to discover the methods of inheritance of each characteristic or trait. After we get precise knowledge of the methods of inheritance of the commoner important traits we shall be in a position to advise, at least in respect to these traits. It would seem a self evident proposition, but it is one too little regarded, that knowledge should precede teaching. In this chapter an attempt will be made to consider many of the traits that are known to run in families and to set forth, so far as known, the laws of their inheritance. We shall begin with some of the general characteristics of man that have been best studied and then pass to a consideration of some human diseases.

In the study of many of these traits I have made use of data that have been furnished by numerous collaborators, chiefly on questionaires known as "Family Records." These are frequently referred to in the following pages, but always anonymously. The Family Records or "Records of Family Traits," as they are also called, are largely derived from professional circles, but not a few from farmers and business men. In respect of several of the special abilities the collaborators have volunteered a numerical grading as follows: 1, poor; 2, medium; 3, exceptionally good. These grades are frequently referred to below.

1. Eye Color

This depends upon the condition of pigmentation of the iris—the colored ring around the pupil. According to Mr. Charles Roberts (1878, p. 134)[1] the iris has on its inner surface "a layer of dark purple called the *uvea* . . . and in brown eyes there is an additional layer of yellow (and, perhaps, brown-red) pigment on its outer surface also, and in some instances there is a deposit of pigment amongst the fibrous structures. In the albino, where the pigment is entirely absent from both surfaces of the iris, the bright red blood is seen through the semi-transparent fibrous tissue of a pink color; and in blue eyes, where the outer layer of pigment is wanting, the various shades are due to the dark inner layer of pigment—the uvea—showing through fibrous structures of different densities or degrees of opacity.

"The eyes of new born infants are dark blue, in consequence of the greater delicacy and transparency of the fibrous portion of the iris; and as these tissues become thickened by use and by advancing age the lighter shades of blue and, finally, gray are produced, the gray, indeed, being chiefly due to the color of the fibrous tissues themselves." Yellow pigment is laid down upon the blue, forming yellow-blue or green eyes. "In the hazel and brown eyes the *uvea* and the fibrous tissues are hidden by increasing deposits of yellow and brown pigment on the anterior surface of the iris, and when this is very dense, black eyes are the result."

While in most races of the globe brown pigment is heavily

[1] For titles of works referred to in text, see Bibliography, at end of book.

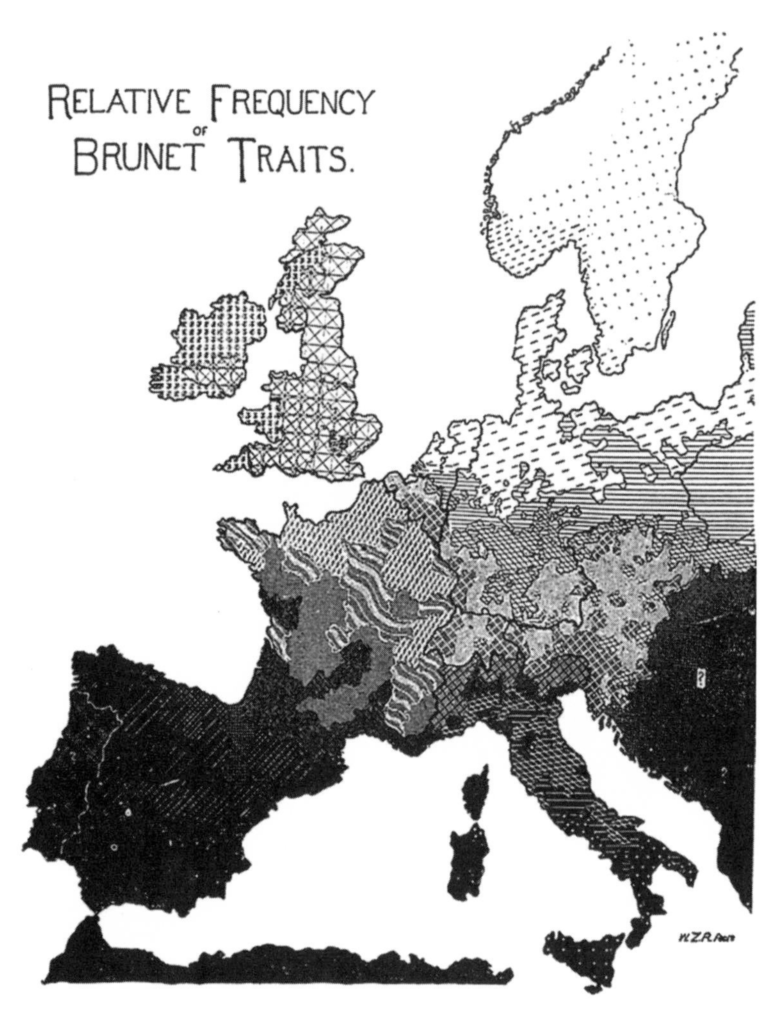

FIG. 8.—Map of southwestern Europe showing the relative frequency of "brunet traits," *e. g.*, brown eye color. On the whole, the darker the shade the greater the proportion of brunet persons in the given area. The lightest areas represent about 20 to 25 per cent brunetness; the darkest European areas over 90 per cent brunetness. At the northern limit of the map "about one third of the people are pure blonds, characterized by light hair and blue eyes;" on the other hand, in the south of Italy the pure blonds have almost entirely disappeared. From W. Z. RIPLEY: "The Races of Europe."

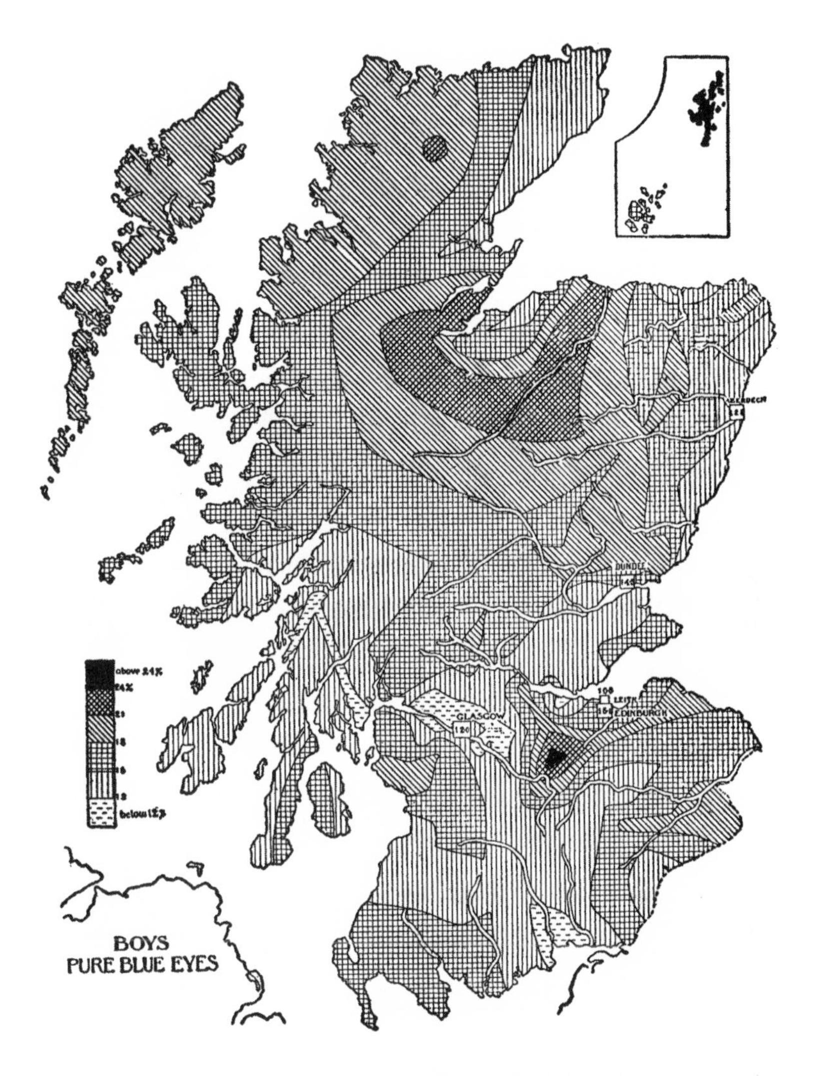

Fig. 9.—Distribution of pure blue eyes among Scottish boys. About 15 per cent of all boys have blue eyes. The relative density is indicated by depth of shading as indicated in the key at the left. A very high density (21 to 24 per cent) occurs in the lower Spey Valley in the northwest. This is the region of the Norse invasion which brought in much protoplasm that was defective in pigmentation. The highest density (over 24 per cent) exists in the coal and iron districts of East Lanarkshire and "this is probably due to the Irish immigrants." J. Gray, 1907.

secreted in the iris, in northwestern Europe blue, gray or yellow-blue eyes are found. It seems probable that, once upon a time, or perhaps at many times, an individual was born without brown pigment in the iris. The offspring of such prospered and spread throughout northwestern Europe and migrated thence to America and Australia (Fig. 8). This defect, lack of eye pigment, has had a wonderful history. By noting its distribution the migrations of peoples can be traced. Thus Gray (1907) has shown that, in Scotland, pure blue eyes are most abundant in the coal and iron districts. "This is probably due to the Irish immigrants, it being well known that blue eyes are very common among the Irish." In the Spey valley of Scotland the density of pure blue eyes is high—probably owing to the Norse invasion at that point. (Fig. 9). So in our country the pigmentation survey that will some day be made will show a high percentage of blue eyes where the Scandinavians and north Germans have settled. Thus eye color, just because it shows no tendency to blend in heredity, is a most valuable aid in history.

Our knowledge of heredity of eye color depends on studies made by Galton, 1899, who noted its alternative nature but otherwise overlooked the true method of its inheritance; more recently, by three studies carried on simultaneously and independently and published by G. C. and C. B. Davenport, in November, 1907; by C. C. Hurst in 1908; and by Holmes and Loomis in December, 1909. Since 1907 the present author has collected additional data. Hurst's data have the advantage of having been collected from personal observation, hence the chance of error due to a diversity of collaborators was eliminated. In the other studies the data were supplied by unprejudiced, if not always critical, recorders.

Applying the test of the 6 (strictly 5) kinds of unions we get the results shown in Table IV.

TABLE IV

		HURST		DAVEN-PORT		HOLMES & LOOMIS		TOTAL		P'ORTION	
One Parent	*Other Parent*	*Blue*	*Pig't*	*Blue*	*Pig't*	*Blue*	*Pig't*	*Blue*	*Pig't*	*Blue*	*Pig't*
pure blue	pure blue	101	0	77	0[1]	51	1	229	1	99.5	0.5
pigmented (Pp)	blue	137	121	428	506	89	85	654	712	48.0	52.0
pigmented (PP)	blue	0	66	0	70			0	136	0	100
pigmented (Pp)	pigmented (Pp)	18	45	98[2]	169	5	34	121	248	33	67
pigmented (PP)	pigmented (Pp)	0	195	0	99			0	294	0	100

Table IV supports the following conclusions:

1. When both parents have pure blue eyes all of the children will have pure blue eyes (the discordant case is probably due to an error).

2. When one parent has pigmented iris while the other has blue, either the fraternity of children will show no blue eyes or else half of them will be blue-eyed. The sum of the latter class, the second case, gives 654:712 or 48 per cent to 52 per cent.

3. When both parents have brown iris either all the children will have brown iris (last case in Table IV) or else about a quarter will lack brown pigment and so will be blue-eyed.

The eugenic value of the inheritance of eye color lies in the consideration, advanced by Major Woodruff, that pigmentation of the eye, skin, etc., better fits a child for life in the tropics or in a country, like the United States, of bright sunlight. Brown-eyed children can be secured from blue-eyed stock by mating with pure brown-eyed stock. We have heard of two blue-eyed parents regretting that they had no brown-eyed children. They wished for the impossible.

[1] Eight hundred and sixty-six additional cases collected subsequently are not included because unchecked.

[2] A number of these blues are doubtless destined to become pigmented in later life.

2. Hair Color

This character is due to the presence of brown granules in the hair and sometimes also to the presence of a diffuse reddish pigment. The study of heredity of hair color is complicated—more than that of eye color—by the fact that the hair grows darker with age, at least until maturity is achieved. If you compare the light browns and the blacks in children under 16 and over 16 you will find twice as many light browns in the younger lot as in the older; but only half as many blacks. In other words, half of the persons who will eventually have black hair still have light to medium brown at 16 years of age.[1] While this tends to obscure the result yet the general fact of segregation in hair color cannot be gainsaid. Let us examine the results of various matings. (Table V).

Table V

The hair-color of the offspring of parents with different classes of hair pigment.

One parent	*Other parent*	*Offspring*
Little brown pigment	Little brown pigment	All with tow, yellow, golden or red hair.
Brown pigment	Little or no brown pigment	Half with light hair, half with brown; in other families all children may eventually gain brown hair
Brown pigment	Brown pigment	Most children have brown hair; some (about one-quarter) have light hair. In some families all children eventually gain brown hair.

The most striking result is that dark-haired children probably never come from flaxen-haired parents. Indeed, a good practical rule is that the children will not acquire hair darker than that of the darker parent.

The inheritance of red-hair color has a certain eugenic importance. There can be little doubt that a young person

[1] Holmes and Loomis, 1909, p. 55.

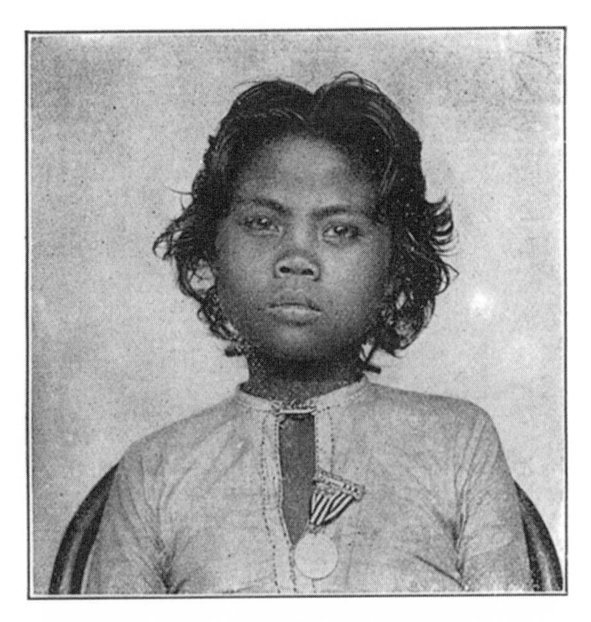

Fig. 10.—Wavy hair; a Segumbar, female, Philippine Islands. (Lent by the American Museum of Natural History.)

who has red hair has a strong antipathy to a red-haired person of the opposite sex. This testimony comes to me from the father of a red-haired daughter. It is confirmed by the fact that, despite prolonged inquiry among thousands of families I have succeeded in obtaining only two cases where both parents had red hair. Though the red was not a clear red in all parents all of the 8 children had red hair. If one parent only forms "red-hair" germ cells exclusively while the other forms exclusively germ cells containing the determiner for black pigment the offspring will show no red; still less will red-haired offspring appear if neither parent forms "red-hair" germ cells. Red-haired offspring may come from two brown or better from glossy black-haired parents provided both form red-hair germ cells. In that case both dark-haired parents will probably

Fig. 11.—Frizzy or kinky hair; a Soudanese male. (Lent by the American Museum of Natural History from a photograph in the Philadelphia Museum.)

have ancestors or other close relatives with red hair. Glossy black hair in the parents is especially apt to produce red hair in the children because the glossiness is usually due to red hidden by black pigment.

3. Hair Form

The form of the hair varies from straight through wavy and curly (Fig. 10) to kinky (Fig. 11) and woolly (Fig. 12), depending largely upon the closeness of the spiral. These different types of hair have a different form on cross-section; *i. e.*, the cut end of a straight hair is nearly circular while

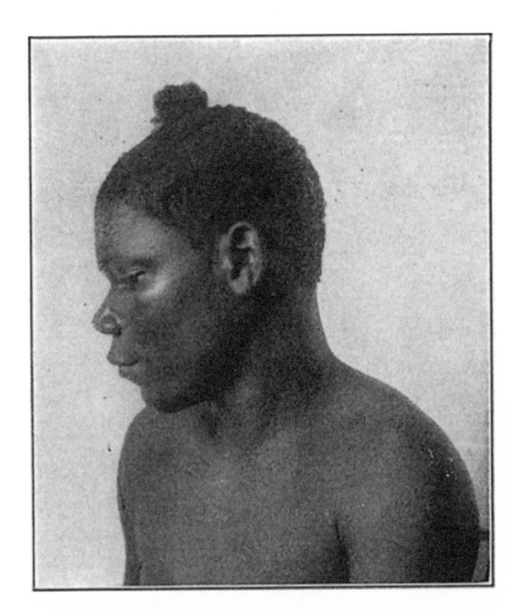

FIG. 12.—Woolly hair; a Congo negro. (Lent by American Museum Natural History.)

that of woolly hair is much flattened, being only half as thick as it is broad. Both the flattening and the curving of hair are due to a modification of the cup or "hair follicle" in which the hair develops. Thus, while straight hair develops in a plain, cylindrical follicle that of the flattened types is curved and inclined in relation to the surface of the skin. Straight hair is the simple condition; curving is due to a special modification. What, now, is the method of inheritance of this special modification?

First, if both parents have hair that from childhood up has been straight, without natural tendency toward curving, then all of the children will have straight hair. There are exceptional cases reported of wavy haired children from straight haired parents, but the exceptions constitute less than 2 per cent.

If one parent has wavy hair while the other has straight hair then, since in wavy haired persons half the germ cells are without the determiner for curved hair, half of the offspring will have straight and half curved hair. If both parents have wavy (simplex) hair about 75 per cent of the children will have curved hair and the others straight hair. But two curly haired parents, both of curly haired stock on both sides, will probably have all curly or wavy haired children. In a word, when either of the germ cells that unite to form the fertilized egg contains the curly determiner the offspring will have curved hair.

4. Skin Color

The pigment of the skin is due to brown granules lying in the deep stratum of the skin. Such granules occur in most people, are common in brunets and still more abundant in negroes. Besides the brown granules a yellow-red pigment is present, but this has been little studied.

Now when both parents are clearly blonds most, if not all, of their offspring are blonds. In 513 offspring reported as derived from this sort of mating 91.4 per cent are recorded as blonds and 6.8 per cent as intermediate, while only 1.8 per cent are stated to be brunet—quite within the limit of error due to inaccuracy of the collaborators. If one person is blond and the other darker, about half of the children will, on the average, be blond and half pigmented but rarely darker than the darker parent. If both parents be dark the percentage of brunets ranges from about 25 to zero. In general, whatever the mating, the children will not be darker than their darker parent.

When one parent is white and the other as dark as a full-blooded negro the offspring are, as is well known, of an intermediate shade (mulatto, mezzotint). If two such mulattoes marry their offspring vary in color. In one fraternity derived from two such mulattoes having 45 per cent and 13 per cent respectively of black in the skin, the proportion of black in the 7 offspring whose color was measured ranged from 46 to 6 (Fig. 13). The lighter limit was as light as most Caucasian skins. In another fraternity whose parents had 29 per cent and 13 per cent of black respectively, the children ranged from 28 per cent to 8.5 per cent of black in the skin color.[1] Here, again, the light-

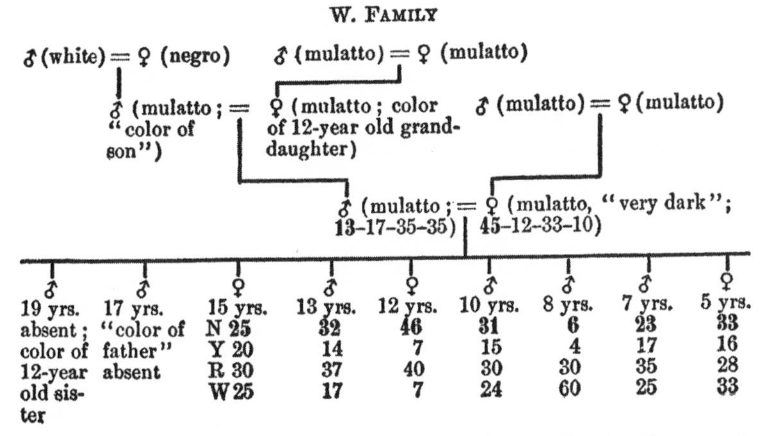

FIG. 13.—Pedigree chart of "W" family of mulattoes, showing the percentages of the four colors; black (*N*), yellow (*Y*), red (*R*) and white (*W*) that combined (as in the color wheel) will give the skin color. ♂, male; ♀, female. For fuller details see DAVENPORT, G. C., and C. B., 1910.

est child has practically a white skin. In the case of the two other families, in which the parents were dark mulattoes (30 to 40 per cent black) none of the children were lighter than 27 per cent black. The germ cells of the parents probably lack the lower grades of pigmentation. On the other hand two very light "colored" parents will have (probably) only light children, some of whom "pass for whites" away from home. So far as skin color goes they are as truly white as their greatgrandparent and it is quite

[1] All colors were determined by means of the Bradley color top.

conceivable that they might have mental and moral qualities as good and typically Caucasian as he had. Just as perfect white skin color can be extracted from the hybrid, so may other Caucasian physical and mental qualities be extracted and a typical Caucasian arise out of the mixture. However, this result will occur only in the third, or later, hybrid generation and the event will not be very common.

Albinism. This is an extreme case of blondness—all pigment being lost from skin, hair and eyes. The method of inheritance resembles that of eye color. When both parents lack pigment all offspring are likewise devoid of pigment. When one parent only is an albino and the other is unrelated the children are all pigmented. Whenever albinos occur from two normals the proportion of these albinos approaches the ideal and expected condition of 25 per cent (Fig. 14).

Albinism is not a desirable peculiarity, despite the beauty of complexion and hair, because the lack of pigment in the retina makes it hard to bear strong light. Albinos may avoid transmitting albinism by marrying *unrelated,* pigmented persons. Pigmented persons belonging to albinic strains must avoid marrying cousins, even pigmented ones, because both parents might, in that case, have albinic germ cells and produce one child in four albinic. Albino communities, of which there are several in the United States are inbred communities; but not all inbred communities contain albinos.[1]

5. Stature

The inheritance of stature has long been a subject of study. It has great interest both because it is easily determined and because it has a great racial range, namely,

[1] This matter is discussed more fully in the "American Naturalist," December, 1910.

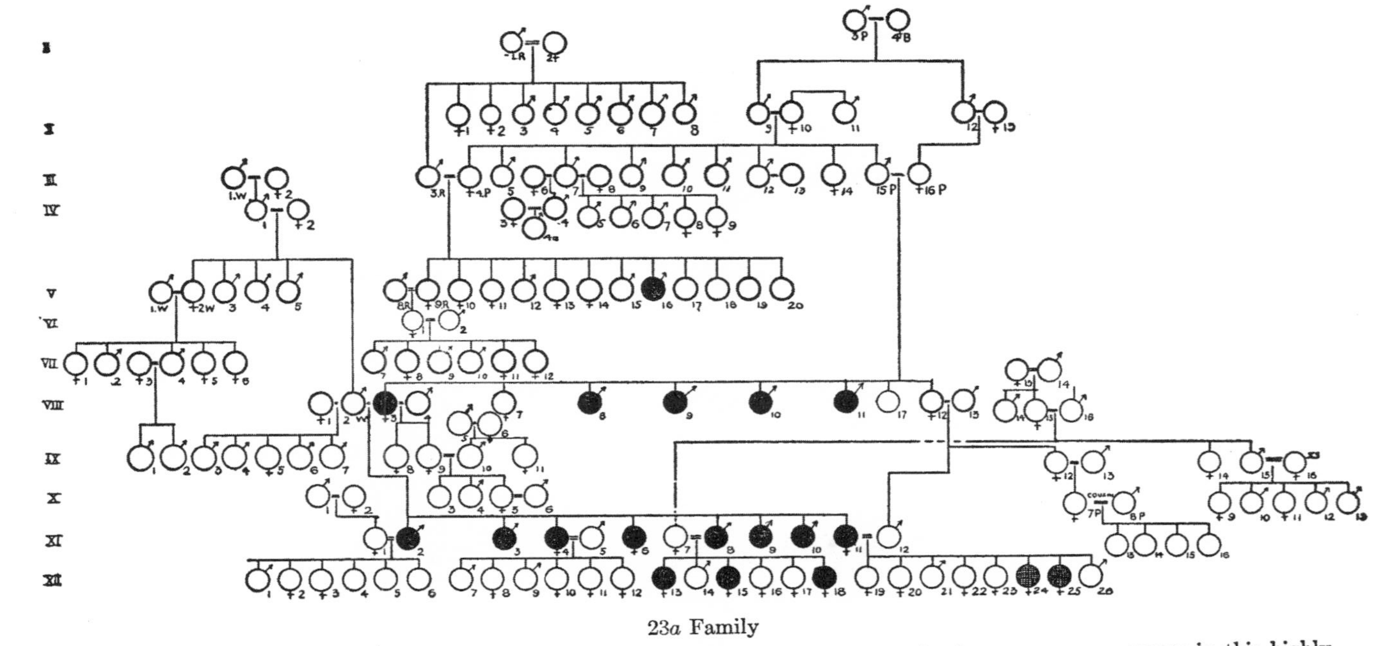

23*a* Family

Fig. 14.—Pedigree chart of an albino family. The letters *B*, *P*, *R*, *W*, represent the four common surnames in this highly inbred community. All black circles represent abinos, ♀ female, ♂, male, O, unknown sex.

from 138 centimeters (or 54 inches) in the negrilloes of Africa to 180 centimeters (or 71 inches) in the Scotch. Among European males, stature ranges from 150 centimeters (60 inches) to 190 centimeters (75 inches), while that of women rarely exceeds 180 centimeters (71 inches).[1]

The importance of stature as a definite character is seen in its distribution in Europe. Apart from the variations ascribed to environment there are clear racial (*i. e.*, inherit-

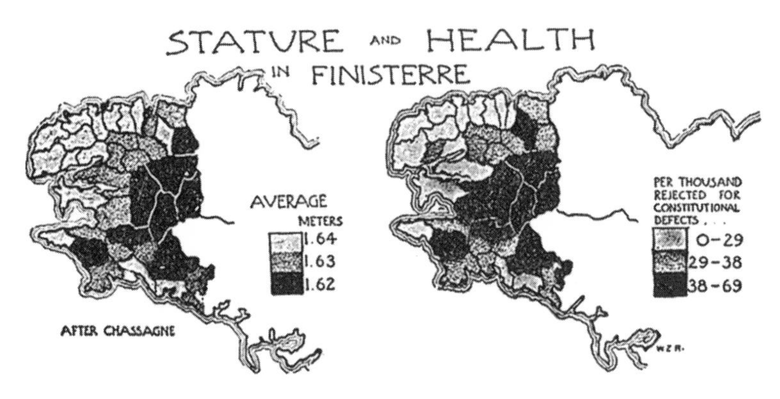

Fig. 15.—Two maps of Brittany, France. On the left is shown the distribution of the various mean statures ranging from 1.62 meters to 1.64 meters. On the right is shown the distribution of rejection of recruits for constitutional defects. Ripley: "The Races of Europe."

able) differences. The rugged hills of Scotland harbor a race that are, relatively, giants; the mild and productive shores of the Gulf of Tarent, Southern Italy, hardly more populous, are inhabited by a people that are, relatively, dwarfs. Conditions of life cannot account for the difference; there is a difference of blood. It is easy to go astray in assigning environmental causes for stature. Thus Ripley (1900, p. 85) referring to a map of Brittany says: "In the interior cantons, shorter on the average by an inch than the population along the sea coast, there is a corresponding

[1] Deniker, "Races of Man," p. 584.

increase of defective or degenerate constitutional types. The character of the environment is largely responsible for this." (Fig. 15). Two maps are given of this territory showing the practical coincidence of the areas of shortest stature and greatest number of rejections of recruits for physical defects. Fifteen pages later, however, practically the same map is used (Fig. 16), the greater height of the

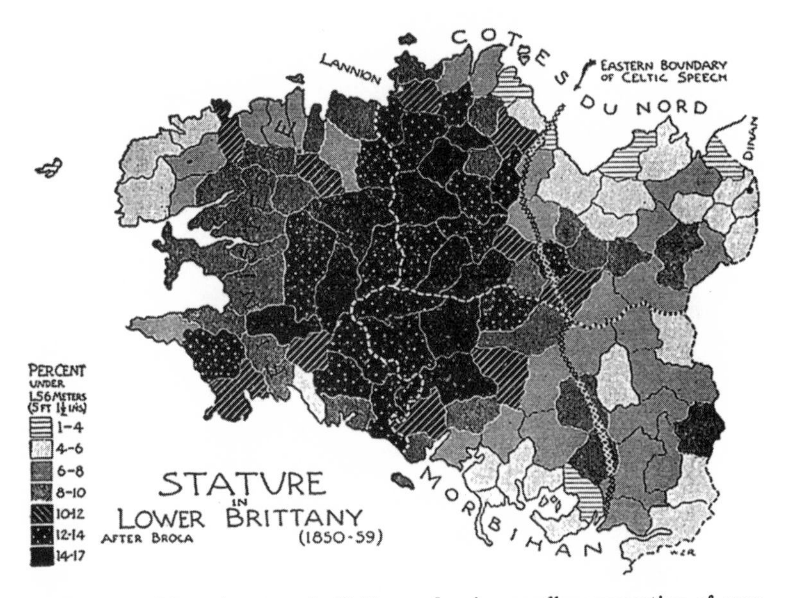

FIG. 16.—Map of stature in Brittany showing smaller proportion of men whose stature is under 1.56 meters in the region subject to Teutonic invasions. RIPLEY: "The Races of Europe."

coastal people referred to, and explained by Teutonic invasions. "The result has been to infuse a new racial element into all the border populations in Brittany, while the original physical traits remain in undisturbed possession of the interior." It appears, then, probable that the greater rejection of recruits in the central country is due less to its unfavorable environment than to its inadequate blood.

Recognizing the inheritable nature of stature it remains

to inquire how it is inherited. First of all it must be conceded that stature is hardly a single unit. It is composed of three elements that would seem to be unrelated, namely, the height of the cranium, the length of the neck and trunk, and the length of the legs. Sitting height is a more significant measure from the standpoint of heredity; but, unfortunately, few persons know their sitting height. A second complication is dependence of stature on age. It increases up to 20 years in the male and about 19 years in the female. Beyond these ages the increase may be neglected. A third complication is that stature is, to a certain degree, dependent on sex. To transmute female measurements to corresponding male measurements Galton (1889) used the method of multiplying them by 1.08 since the mean of male stature is that much greater than the mean of female stature. We can avoid this complication by using, in place of the absolute or transmuted measures, the deviation in each sex from its own mean. The mean stature for the adult males of the white population of the United States may be taken at 69 inches (175 cm); that of females at 64 inches (163 cm). Despite all these complications, which tend to obscure the result, we can still seek an answer to the question: What general laws are there of inheritance of stature?

The first general law is that, in case the four grandparents are very unlike, the adult children will vary greatly in stature, whereas when the grandparental statures are closely alike those of the children will be also. This is shown in the following Table:

	Inches					
Difference between the shortest and the tallest child:	3	4	5	6	7	8
Difference between the shortest and the tallest grandparent:	4.6	5.0	6.0	6.5	[illegible]	7.2

This law seems to indicate that the reason why in some

families the children vary greatly in stature while in others they vary little is because more diverse elements have entered into the make-up of the children in the first case than in the second. In the first case long and short blood are commingled in the ancestry while in the second case exclusively long or exclusively short ancestry as the case may be.

The second general law is that when both parents are tall all of the children tend to be tall; but, on the contrary, if both parents are short some of the children will be short and some tall in ratios varying from 1:1 up to 2:1. If all of the grandparents are short then there tend to be twice as many short children as tall; but if one grandparent on each side be tall there will tend to be an equality of short and tall offspring.

The evidence for the foregoing is found in the study of 104 families which furnished quantitative data as to stature for children, parents and grandparents.

To illustrate the inheritance of extreme short stature in a family I may quote from C. F. Swift (1888). He says (p. 473) "I am unable to give a particular account of the Little Hatches of Falmouth. [Mass.] They were children of Barnabas, who married in 1776 his relative Abigail Hatch and had two sons and seven daughters. Six daughters were less than 4 feet in height. None married. The seventh daughter Rebecca was of common size and married Robert Hammond. The two sons, Barnabas, born in 1788, and Robinson, b. 1790, were both of low stature, one, scarce 4 feet in height, was a portly gentleman almost as broad as long." It may be predicted that the tall daughter who married had only tall children.

6. Total Body Weight

Adult weight (assuming density to be constant) depends upon stature and circumference. It is, therefore, still more complicated than stature and still further removed from any semblance of a unit character. Moreover, it is much more dependent upon conditions of life, for, as is well known, a sedentary life with overfeeding and drinking tends, *in persons so disposed*, to increase weight, even as strenuous activity and dieting favor the reduction, within certain limits, of weight.

Despite this dependence of weight on environment we may attempt to learn if it shows any trace of heredity. First, it is necessary to avoid the use of absolute weights on account of sex differences. So we find the mean weight of American fathers and mothers and calculate our weights as deviations from these means. The mean weight of fathers in our data is 162 pounds; of mothers 131 pounds. The range in weight of fathers is from 110 to 250 pounds. The range in weight of mothers is from 90 pounds to 360 pounds.[1] In our study we are, however, concerned less with absolute deviations in weight from the average than in the deviations in corpulency and so we make our starting point the weight for a given stature and calculate in each case the deviation from the weight that is normal for the given stature. The table of normal weight that we employ is Table VI.

TABLE VI

NORMAL WEIGHT, IN POUNDS, FOR EACH INCH OF STATURE AND EACH SEX

Inches of stature		59	60	61	62	63	64	65	66	67
Normal weight in pounds for	male		131	132	134	137	140	143	147	152
	female	107	112	117	122	126	131	136	139	141

Inches of stature		68	69	70	71	72	73	74	75
Normal weight in pounds for	male	157	162	167	172	177	182	190	198
	female	144	150	155	160	165	170		

The first result is that when both parents are slender in build or of relatively light weight the children will tend all to be slender.

[1] This maximum occurred in a single case of our records; the next lower weight is 225 pounds.

The evidence for this has never been fully set forth. It rests on five fraternities in which the ten parents diverged (in pounds) from the normal as follows: 1, 1, -2, -7, -7, -9, -11, -12, -33, -47. Every grandparent was below normal in weight except one who was just normal. Of 23 children only 3 are above normal. Their total excess weight amounts to 25 pounds, while the total deficiency of the 20 remaining children is 374 pounds—an average deficiency for the 23 children of 15 pounds. Truly, a slender population.

If both parents are heavy and of heavy ancestry their children tend, on the whole, to be heavy (Fig. 17).

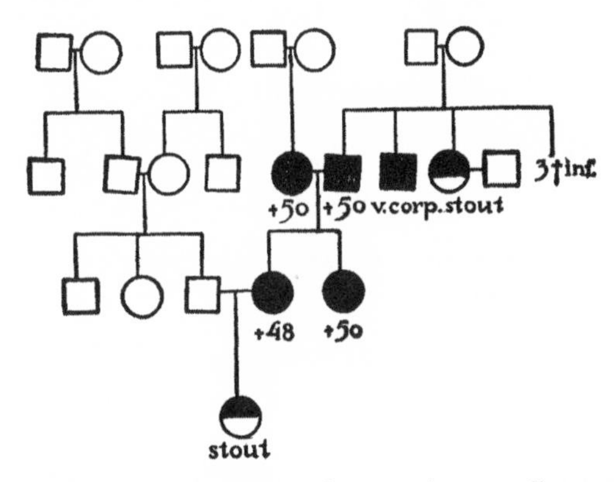

Fig. 17.—Pedigree of family with corpulency. Great-grandparents, grandparents and one of the parents are much above normal weight for their stature and the same tendency is found throughout the fraternities to which they belong. The father is slender. His daughter is, at an early age, inclined to stoutness. F. R.; Hal. 3.

I have data on four families that meet these conditions and give in Table VII all the data concerning their deviations in weight from the normal.

Table VII

THE DEVIATIONS FROM NORMAL STOUTNESS (WEIGHT ÷ STATURE) OF THE ANCESTORS AND CHILDREN WHEN BOTH PARENTS ARE HEAVY

Reference letters	FF	FM	MF	MM	F	M	C1	C2	C3	C4	C5	C6
Ave.—2	23	25	24	28	27	35	—10	—6	23			
Gan.—1	1	23	3	9	18	21	— 6	8	8	9	13	18
Eld.—1	8	11	21	33	33	5	—12	32	38	53		
Elt.—1	3	11	3	44	8	18	—22	—2				

C, child; F, father or father's; M, mother or mother's.

It is to be kept in mind that the children are mostly young, 18 to 25, and consequently do not show their potentialities in weight. Nevertheless, while there are 6 children below the normal in weight, giving a deficiency of 58 pounds, there are 9 above the normal with an excess of 202 pounds.

When both parents are heavy (disregarding grandparents) the numbers of light and heavy children are practically equal (39 light to 34 heavy or 465 pounds total deficiency to 490 pounds total excess).

When one of the parents is heavy and the other slender both heavy and slight offspring occur and, in youth at least, the slight are more numerous than the heavy. Table VIII gives the data on this mating.

Table VIII

THE DEVIATIONS FROM NORMAL STOUTNESS (WEIGHT ÷ STATURE) OF THE ANCESTORS AND CHILDREN IN SIX FAMILIES WITH ONE SLENDER AND ONE HEAVY PARENT

Reference letters	FF	FM	MF	MM	F	M	C^1	C^2	C^3	C^4	C^5	C^6
Bab.	21	44	—32	29	10	— 7	—10	—6	23			
Bra.—3	—2	—6	8	44	—17	9	— 8	—16	—16	—33	7	7
Cro.—2	3	33	—43	3	58	—26	3	— 7	—17	—25	8	—28
Elk.—1	8	48	—20	2	33	—14	—13	—26	—10	—13		
How.—1	—32	—17	63	78	—45	78	—27	—26	—10	—12	19	
Ran.—1	17	—11			—40	44	13	—17	— 4			

In Table VIII are included 27 children, 7 above the normal stoutness and 20 below, or a total of 30 pounds excess to 324 pounds deficiency.

A pedigree of a family with hereditary obesity is described by Rose (1907). A girl of 15 with a stature of 145 centimeters (57 inches) weighed 75 kilograms (165 pounds). The father and his parents were not obese.[1] The mother, on the other hand weighed 88 kilograms and her father 99 kilograms, while the mother's mother is slender. Of the four children

[1] There is no evidence that they did not carry the factor that favors obesity or that they were wholly unrelated to the maternal side.

two (including the girl of 15) are very obese, one normal and one under weight. This result accords with the hypothesis that obesity is due to a defect. It is noted that the mother's mother had a goitre; and it is probable that in this family there is an hereditary deficiency in growth control.

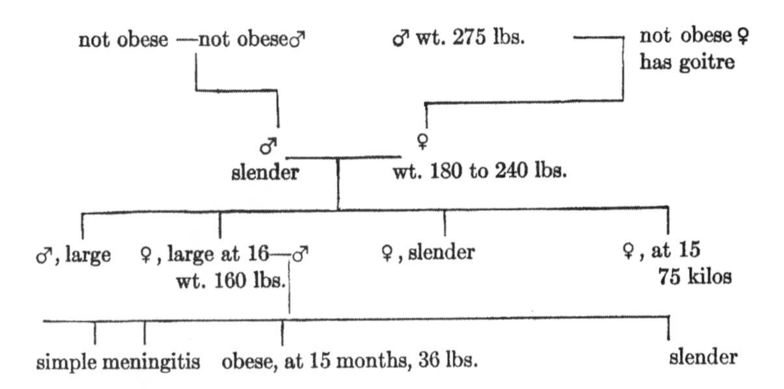

Longevity. When Dr. O. W. Holmes was asked for specifications for a long life he advised, in effect, first to select long-lived grandparents. This advice accords with a widespread opinion that longevity is inheritable. But length of life is not a unit character. It is a resultant of many factors; especially

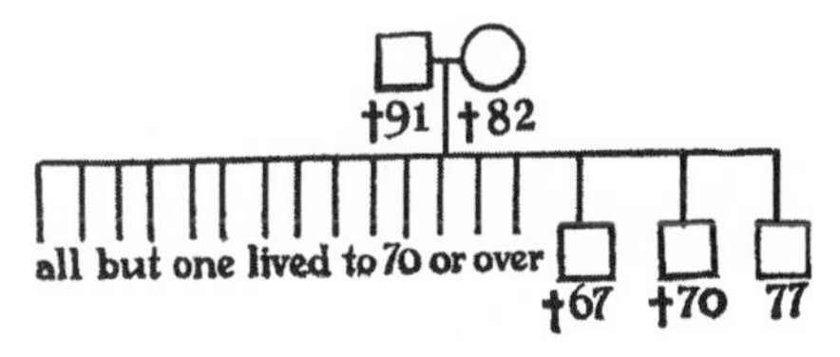

FIG. 18.—A short pedigree (early 19th century in United States) illustrating "inheritance" of longevity. F. R.; Att. 1.

of those factors that resist causes of death. Such factors are absence of defects of bodily structure, resistance to the commoner virulent germs of disease, and environmental conditions that maintain at its highest point internal resistance. The first two factors are "inheritable" and the last remains tolerably uniform for the people of a certain social class such as the members of one and the same family belong to; so it is not strange that some families with perfect structure and high resistance should be long lived (Fig. 18) and others, with organic defects and low resistance, should be short lived

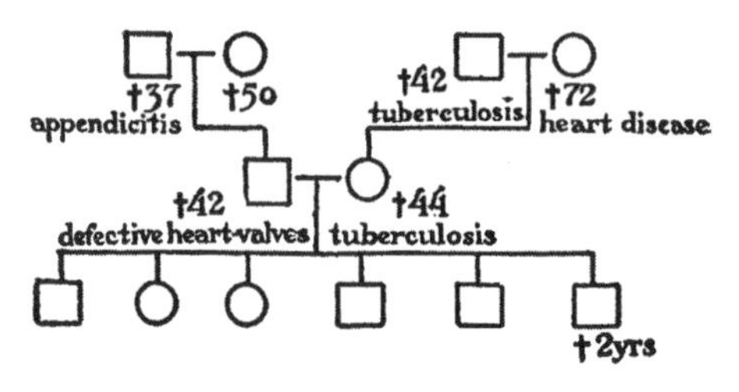

Fig. 19.—Fragment of pedigree of a high class family with slight longevity due in part to heart defects and non-resistance to tuberculosis. The latest generation comprises only young children. F. R.; Pyn. 1.

(Fig. 19). Thus, while longevity is not a biological unit of inheritance a person belonging to a long lived family is a better "risk" for a life insurance company than a person belonging to a short lived family.

7. Musical Ability

This quality is one that develops so early in the most marked cases that its innateness cannot be questioned. A Bach, matured at 22; a Beethoven, publishing his compositions at 13 and a Mendelssohn at 15; a Mozart, composing at 5 years, are the product of a peculiar protoplasm of whose tenacious qualities we get some notion when we learn that the Bach family comprised 20 *eminent* musicians and two score others less eminent. The exact method of inheritance of musical ability has not been sufficiently analyzed. Hurst (1908) suggests that it behaves as a recessive, as though it depended on the absence of something. The "Family Records" afford some data on this subject. A statement of the grade of musical ability of each person, whether poor,

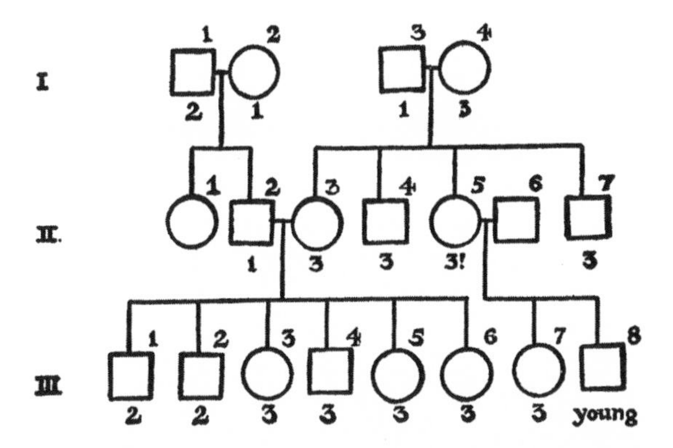

Fig. 20.—Pedigree of an American family of singers. Numbers below symbols designate grades; thus: 1, little or no musical ability; 2, medium ability; 3, exceptionally high ability. Numbers above the individual symbols are for reference.

I, 1. Extremely fond of music, had organ and piano in his home; a very cultivated man of artistic tastes. Married I, 2, non-musical, belonging to an utterly non-musical family. Their son, II, 2, is not musical.

I, 3. Fond of music, could "carry a tune" easily. A mathematician and astronomer. His wife, I, 4, was sufficiently musical to sing in such a simple church choir as was to be found in the State of Maine in the middle of the nineteenth century and her mother and mother's sisters were singers. All of their four children were musical. One son, II, 7, who died unmarried had a fine voice and was a good singer. The other son, II, 4, had a musical ear and a fine voice; he sang much without having taken lessons. His wife is non-musical and their 14-year old daughter is as non-musical as her mother. One of the daughters, II, 5, had a fine voice and still keeps up her music; she married an utterly non-musical man and they have one son who cannot even "carry a tune" and one daughter who is *a famous opera singer*. The other daughter, II, 3, is a fine singer, and plays the piano, organ and guitar. She married the above-mentioned non-musical man, II, 2. They had six children all of whom have fine voices; III, 1, has a fair baritone voice; III, 2, has an unusually deep bass voice; III, 3, died at 27 years. Her voice was said by good judges, such as the De Reszkes, Anton Seidl, etc., to be more beautiful even than that of III, 8. III, 4, is organist and choir master in a large church in New York City. III, 5, is very musical; III, 6, died young but had already developed much musical talent and could read music with wonderful ease. F. R.; H.

medium or exceptionally good was asked for. Altogether data were obtained for 1008 children, their parents and most of their grandparents. The following rules are deduced from these data.

When both parents are exceptionally good in music

(whether vocal or instrumental) all the children are medium to exceptionally good.

There were 48 cases where both parents showed exceptional musical ability. Of the 202 children 81 had exceptional ability and 120 fair musical ability. Only one is returned as being poor in music; and this case may be cast aside as quite within the probability of an error due to carelessness in making the returns or to bad classification. These results come out so smoothly as to indicate that high attainment in vocal and instrumental music are due to the same defect in the protoplasm.

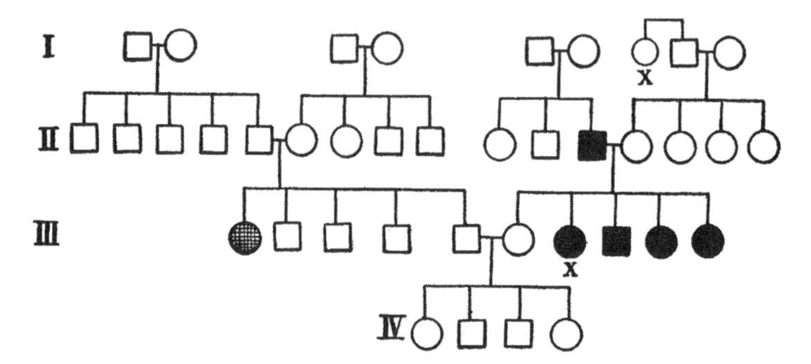

FIG. 21.—Pedigree of singing ability and peculiar form of toes. I, 7. (X) has bones of both fifth toes cartilaginous and toe crossed over upon fourth toe; and her granddaughter III, 7, has exactly the same peculiarity; II, 12, has an exceptionally good bass voice; his daughter III, 6, cannot sing; but III, 7, has a beautiful soprano voice; III, 8, has an exceptionally good baritone voice; III, 9, has a 'beautiful contralto voice' and III, 10, has great musical ability. On the other side of the house, III, 1, has good musical ability. But in the fourth generation there is no musical ability. F. R.; Ait. 1.

To illustrate inheritance of musical ability by a concrete example the pedigree of a noted New England singer is appended (Fig. 20). This particular example alone could not be used to demonstrate either the hypothesis that musical ability is due to a new unit or that it is due to a defect.

When both parents are poor in musical ability and come of ancestry that lacks on one or both sides such ability the children will all be non-musical.

Four families of this sort are given in the *Records*. All 29 children are poor in music. Compare Fig. 21.

When one parent has high musical ability and the other has little the children will vary much in this respect.

Thus of 257 offspring of such matings 45 are without musical ability, 84 are exceptionally good at music while 128 are intermediate. The result indicates a partial blend in the musical ability of the offspring of mixed origin.

As an example that illustrates the law approximately may be cited the Hutchinson family (Hutchinson, 1876). According to the statement of Wm. Lloyd Garrison, Jesse and Mary L. Hutchinson, progenitors of the tribe, lived in Milford, N. H., 1777–1863. The father possessed a rare baritone, the mother a sweet and mellow contralto voice. Of the sixteen children, three died young. The remaining thirteen are described as follows: David, deep bass voice; Noah, tenor voice; Andrew, baritone and bass voice, deeply interested in music; Zephaniah, passionately fond of music; Cabel, baritone voice; Joshua, very musical, sang; Jesse, editorial work; Benjamin, not gifted musically; Judson, musical genius; Rhoda, high contralto; John, most commanding vocal talents of all; Asa, inherited a large share of musical gift; Abbe, contralto voice, one of quartette. Details are lacking concerning the voice of Jesse, and the description of Benjamin is all too vague, considering the importance of this case, and so too much emphasis cannot be laid on these two cases; but aside from them the uniformity of testimony as to vocal talent of the family is striking.

8. Ability in Artistic Composition

Like musical ability, artistic talent shows itself so early as to demonstrate its innateness. Thus extraordinary talent was recognized in Francesco Mazzuoli (though ill taught) at 16, in Paul Potter at 15, in Jacob Ruysdael at 14, in Titian Vecelli at 13. Galton gives the following pedigree of the Vecellis. All the persons named were painters. "The connecting links indicated by crosses are, singularly enough, every one of them lawyers" (Fig. 22).

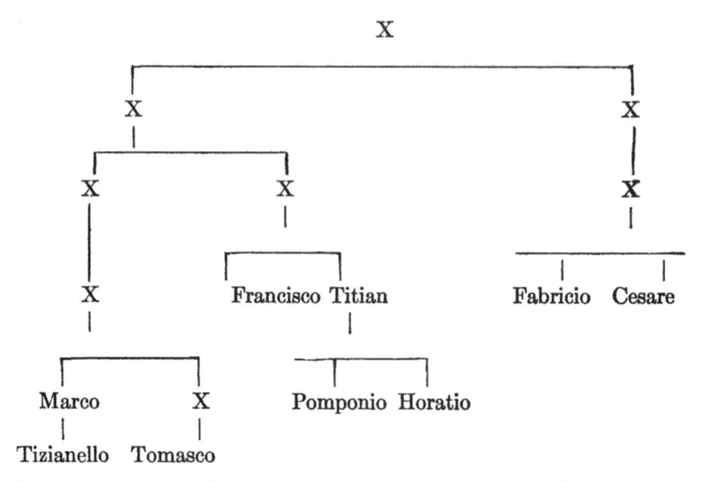

FIG. 22.—Pedigree of the painter family Vecelli.×, father (always a lawyer).—GALTON, 1869.

The data furnished by the Family Records seem to justify the following conclusions.

When both parents have exceptional artistic ability their children will, in most cases, all have high artistic ability (Fig. 23).

The data for this generalization are sparse. Four matings of this sort furnished 13 children of whom 10 had a high grade, 1 is recorded as medium and two as poor; but both of the latter occur in one record that gives internal evidence that the question was not clearly understood.

When both parents are devoid of artistic talent and come from an unartistic ancestry none of the children show exceptional ability in art.

From 103 such matings (grade 1) there were derived 391 children of whom 185 are given as of grade 1 and 206 as of grade 2, while to none was ascribed grade 3.

When one parent is artistic and the other neither himself artistic nor of artistic ancestry then probably none of the

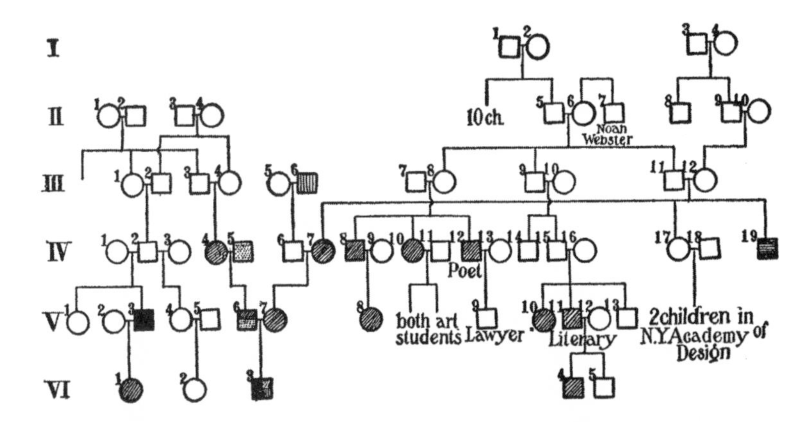

Fig. 23.—Pedigree of artistic ability (solid black for high talent, oblique shading for talent of a less degree). The family shows also the traits of taste for history (dots), of mechanical talent (vertical lines), and of wood carving (horizontal lines). II, 3, Nathan P, had son Wm. F. (III, 2) who was grandfather of an artist, V, 3; and a daughter Mary (III, 4) who was the great grandmother of artist J. W. F. (VI, 3). This brother and sister (III, 2, and III, 4) married a brother and sister, (III, 1 and III, 3) and it is in this stock that we first find the inheritance of artistic ability. IV, 4, married John E. F. (IV, 5) a man who through life had a love of historic research. This love of history appeared again in George E. F. (V, 6) who became a journalist and subsequently author of several valuable works on Indian history. In his son (VI, 3) in turn this love of history cropped out, as shown both in his Art History researches and as a painter of Indian history scenes. On his father's side, the lineage of VI, 3, has been traced back to 1630. No artistic genius was found in the male line except in V, 6 and VI, 3. His grandmother (IV, 4) displayed artistic tendencies, painting notable pictures throughout life.

We turn now to the mother of VI, 3, and her family. Her great-grandfather, Joel L., II, 5, married Jerusha, sister of Noah Webster, II, 7. Their son Chester's second son, Edward, IV, 15, a distinguished clergyman married Mary J. S., IV, 16, an educated lady and great lover of art. Their son, V, 11, was editor of the N. Y. "Sun," educator, Regent of the State of New York and fond of drawing and painting in an amateurish way. Artistic gift exists in his sister Anna and his older son, Kenneth.

III, 8, married Rev. S. P., graduate Andover Theological Seminary, first Presbyterian missionary to Oregon. Their first son, IV, 8, entered the ministry and was afterward a physician, also having marked artistic genius. His daughter Florence, V, 8, had marked artistic ability. His sister, IV, 10, was also a natural artist and this talent developed in her children and grandchildren to some extent. A brother, IV, 12, was clergyman, author-poet and professor in art. His son, V, 9, was a lawyer. Of children of III, 11–12, IV, 19, was gifted as a wood-carver, a trait which appeared in his great-nephew, VI, 3. IV, 17, married, and two children were proficient in the N. Y. Academy of Design. IV, 7, had an artistic turn of mind and her daugh-

children will have high artistic talent. But if the unartistic parent have artistic ancestry there will be artistic children.

From 15 such matings there were derived 37 children of whom 15 were poor in artistic ability and 22 medium. Among the 120 children derived from the mating: non-artistic parent having some artistic ancestors × artistic parent, there were 43 with exceptional artistic ability.

9. Ability in Literary Composition

The inheritance of the ability to express oneself in literary form is commonly recognized. "Poets are born; not made." Many literary men show their talent very early, before they had received much training in expression. Burns, the plowboy, was celebrated as a poet at 16, Calderon at 14, Goldoni produced comedy at 8, Charlotte Brontë published "Jane Eyre" at 22, Fénélon was known at 15, Sir Philip Sidney was famous at 21. As illustrations of heredity we have two of Charlotte Brontë's sisters writing a famous book, besides a brother Patrick said to be the greatest genius of them all. The father and the father's father of T. B. Macaulay, two uncles, a cousin and a nephew were all writers of note. Four generations of Taylors in England were authors of an "evangelist disposition."

The precise method of inheritance of literary ability has not hitherto been made clear; but a study of the Family Records seems to justify the following conclusions.

When both parents have high to good literary ability all (or nearly all) of the children will have likewise good literary ability.

There are 643 offspring of such matings in the Family Records and of them 93 per cent have medium to high literary capacity. No doubt these terms are used somewhat loosely and this may account for the exceptional cases.

---

ter, Mary L. B., had a decided artistic talent which she inherited from her father's family as well as her mother's.

It may be of interest to state further that VI, 3, possessed a mechanical genius, as did his great-grandfather, Joseph B., III, 6, a skilled jeweler, many of whose descendants to the fourth generation were also skillful jewelers.

When both parents have poor literary ability and come from a strain devoid of it the children will, typically, have poor literary capacity. This generalization is based on the 19 children, all non-literary, of four matings of this sort. But when literary ability appears in remoter ancestry it will occur in some of the children. Thus in 23 matings of this sort only 25 per cent of the children are without literary capacity.

10. Mechanical Skill

There can be little doubt of the inheritance of some of the elements of mechanical ability. The case of John Roebling and his sons, builders of the first great suspension bridge over the East River, New York City, and of Charles Martin, long chief engineer of that bridge, and his son, Kingsley Martin, for some years chief engineer of the bridges of New York City, are examples familiar to modern Americans. Not less striking is the family of boat designers whose pedigree is shown in Fig. 24. Five of the seven sons of the illustrious head of the family were inventors and boat designers, and high technical ability has appeared also in the third generation.

The Pomeroys are another American family that illustrates the inheritance of mechanical skill. The first of the family in America was Eltweed Pomeroy at Dorchester in 1630 and later at Windsor, Connecticut. He was by trade a blacksmith, which in those days comprehended practically all mechanical trades. His sons and grandsons, with few exceptions, followed this trade. "In the settlement of new towns in Massachusetts and Connecticut the Pomeroys were welcome artisans. Large grants of land were awarded to them to induce them to settle and carry on their business." "The peculiar faculty of the Pomeroys is not the result of training and hardly of perceptible volun-

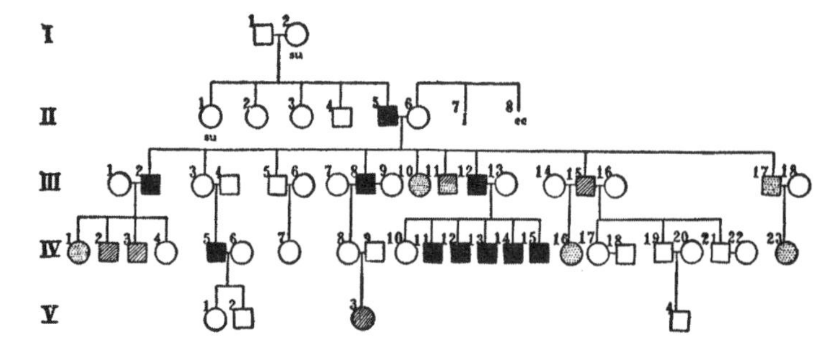

Fig. 24.—Pedigree of family with mechanical and inventive ability, particularly in respect to boat-building. I, 2, a suicide: II, 1, a suicide. His brother, II, 5, a builder of swift boats and yachts, II, 7, insane; II, 8, eccentric. The union of these two strains with evidence of nervous instability resulted in a family of 9 children and 18 grandchildren. Four of the sons show a high degree of inventive ability and 2 of these III, 8–12, developed the genius of their father in designing and building swift and beautiful boats. Three are musicians, III, 10, 11, 17, and one of them, III, 11, shows also mechanical ability. In the next generation these traits reappear in the various fraternities. IV, 1, is a musician; 2 has much mechanical skill and 3 is inventive; 5, is a builder of fine boats; IV, 11–15 represent 5 boys, none over 22, but already designing boats; two other daughters of this generation show artistic and musicial talent and, finally, in the next generation we have a girl of 14, V, 3, designing boats. F. R.; H.

tary effort in the individual. Their powers are due to an inherited capacity from ancestry more or less remote, developed for generations under some unconscious cerebration." There was Seth Pomeroy (1706–1777) an ingenious and skillful mechanic who followed the trade of gunsmith. At the capture of Louisburg in 1745 he was a major and had charge of more than twenty smiths who were engaged in drilling captured cannon. Other members of the family manufactured guns which in the French and Indian wars were in great demand and in the Revolution, also, the Pomeroy guns were indispensable. "Long before the United States had a national armory, the private armories of the Pomeroys were famous. There was Lemuel Pomeroy, the pioneer manufacturer of Pittsburg, stubborn but clear headed, of whom a friend said: There would at times be no living with him if he were not always right." There was also Elisha M. Pomeroy of Wallingford a tinner by trade. He invented the razor strop and profited much by its success. [C. H. S. Davis, 1870, History of Wallingford.] In the sixth generation we find Benjamin Pomeroy a successful lawyer entrusted with important public offices. "But he was conscious of powers for which his law practice gave him no scope. He had a taste for mechanical execution, and as a pastime between his professional duties undertook the construction of difficult public works—the more difficult the better he liked them. The chief of the United States Topographical Engineers was a friend of Mr. Pomeroy and repeatedly consulted him in emergencies wherein his extraordinary capacity was made useful to the government. By him were constructed on the Atlantic coast beacons and various structures in circumstances that had baffled previous attempts." The value to this country of the mechanical trait in this one germ plasm can hardly be estimated. Especially is it to be noted that, despite constant out-marriages, it goes its course unreduced and unmodified through the generations.

The Fairbanks family of St. Johnsbury, Vermont, illustrates the inheritance of inventiveness combined with executive ability, specialized in the iron trade. The inventor of the "platform scales" belonged to a family not merely of iron workers but to one with imagination such as made other members literary men (Fig. 25).

The Family Records give rather definite information as to the method of inheritance of mechanical skill.

When both parents have good or exceptional mechanical skill all of their children will have it also.

Out of 413 children of such matings (including both sexes) all but 7 show some mechanical ability, and 118 of them ability of an exceptional order. Indeed, most persons of exceptional skill come from this mating.

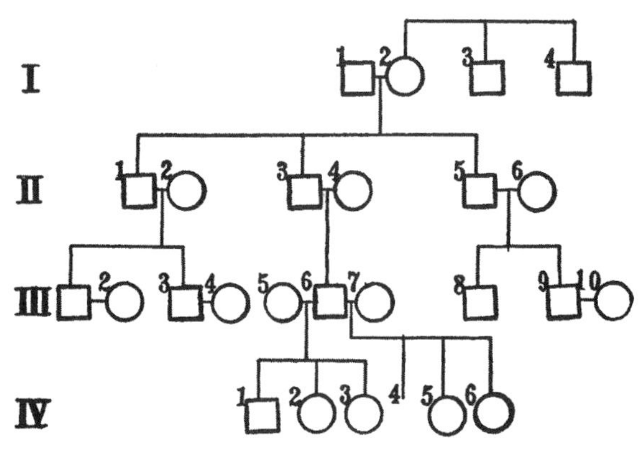

Fig. 25.—Pedigree illustrating inheritance of special ability in the Fairbanks family of Vermont. I, James Fairbanks; I, 2, Phoebe Paddock. Her two brothers, I, 3 and 4, were *iron workers*, II, 1, Erastus Fairbanks moved at 19 years to St. Johnsbury, Vermont and began to manufacture stoves, plows, etc.; II, 2, Lois Crossman; II, 3, Thaddeus, a natural mechanic, invented the platform scales; II, 4, Lucy Barker; II, 5, the third brother, Joseph P. Fairbanks was a lawyer, with literary tastes.

Erastus and Lois had two sons of whom the elder, III, 1, went into the scale business, showed much inventive ability and a strong taste for natural history. His brother Horace, III, 3, was an excellent administrator and became Governor of Vermont. Dr. Henry Fairbanks, III, 6, son of Thaddeus went into the ministry, but his love of invention drew him into the iron business. He combined mechanical and literary gifts. III, 8, was a minister and III, 9, a sagacious and exact man, was secretary and treasurer of the Fairbanks Company.

If both parents lack mechanical skill and come from an ancestry that lacks it no offspring will have mechanical ability. Even if mechanical skill is found in the ancestry of one side, but not of the other, still there will be no marked mechanical ability in the children.

If one parent has mechanical ability and the other belongs to a strain that lacks it then exceptional mechanical ability will be absent or uncommon. But if the parent that lacks mechanical ability comes from an ancestry that possessed it a large proportion of the children will have such ability. Also when both parents that have slight mechanical ability are descended, on one side, from persons with skill, such skill will reappear in approximately one child in four.

11. Calculating Ability

The inheritance of great mathematical ability cannot be denied and is well illustrated in the case of Bernoulli: Jacques, his nephews Nicholas and Jean, and three nephew's sons were mathematicians of high rank.

Our Family Records afford a limited amount of data on the subject of inheritance of mathematical ability. They do give information concerning the inheritableness of the ability to calculate—a broader phenomenon. The following rule seems justified: When both parents are good at calculating all (or nearly all) of their children will be so likewise.

Of 728 offspring of this class of matings all but 48 (or 8 per cent) were good at calculating. In no case were both parents returned as poor at calculating; but in 47 matings both parents were only medium and 13 per cent of their children were poor at calculating.

12. Memory

There is no doubt that people vary in their ability to remember and there is no question that a good memory is an innate quality. Phenomenal memories are often associated with mental defect in which case it is clear they are independent of training. In other cases they are associated with high scholarship. Thus Galton cites the case of Richard Porson, an eminent Greek scholar, whose memory became stupendous. His mother had a remarkable memory and so did his sister.

The Family Records throw some light on the inheritance of a good memory; although the term is a relative one and lacks in precision. Nevertheless for a preliminary study the data are not to be despised although there are not a few exceptions to any generalizations one may hazard.

When both parents have an exceptionally good memory

most, if not all, of the children have a memory that is medium to exceptional.[1]

When both parents have a poor memory and come from ancestry so characterized few if any of the children have an excellent memory.

Two "poor" parents (with "poor" grandparents) have 10 children all with poor memory.

When one parent has a memory that is either excellent or fair and the other has one that is "poor" all children have a medium memory; and, conversely, parents with medium memory may have 20 to 25 per cent of children with excellent and as many with poor memory.

13. Combined Talents and Summary of Special Abilities

While the separate talents may, for purposes of analysis, be considered separately they usually, as our illustrations suggest, occur in combination in a single family. And such talents are frequently enough associated with insanity or mental defect in some of its members as apparently to justify the poet's conclusion: "Great wits are sure to madness near allied" (Fig. 26).

In many cases artistic, literary and musical talent are found in the same family—two or all three of them are occasionally found in the same person (Fig. 27, Fig. 28). The conclusion seems justified that artistic, literary and musical skill are unit characters that may occur in any combinations—the common inherited factor may be only a highly developed imagination.

In the foregoing cases the method of inheritance of many of the elements of the mental makeup have been considered

[1] The Family Records give 4 per cent of children of such matings as having a poor memory.

and the remarkable result has been deduced that the higher grades of all these qualities act, in inheritance, as though they were due to the absence of something that is present

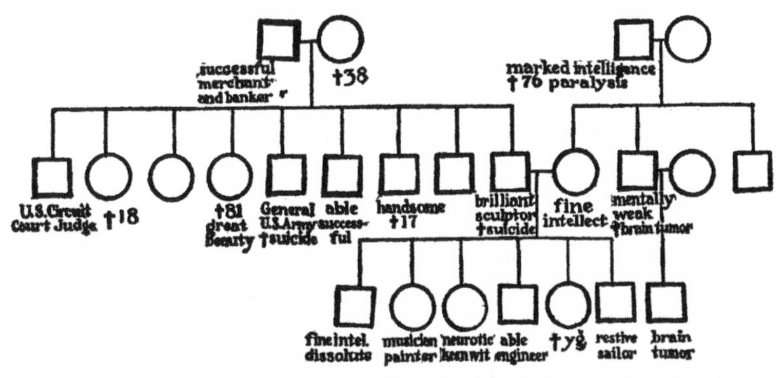

Fig. 26.—Pedigree of brilliancy combined with defect and melancholia. F. R.

in persons of poorer ability. It is as if the difference between a person of high ability and one of low ability in respect to any mental trait is that the person of high ability has

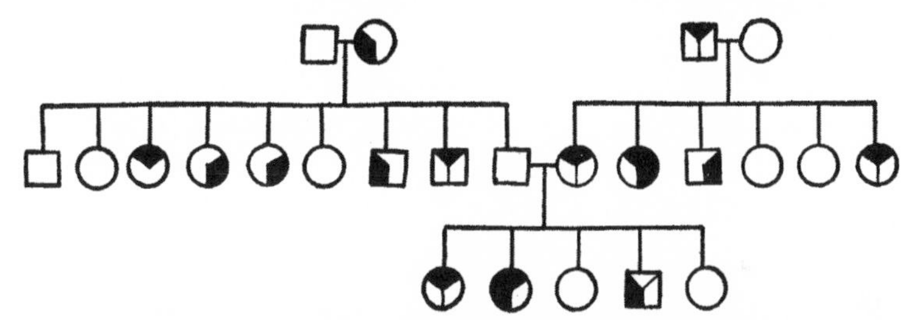

Fig. 27.—Pedigree of family with artistic (dark upper section), literary (right section) and musical (left section) ability.

got rid of a something possessed by the person of lower ability that prevents the latter from fully exercising his faculties;—he has sloughed off one or more inhibitors.

14. Temperament

Two contrasted temperaments are usually recognized. One phlegmatic, slow, rarely depressed; the opposite ner-

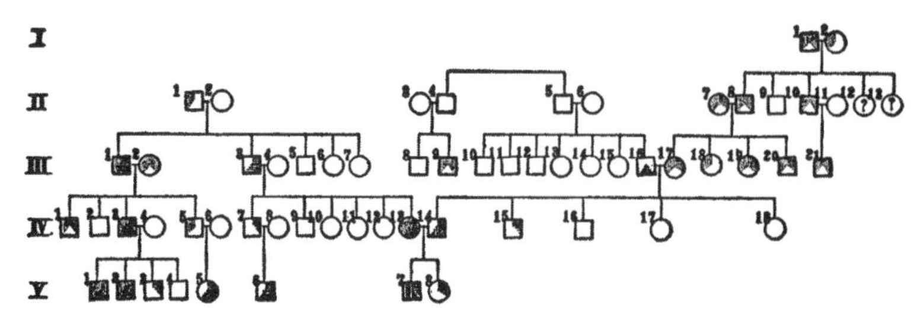

Fig. 28.—Pedigree of a pedigree-complex (Abbott-Buck-Wolff) showing inheritance of musical (dots), literary (horizontal lines) and inventive (vertical lines) ability. Variations in the area covered by each symbol indicate roughly a variation in degree of ability of the given kind. I, 1, a musician of the eighteenth century and I, 2, his wife, the daughter of a professor of music. One of his sons adopted a seafaring life and died in Mozambique. Two sons, II, 8, 11, were instructors in the Geneva Conservatory of Music. The son, III, 21, of one of these was a professor of music and a composer. The other married a woman, II, 7, with literary and musical ability and had four children of whom III, 19, was a literary composer; III, 18, had good musical ability; III, 20, was brilliant piano player with a fine baritone voice and literary; and III, 17, both literary and musical, married a man with inventive ability whose first cousin, III, 9, was an organist and musical composer of high rank. Two of their children, IV, 14, 15, show literary ability and IV, 14, inventive ability also. He married into a family famous in American literature and with much musical ability and the product was two children both literary and one, V, 7, an inventor of high rank. III, 1 and 3, derived from a musical father, have literary ability of a high order. One who has also some musical ability married a very musical wife and of the 4 sons at least 3 have musical ability. One of these, IV, 3, combined with the musical also literary ability, married a woman with some literary ability and had 4 sons of whom 3 at least are litterateurs and two have much musical ability. V, 5, is a well-known authoress.

vous, quick, often elated or alternately elated and depressed. Between the extremes lie, as is usually the case, many intermediates. While it is clear that there are no sharp lines to be drawn between these conditions, some insight into their hereditary behavior may be gained by an examination of the opinions furnished by collaborators in the Family Records.

When phlegmatic is assumed to be a condition recessive to the "intermediate" and nervous conditions we find that in three families with 13 offspring, 10 or 77 per cent, are likewise phlegmatic. On the other hand, when nervous is assumed to be recessive to intermediate and phlegmatic in 130 offspring of nervous parents 64 or 49 per cent were nervous.

So far as the data go they support the following conclusions. The offspring of two phlegmatic parents tend to be phlegmatic and the nervous parents of purely nervous origin have nervous children. But one phlegmatic parent mated to a nervous one will produce chiefly nervous children and many who are intermediate. When both parents are nervous with phlegmatic ancestry a fairly large proportion (up to about a quarter?) will be phlegmatic.

15. Handwriting

Inheritance of peculiarities of handwriting is often alleged (Darwin, 1894, p. 449), but it is difficult to get satisfactory evidence about it. A correspondent (Hal-2) writes:—"We belong to a family of penmen. My four brothers and myself inherited our handwriting (the English legal copyist's handwriting) from my father. Two of our uncles and two cousins also wrote the family hand. I believe it was asserted that our paternal grandfather wrote the same. We could distinguish the writing of each, but the general family resemblance was there, especially when we were all young men and my father was not old. . . . We descended from a family that included officemen, lawyers, recorders to whom expert penmanship was necessary."

16. General Bodily Energy

Of the inheritance of this quality there can be no doubt. If we take the class of commanders as one characterized above all by bodily energy we see the intensity of its heredity. It is exemplified in the family of Alexander the Great from Philip of Macedon down, the family of Charlemagne including Pepin le Gros and Charles Martel, of Gustavus Adolphus, and of Scipio Africanus.

Can we discover how bodily energy, which reaches its highest degree in such commanders, is inherited? Here again I appeal to the Family Records in which energy is recorded in the three grades: below average, medium, decidedly above average. The following principles seem established.

When both parents have bodily energy that is regarded as "decidedly above average" all of their children will have either exceptional or at least medium energy.

The mating of two energetic parents in 192 families produced 413 offspring (or 2.2 children to the family). Of these 301 (73 per cent) are placed in the highest grade; 100 (24 per cent) in the middle grade and only 12 (3 per cent) in the low grade. Considering the probability of errors this lowest grade is negligible.

When both parents have medium to low energy and come from ancestry of this sort all offspring have medium to low energy.

There are 54 matings of this sort, with 219 children (or 4.1 children to the fraternity). All but 4 are in the medium class.

When one parent has great bodily energy while the other has no great energy in himself or his ancestry all of the children (86) have medium (82) or low energy (4). But if there be energy in the grandparents on the low side about half of the children will have energy that is decidedly above the average.

There are 105 matings of the latter sort, producing 456 children (or 4.3 children to the fraternity) of whom 226 were classed as of great energy, 208 of medium and only 22 as low.

On the whole the facts support the hypothesis that excessive bodily energy is due to a loss of something—perhaps an inhibitor that prevents persons from achieving the best that is in them. However, the whole subject deserves a more thorough investigation.

17. General Bodily Strength

Like other bodily traits general strength is clearly inherited. This appears repeatedly in our records. An example is given in Fig. 29.

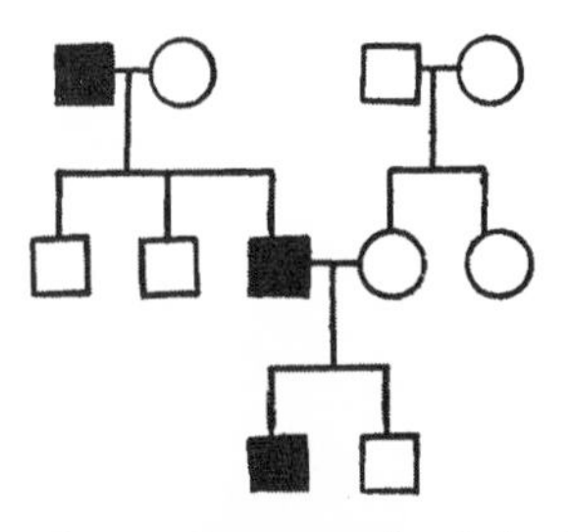

Fig. 29.—Inheritance of muscular strength. I, 1, of great physical strength. His son II, 3, was likewise possessed of unusual strength. His elder son in turn was athletic but became dissipated. F. R.; St. 1.

18. General Mental Ability

The general mental ability of a person is a vague concept which is, however, in common use. We speak of a man as weak minded, as of mediocre ability, as exceptionally able without attempting a closer analysis of the subject.

General mental ability, like stature and weight, undergoes a progressive development so that in studying its heredity we must compare it in adult persons or else measure it by the deviation the person shows from the normal of his age. Thus we may call "weak mindedness" such a defect as would keep a child of 10 in a school grade where the other children are 6 or 7; a child of "mediocre" ability is not more than two years behind the average grade for his age; "exceptionally able" would imply, say, two years in advance of children of his age. A series of tests (the Binet-Simon tests) have been devised to gauge mental ability by gauging a variety of capacities such as general information, ability to count and to repeat phrases, to recognize names and describe common things and to make fine sense discriminations. Such tests show that there are all grades of mental ability. At one extreme is the idiot, without language and incapable of attending to his bodily needs. He may retain to maturity the mentality of a child of a few months. In a higher grade mentality of a child of 3 to 5 years is retained throughout life; such are the imbeciles; then come the merely backward children who make dull adults of all

Fig. 30.—Family group from a long-settled valley where much consanguineous marriage has taken place.

grades to the normal condition (Fig. 30). Finally, there are the exceptionally bright, quick children some of whom at least, become superior adults. It is hard to recognize a unit character in such a series any more than in human hair color. Nevertheless there are laws of inheritance of general mental ability that can be sharply expressed. Low mentality is due to the *absence* of some factor, and if this factor that determines normal development is lacking in both parents it will be lacking in all of their offspring.

Two mentally defective parents will produce only mentally defective offspring. This is the first law of inheritance of mental ability. It has now been demonstrated by the study of scores of families at the Vineland (N. J.) Training School for defectives by Dr. H. H. Goddard. Some pedigrees illustrating this law, and those that follow, are given in Figs. 31-35.

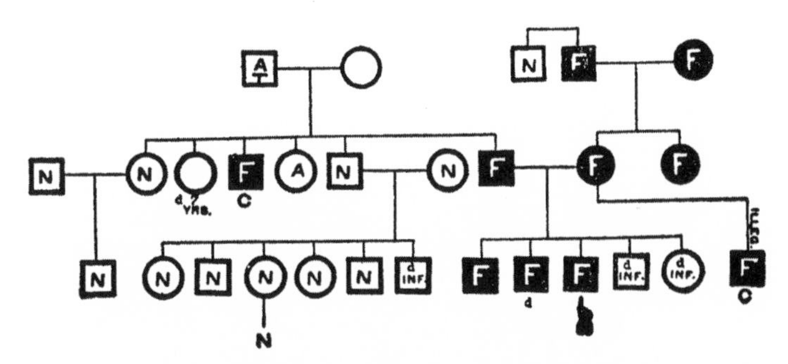

FIG. 31.—Pedigree chart illustrating the law that two defective parents have only defective children. *A*, Alcoholic; *C*, criminalistic; *D*, inf., died in infancy; *F*, feeble-minded; *N*, normal, *T*, tubercular. GODDARD, 1910.

The second law of heredity of mentality is that, aside from "mongolians," probably no imbecile is born except of parents who, if not mentally defective themselves, both carry mental defect in their germ plasm. Fig. 36 (left side of chart). Many a person of strong mentality may carry defective germ cells and, whenever two such persons marry, expectation is that one-fourth of their offspring will be defective. If a person that belongs to a strain in which defect is present (and who, consequently, may be carrying the defect in his germ plasm) marry a cousin or other near relative (in whom the chance is large that the same defective germ plasm is carried) the opportunity for *two* defective germ cells to unite is enhanced. Such consanguineous marriages are fraught with grave danger.

In view of the certainty that all of the children of two feeble-minded parents will be defective how great is the folly, yes, the crime, of letting two such persons marry. It

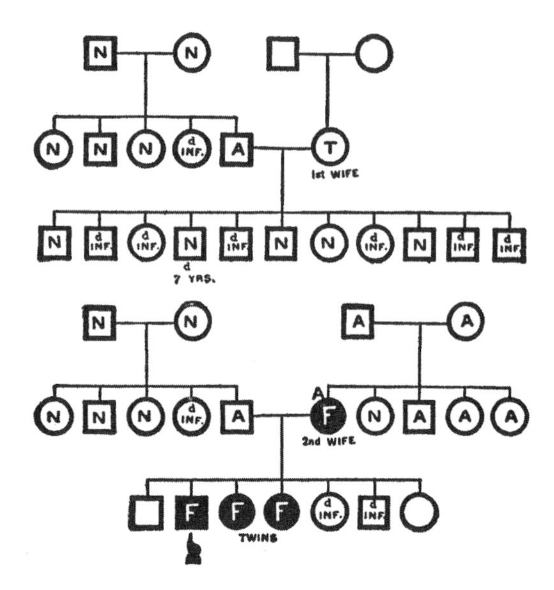

Fig. 32.—Pedigree chart illustrating the inheritance of feeble-mindedness. In chart *A*, the central mating is of an alcoholic man with a normal woman who died of tuberculosis. Of their 11 children, 5 are known to be normal, the others died early. Then (B) this man married a feeble-minded woman and of 7 children 3 are certainly feeble-minded, and 2 were, as young children, killed at play, in a fashion indicating a lack of ability to avoid ordinary dangers. Goddard, 1910.

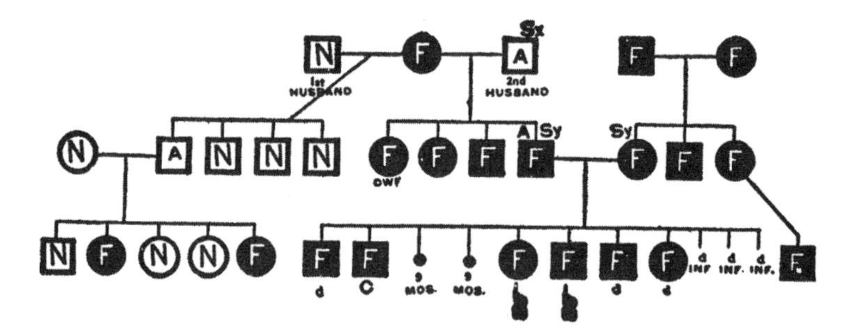

Fig. 33.—Here a feeble-minded woman (of the first generation) has married a normal man and has 4 normal children (except that 1 is alcoholic); then she marries an alcoholic sex-offender (who is probably also feeble-minded) and has 4 feeble-minded children. Here the mental strength of the first husband brought the required strength into the combination, so as to give good children. Goddard, 1910.

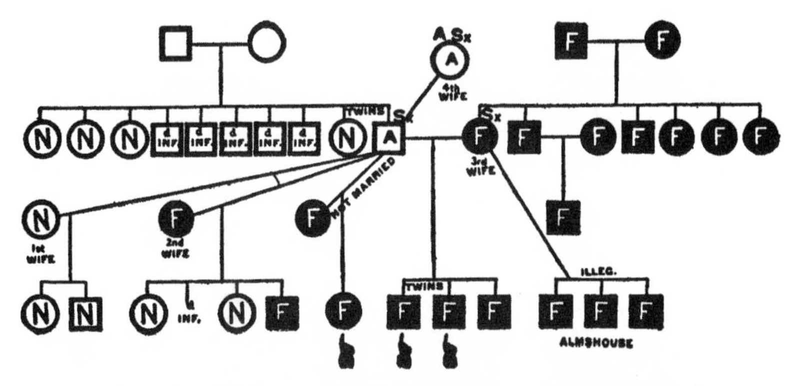

FIG. 34.—An alcoholic man of good family but probably simplex in mentality has by a normal woman 2 normal children and by a feeble-minded woman 2 normals and 1 feeble-minded. He has had 4 other children by feeble-minded women, all feeble-minded. *Sx*, sex-offender. GODDARD, 1910.

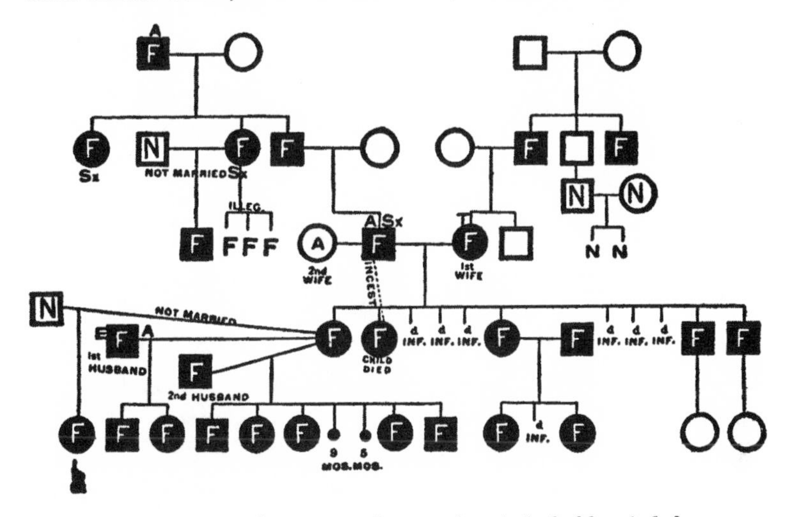

FIG. 35.—This chart shows several cases of entirely feeble-minded progeny from two defective parents. GODDARD, 1910.

has happened many times that keepers of poorhouses have let feeble-minded women in their charge go to marry a half-witted farmer in order to relieve the town of the burden of maintaining her. Some years later both she and her husband come to the poorhouse as permanent inhabitants and

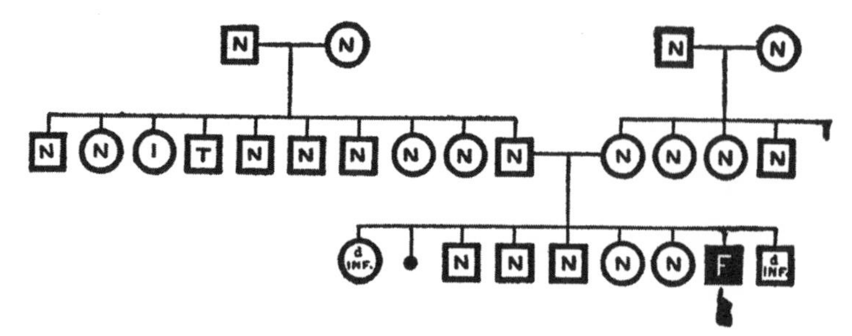

FIG. 36.—Pedigree of a "mongolian" imbecile. Except for an insane uncle (1) there is no evidence of a psychopathic condition in the parental germ plasms. GODDARD, 1910.

bring half a dozen imbecile children to be a permanent charge on the community. Surely there is no economy in this.

A still more appalling piece of testimony is given by a delegate from Alabama to the 26th National Conference of Charities and Correction. He said: "In our poor institutions the males and the females are allowed to run together and, so long as that is allowed, you cannot cut off the increase. It is perfectly appalling how the children accumulate in institutions."

Anyone acquainted with rural poorhouses (Fig. 37), particularly in the South, will appreciate that the people housed in them are mostly mentally inferior. By bringing together defective men and women, without proper segregation of the sexes, and by protecting and nursing the defective offspring of defective parents and then turning them out upon the community, the improperly conducted county poorhouses constitute one of the country's worst dangers. What is the state of your county poorhouse, reader?

An apparent paradox may well have occurred to the reader, and that is that mental defect and the elements of exceptional ability are inherited in the same way. This certainly looks like a self-contradiction. Are not the feeble-

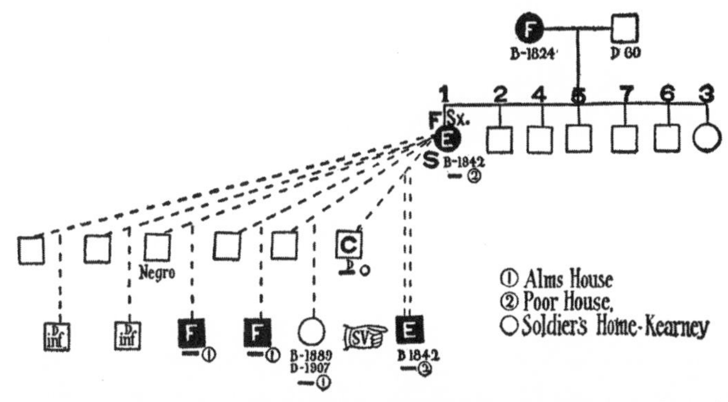

Fig. 37.—The "*poorhouse*" type of reproduction of the feeble-minded and epileptic. A lewd, feeble-minded and epileptic woman whose mother was certainly feeble-minded (but of whose father, brothers and sisters nothing is known) was the inmate of a county poorhouse. While there she had 6 children, of whom 2 died in infancy, 1 died at 18 in the almshouse, 2 were feeble-minded and are now living in the almshouse (1 the son of a negro) and 1 was epileptic, the son of a man with a criminal record. *C*, criminalistic; *D*, dead; *E*, epileptic; *F*, feeble-minded; *S*, syphilitic; *Sx*, sexually immoral.

minded and the talented at opposite extremes of the mental series? Why, then, this resemblance in the inheritance of their traits? Improbable as the result may appear it is precisely that to which students of hereditary genius have come. Says Havelock Ellis: "We may regard it (genius) as a highly sensitive and complexly developed adjustment of the nervous system along special lines, with concomitant tendency to defect along other lines. Its elaborate organization along special lines is often built up on a basis even less highly organized than that of the ordinary average man. It is no paradox to say that the real affinity of genius is with congenital imbecility rather than insanity." Ellis notes that eminent men are more apt to be eldest or youngest sons. Now this fact is in agreement with the observation that feeble-minded persons of certain types ("mongolians,") are more apt to be eldest or youngest children than intermediates. This type seems to be caused solely by the defects in development due to imperfect nutrition of the child born of parents (particularly mothers) that are immature or too old. The contention that geniuses and some defectives are born chiefly at the extremes of the reproductive period supports the view of their relationship.

19. Epilepsy

This term is believed by many professional men to cover a number of distinct brain disorders that have in common the symptoms known as convulsions or "fits." All too little is known about the physiology of the forced movements of convulsions, accompanied as they typically are by temporary loss of consciousness. It is known that convulsions may sometimes be induced in guinea pigs by a heavy blow on the brain case, and similar injuries are stated to have produced epilepsy in man. In other cases the "cause" is stated to be disturbance in the cerebral circulation due to a local stoppage in the blood vessels. However, it may well be questioned whether such causes are sufficient and not merely *inciting*, whether an inherent weakness did not first exist, which was only disclosed by the blow or disturbance in the circulation. A fall on the ice may result in a child's first epileptic fit but thousands fall on the ice without more than temporary discomfort; it was not the fall merely but the fall plus the too delicate nervous organization.

The hereditary basis of epilepsy has been studied and, rather remarkably, it follows the same laws as feeble-mindedness. Two epileptic parents probably produce only defective offspring, and the defect sometimes takes the form of epilepsy, sometimes that of feeble-mindedness. It does not seem necessary to repeat the laws of heredity for epilepsy since in them the words epilepsy and feeble-mindedness are almost interchangeable (Figs. 38–43).

The warning against the evils of poorhouses as breeding

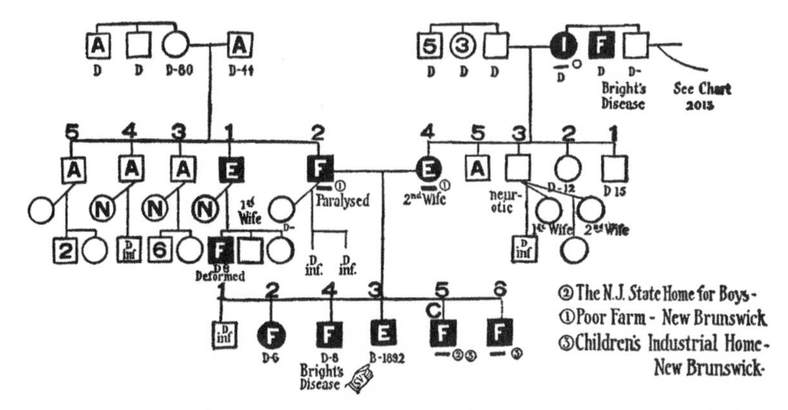

Fig. 38.—The product of a feeble-minded man (who has an epileptic brother) and his epileptic wife (whose father was insane and uncle feeble-minded); the first child died in infancy, the next two were feeble-minded and died young, the next is an epileptic at the New Jersey State Village; the next is feeble-minded, has a criminal record and is in the State Home for Boys; the last is feeble-minded and is in the Children's Industrial Home. Six in this family have been or are wards of the State. *A*, alcoholic; *C*, criminalistic; *D*, deaf; *E*, epileptic; *F*, feeble-minded; *I*, insane; *N*, normal. *SV* in the ☞ means an inmate of a State Village for Epileptics.

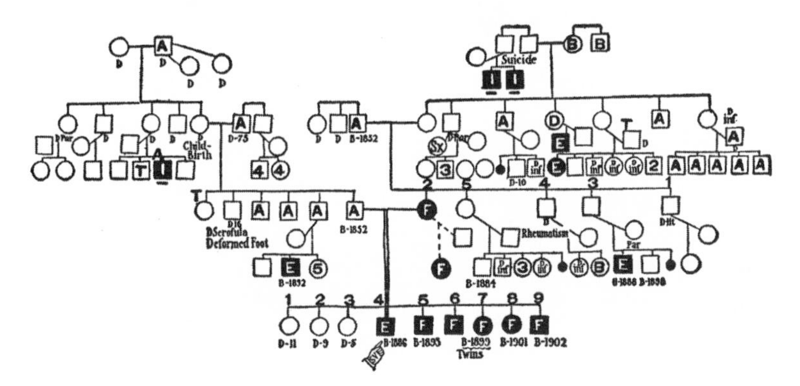

Fig. 39.—The central mating is that of a feeble-minded woman of an intensely neuropathic strain and an alcoholic man, who has 3 alcoholic brothers, father and grandfather alcoholic, an insane cousin and an epileptic nephew. The husband, though recorded as alcoholic, is probably also feeble-minded, at least all (6) of his children who survived were feeble-minded or epileptic. This chart shows 4 wards of the State and many others who should have been segregated. *A*, alcoholic; *B*, blind; *B*, (below), born; *D*, deaf; *D*, (below), died; *E*, epileptic; *F*, feeble-minded; *Ht*, heart-disease; *I*, insane; *Par*, paralysis, *Sx*, sex-offense; *T*, tubercular.

places of feeble-mindedness needs to be repeated for epilepsy and the dangers of consanguineous marriage are equally great (Fig. 43). If these two sources of epileptics—namely the poorhouse and the hovel (Fig. 44)—were cut off the supply of epileptics would be markedly reduced. And it is

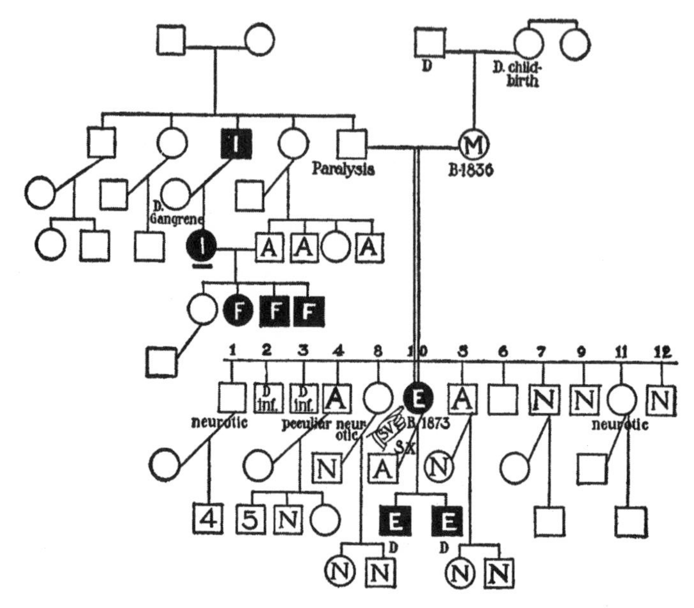

FIG. 40.—This mating illustrates the principle that migraine (*M*) and paralysis frequently indicate the presence of defective germ cells, as well as normal. In the central mating the paralytic father has an insane brother, an insane niece and 3 feeble-minded grandnephews, besides a grandniece, who died in convulsions. By his migrainous wife he had 12 children about 9 of whom something is known. One is epileptic, 3 "neurotic" or very nervous, 1 "peculiar" and alcoholic, while 3 are normal. The epileptic child has by an alcoholic husband 2 epileptic sons. Abbreviations as in Figs. 38, 39.

to be observed that these two sources of supply are quite within the control of society. A little larger appropriation to provide for the complete segregation of the sexes and a better superintendence will shut off the poorhouse supply and the inmates of the hovels should be brought under surveillance,—if necessary under public care.

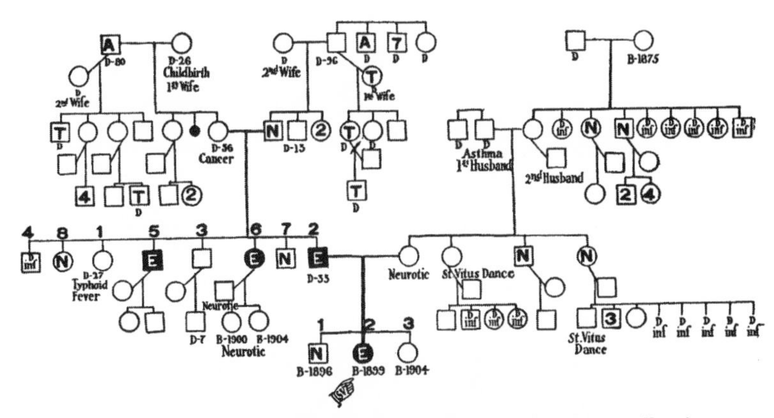

FIG. 41.—The central mating in this chart is that of an epileptic man, of a highly neuropathic strain, and a neurotic woman, whose sister and nephew have had chorea or St. Vitus' dance. The product is 1 normal child, 1 epileptic, and 1 as yet only 7 years old. Abbreviations as in Figs. 38, 39.

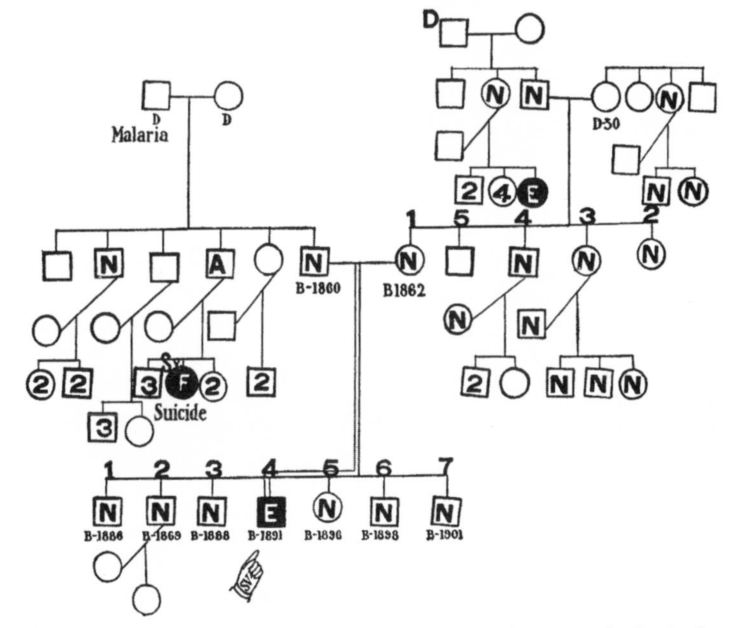

FIG. 42.—The central mating is that of 2 normal parents, both of whom belong to stock that shows evidence of being neuropathic. Doubtless some of the germ cells of both parents are defective in mental strength. Along with 6 normal children appears 1 epileptic. Abbreviations as in Figs. 38, 39. Figs. 37–43, are contributed by Dr. DAVID F. WEEKS.

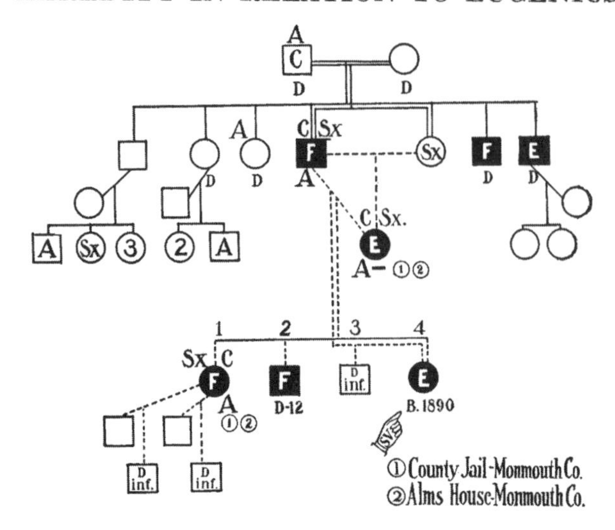

FIG. 43.—The *"Hovel"* type of reproductions of defectives. In a hut in the woods there was brought up a family of defectives. One of the boys, who is a drunken, feeble-minded fellow with criminalistic tendencies, had by his own sister a daughter who is a drunken epileptic, who has been the inmate both of the county jail and the county poorhouse. By her father she had 4 children of whom 1 is epileptic, 2 are feeble-minded (the girl has a very bad record of drunkenness, crime and immorality) and the other one was an idiot monster who died directly after being born. Close inbreeding of such a strain results only in this imperfect fruit. Abbreviations as above.

FIG. 44.—A hovel in a rural district, removed from social influences and liable to become the scene of anti-social acts. F. W.; 5, 1911.

20. Insanity

If the word epilepsy is a wardrobe then the word insanity is a veritable lumber room, including a great variety of mental diseases which have this in common that they render their victim incompetent and irresponsible before the law. Two great classes of insanity are distinguished: the "organic" and the "functional." The first group includes cases of mental deterioration associated with venereal diseases, alcoholism, degeneration of the blood vessels and trauma; the second includes cases of distinct neuropathic taint which shows itself in the slighter forms as melancholia or manic depressive insanity and in the profounder forms as dementia precox. Concerning heredity in the functional forms there is no doubt. Berze (1910) gives a case of dementia precox in a father and three sons; another of two children, their mother and her father; and numerous other cases with two or three to the family—all with a more or less typical form of dementia precox. But the mental defect that is "inherited" is not always of the same type. Thus in the same family may be found cases of manic depressive insanity, of senile dementia, of alcoholism and of feeble-mindedness. It would seem to be the neuropathic taint that is inherited.

This is the conclusion to which Cannon and Rosanoff (1911) have come in their study based on house to house investigations of the families of patients at a State Hospital. They omit from consideration the "organic" class of cases as "probably purely exogenous in origin." Aside from these they find that when both parents have any form of insanity all of their children will "go insane." If one parent is insane and the other normal but of insane stock half of the children tend to become insane; when both parents, though normal, belong to an insane stock about one-fourth of the

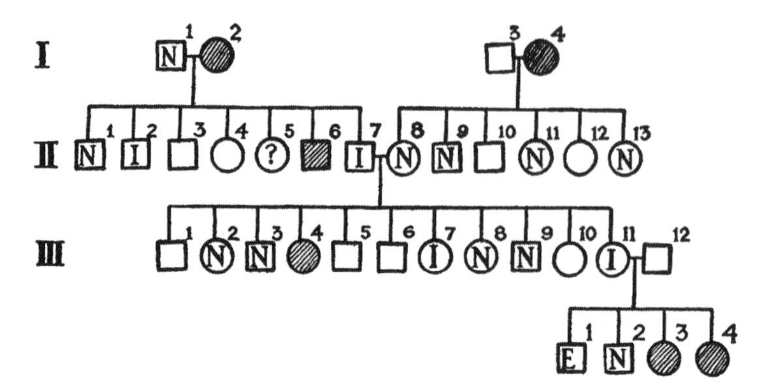

Fig. 45.—The central mating, II, 7, II, 8, is that of a man, II, 7, who is subject to melancholia and has an insane brother and another who is neuropathic. His wife is normal but her mother was neuropathic. The product of this union is 11 children of whom 3 are neuropathic. One of these insane children marries a normal person (probably of neuropathic ancestry), and has 2 neuropathic children besides 1 that is epileptic, IV, 1. *E*, epileptic; *I*, insane; *N*, normal; shaded symbols imply some neuropathic condition other than insanity. Cannon and Rosanoff, 1911.

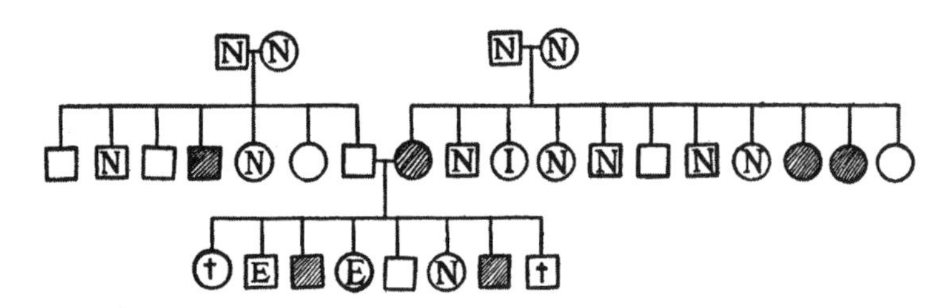

Fig. 46.—The central mating is that of a normal man of neuropathic stock with a neuropathic woman who has an insane sister. Since by hypothesis all of her germ cells and half of his are "neuropathic" it is to be expected that half of their offspring will be neuropathic in some degree. Actually, of 6 surviving children 2 are epileptic, 2 highly nervous and 2 normal so far as known. There is a slight, but not unreasonable deficiency of normals, namely, 1. The shaded symbols represent nervous subjects. Cannon and Rosanoff, 1911.

children become insane. The typical laws of heredity are followed here (Figs. 45–47).

But is it so certain that alcoholic, traumatic, even syphilitic dementia have no hereditary basis? On the contrary

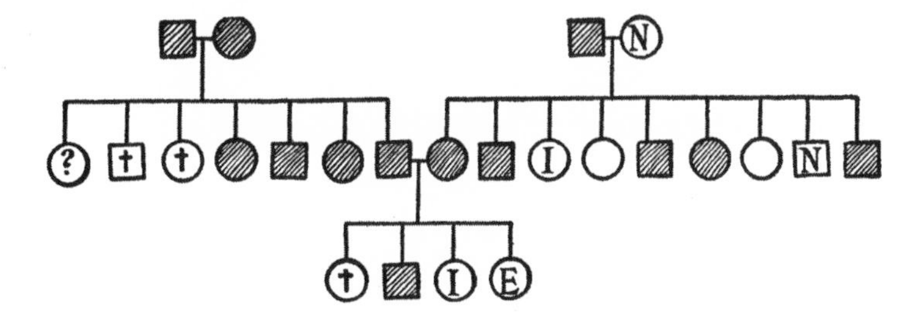

Fig. 47.—The central mating is that of a pair who, though not insane, have pronounced neuropathic manifestations. The mother has an insane sister and the father comes of neuropathic stock. Of the 3 surviving children 1 is neurotic, 1 insane and 1 epileptic. A similar mating of 2 neuropathic persons is seen in the parents of the father—all of their offspring are neuropathic. The shaded symbols represent neuropathic individuals. Cannon and Rosanoff, 1911.

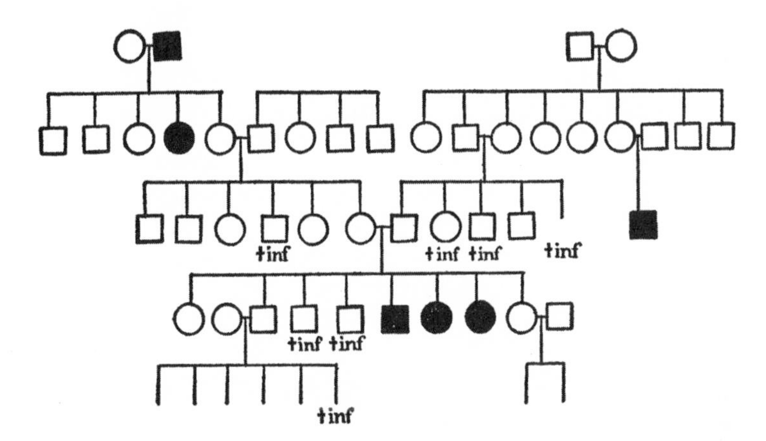

Fig. 47a.—Inheritance of "insanity." From the central mating of 2 normal persons there are derived 8 children, 3 insane. But there is the hereditary tendency in the germ plasm of *both* parents. Mott, 1905.

it is fairly open to debate whether alcoholics are not usually mentally defective and the delirium tremens that develops is a symptom of their mental weakness. Similarly a blow is often just the stress that reveals the mental weakness;

the syphilitic poison in some, if not most cases, likewise acts most disastrously on the neuropathic constitution. Thus, probably an hereditary predisposition lies at the basis of most cases of insanity; and this predisposition behaves in heredity like a defect.

21. Pauperism

Pauperism is a result of a complex of causes. On one side it is mainly environmental in origin as, for instance, in the case when a sudden accident, like death of the father, leaves a widow and family of children without means of livelihood, or a prolonged disease of the wage earner exhausts savings. But it is easy to see that in these cases heredity also plays a part; for the effective worker will be able to save enough money to care for his family in case of accident; and the man of strong stock will not suffer from prolonged disease. Barring a few highly exceptional conditions poverty means relative inefficiency and this in turn usually means mental inferiority. This is the conclusion that social workers in many places have reached. Thus from Harrisburg, Pa., come these cases: (a) Mr. and Mrs. R., applicants for relief and living in a slum district, are parents of 14 children of whom 10 are living. These parents are both epileptic and feeble-minded. (b) Mother and father are both feeble-minded. There are 6 children, all of marriageable age, all unfit to earn in any case more than $1.50 per week, and all recipients of public alms. Such cases might be multiplied indefinitely.

In the larger pedigrees of the Jukes and Zero families more definite data as to inheritance of some of the elements of poverty can be gained. Let us take "shiftlessness" as an important element in poverty. Then classifying all persons in these two families as very shiftless, somewhat shiftless, and industrious the following conclusions are reached. When

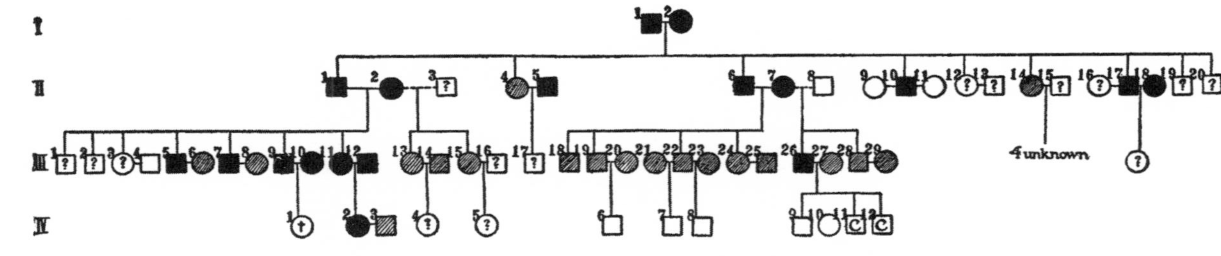

FIG. 48.—A fragment of the Jukes pedigree, being descendants of the elder daughter of "Ada Jukes." Showing occurrence of *shiftlessness* (black symbols) and partial shiftlessness (striated symbols).

I, 1, a lazy mulatto; I, 2, a non-industrious harlot, but temperate; II, 1, a frequent recipient of out door relief, in jail for assault; II, 2, lazy and a harlot; II, 4, twice recipient of out relief; II, 5, laziness, assault, vagrancy; II, 6, vagrant, out relief, jail; II, 7, vagrancy; II, 10, lazy, in poorhouse; II, 14, a harlot recipient of a little out relief; II, 17, lazy, licentious; II, 18, a harlot; II, 19, died young; II, 20, unknown; III, 1, 2, 3, little known; III, 5, 6, received out relief; III, 7, vagrant in jail; III, 8, received a little out relief; III, 9, soldier, pauper; III, 10, harlot, poorhouse inmate; III, 11, harlot, jail, out relief, III, 12, out relief; III, 18, bad boy; III, 19, licentious, pauper; III, 20, harlot; III, 22, licentious; III, 24, harlot, pauper; III, 26, pauper and drunkard; III, 28, basket-maker and pauper who later acquired some property; IV, 2, harlot; IV, 11, 12; criminalistic. DUGDALE, 1902, Chart II.

both parents are *very* shiftless practically all children are "very shiftless" or "somewhat shiftless." Out of 62 offspring, 3 are given as "industrious" or about 5 per cent (Fig. 48). When both parents are shiftless in some degree about 15 per cent of the known offspring are recorded as industrious. When one parent is more or less shiftless while the other is industrious only about 10 per cent of the children are "very shiftless." It is probable that both shiftlessness and lack of physical energy are due to the absence of something which can be got back into the offspring only by mating with industry.

22. Narcotism

The love of alcoholic drink, opium, etc., is commonly regarded as due solely to its use. It has even been asserted that the "taste" is usually an acquired one; and we are assured that drunkenness results from bad associates and imitation of bad habits. Cases are cited of persons who, after an exemplary youth, have suddenly through drink been started on the downward road. On the other hand there are those who maintain that the desire for narcotics is a symptom of a neurasthenic tendency. "So long as there is a call for these narcotics must our race be stamped as degenerate" (Gaupp quoted by Mason, 1910). Says Lydston (1904, p. 200) "Practically, then, inebriety means degeneracy, the subject being usually primarily defective in nervous structure and will-power. It is a noteworthy fact that the family histories of dipsomaniacs are largely tinctured with nerve disorders. Hysteria, epilepsy, migraine and even insanity are found all along the line. In such cases inebriety is but one of the varying manifestations of bad heredity." Each of these contrasted views is partial. Whether a person who has taken a first glass of alcoholic liquor shall take another is determined largely by the effect upon him of the first. If the alcohol is very distasteful he will probably not continue to drink; if it wakens a strong desire for more he will probably become (or is) a dipsomaniac.[1] The result in these extreme cases is determined by innate tastes which are doubtless hereditary. But in most cases the person who takes a first glass finds it indifferent. His subsequent relation to alcohol depends largely upon his associates; but his selection of associates again depends on innate tastes. Some like the steady, quiet, serious youth for their companions; others select the reckless, jolly fellows, careless of the proprieties and—"birds of a feather flock together." The influence of precept is not to be overlooked; this is, however, most important in determining the first drink. No doubt a strong susceptibility to social sentiment restrains many of the border line cases.

A strong hereditary bias toward alcohol runs through not a few families of the United States. A pedigree of one such is given in Fig. 49. The neighbors say: "It is a family of drunkards," yet some of the individuals never touch liquor. The bad environment has its result first and chiefly on those individuals with an hereditary predisposition toward narcotics and this hereditary bias is stronger in some families than others, depending on the nature of the family trait, and it occurs in a larger proportion of the cases in some families than others, depending on the nature of the matings that have occurred in that family.

23. Criminality

In connection with the subject of nervous defect and disease the topic of an hereditary tendency to crime must be

[1] Dr. L. D. Mason, head of the Inebriates' Home for Kings County (N. Y.) tells this story from his experience. He knew of a young man of such ancestry that a dipsomaniac was predicted. For years the youth refrained from drink, and led an exemplary life. Finally, he was operated on for appendicitis and, to hasten recovery, the surgeon gave him some brandy. An uncontrollable appetite was awakened and the man soon died from alcoholism.

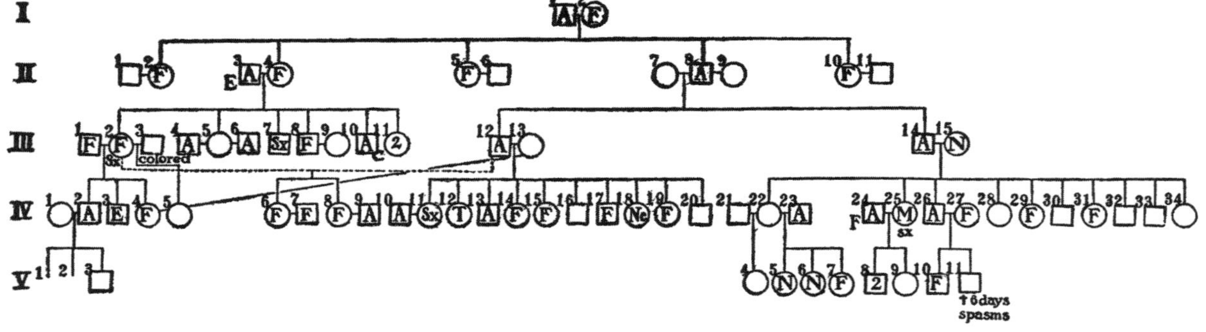

Fig. 49.—Pedigree of a Massachusetts family comprising much feeble-mindedness or imbecility, F; associated with alcoholism, *A*; criminality, *C*; sex immorality, *Sx*; epilepsy, *E*; *M*, migrainous; *Ne*, neurotic; *T*, tubercular. Note the association of alcoholism with imbecility.

I, 1, a basket-maker, alcoholic, married a feeble-minded woman. Of their 5 children (II, 2–10) 4 were feeble-minded and the other, II, 8, shiftless and an alcoholic. II, 4, married an epileptic alcoholic and they had 7 children. The oldest Amanda, III, 2, feeble-minded and sexually immoral, married, first, a feeble-minded man, III, 1, by whom she had 3 children, 1 alcoholic and immoral, 1 epileptic and a cripple, 1 feeble-minded; secondly, by a colored man, III, 3, she had 1 illegitimate colored child, IV, 5; thirdly, by an alcoholic, she had 2 other feeble-minded illegitimate offspring, both of whom married, the first a feeble-minded man, the second an alcoholic consumptive. The second daughter, III, 5, married twice, both alcoholics, but had no offspring; III, 7, was sexually immoral as was also the imbecile son, III, 8. III, 10, was alcoholic and criminal and two daughters, III, 11, were normal. II, 8, a shiftless alcoholic known as "Woodchuck Pete" married twice. By his first wife, a normal woman, he had two sons, both alcoholic. The oldest, III, 12, we have met above as the illegitimate husband of III, 2. He married later her daughter, IV, 4, they had 11 children. The oldest daughter, sexually immoral, IV, 11, married an alcoholic but had no children. IV, 12, was a consumptive; IV, 13, an alcoholic, and of the others, 4 are imbeciles and 1 neurotic, 4 being state wards. III, 14 married a normal woman and they had 10 children. IV, 22, married first a normal but shiftless man whom she left for non-support; her second husband was an alcoholic wanderer, by whom she had 2 normal and 1 feeble-minded child. IV, 25, migraine and immoral, married an alcoholic imbecile by whom she had 3 normal but scrofulous children. IV, 26, an alcoholic, married a feeble-minded woman and had 2 children, 1 feeble-minded, 1 died at 6 days of spasms. Of remaining 7 children, 2 are feeble-minded. F. W., 1.

alluded to. Despite the conservatism of the courts, despite the fact that scientifically ascertained general principles usually weigh less than precedent, the treatment of the criminal has made progress during the past century. It is stated that "Mackintosh speaking in the English House of Commons so late as March 2, 1819 said 'I hold in my hand a list of those offenses which at this moment are capital, in number two hundred and twenty three' " (Johnston, 1887, p. 106). Physical severity, frequent floggings, chaining to the floor, unsanitary surroundings, insufficient and improper food were the elements of a treatment by a society that was exasperated into severity by the realization of its impotent ignorance. Only slowly has the idea of hospitals for insane criminals spread; but though several states maintain great institutions of this sort they still receive a quite insufficient proportion of those convicted of crime.

A few pictures of the youth with hereditary criminal instincts may properly be quoted here.

1. O. L., female, father and mother both intemperate and degenerate, and always on the verge of pauperism. The patient is cruel to animals and children; thus, she put a cat on a red hot stove, threw knives and stones at playmates, wished to have a small baby to strike and kick; and helped drown a comrade in a bath tub. She is very untruthful and a chronic thief; has fits of temper when she screams, tears clothing, and pulls out her hair; is in a state of chronic rebellion against the constituted authorities, a trouble maker and inciter of mischief. She talks fluently, is sly and cunning, vain as to her personal appearance and boastful to attract attention. Age 16. This person has committed the crimes of wanton cruelty to animals, petty larceny, truancy, assault and murder. She is a moral imbecile.

2. O. K., male, entered a school for feeble-minded at 9, at the time of the description is 11. He has a bright, knowing,

intelligent manner, has a fund of general information and is very talkative. He is very cruel to younger children, has an ungovernable temper, is an inciter of discontent and rebellion among the other patients, lies maliciously, ingeniously and convincingly, and steals inveterately and without motive.

This child, removed into an excellent school with the best of surroundings, at the tender age of nine reveals striking criminalistic traits which no care can correct. In this case the hereditary history is unknown. In those that follow it has been precisely ascertained.

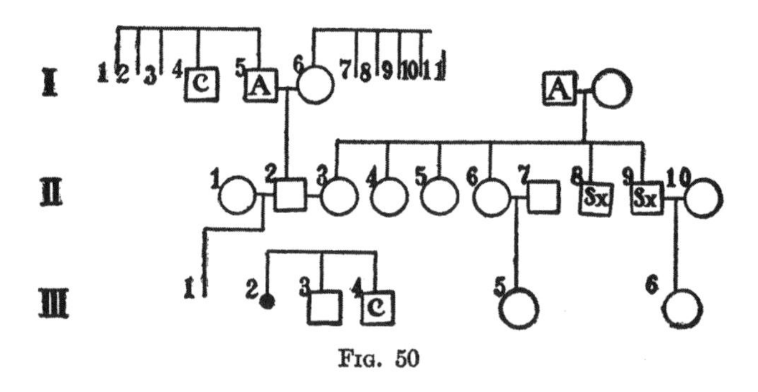

FIG. 50

3. Figure 50, III, 4 is an eleven year old boy who began to steal at 3 years; at 4 set fire to a pantry resulting in an explosion that caused his mother's death; and at 8 set fire to a mattress. He is physically sound, able and well informed, polite, gentlemanly and very smooth, but he is an inveterate thief and has a court record. His older brother, 14, has been full of deviltry, has stolen and set fires but is now settled down and is earning a living. Their father is an unusually fine, thoughtful intelligent man, a grocer, for a time sang on the vaudeville stage; his mother, who died at 32, is said to have been a normal woman of excellent character. There is however a *taint* on both sides. The father's father was wild and drank when young and had a brother who was an inveterate thief. The mother's father was alcoholic and when

drunk mean and vicious. Some of the mother's brothers stole or were sexually immoral.

4. A healthy man (Fig. 51, II,1) employed on a railroad as a fireman and using neither alcohol nor tobacco married a woman who was born in the mountains of West Virginia near the Kentucky line and who shows many symptoms of defectiveness. She has epileptic convulsions as often as 2 or 3 times a week, has an ungovernable temper, smokes, chews and drinks, is illiterate and sexually immoral. There

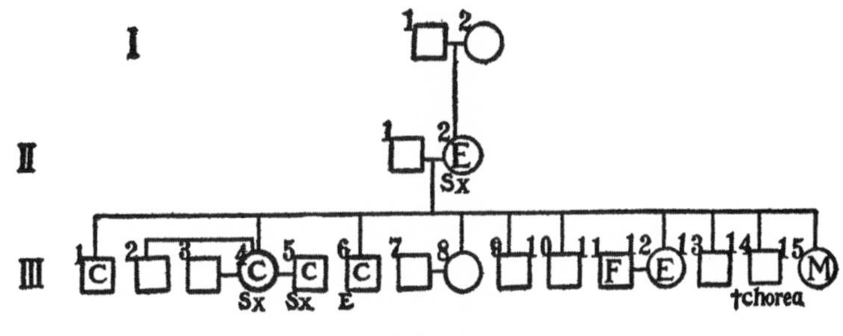

Fig. 51

are 10 children, of whom something is known about 7. One died early of chorea, one of the others (III, 8) seems normal; III, 1 has killed two men including a policeman; III, 4 had her husband killed and lives with his slayer; III, 6, an epileptic and cigarette fiend, convicted of assault; III, 12 has hysterical convulsions and is afraid in sleep; III, 15 has migraine. The combination in the fraternity of migraine, chorea, hysteria, epilepsy and sexual immorality and tendency to assault is striking and appalling.

5. A 10 year old boy (Fig 52, IV, 4) who was precocious as a raconteur at 22 months, does well at school except for inattention; is fond of reading and athletics, cheerful, and polite. But he prefers the companionship of older, wild boys and cannot be weaned from them. He lies, runs up accounts in his parents' name, is acquiring bad sexual habits, and runs

away from home. He has two fine, studious brothers. His father is a strong character and a successful lawyer, his mother an excellent woman, intelligent and firm. She has a brother who left home at 14 to seek a life of adventure. He finally settled down to a steady life. Their father's father was erratic. He loved Indian outdoor life, always used an Indian blanket and at over 70 years swam the Mississippi River. He traced back his ancestry to Pocahontas. He has another grandson, III, 2, who is an unruly character with a

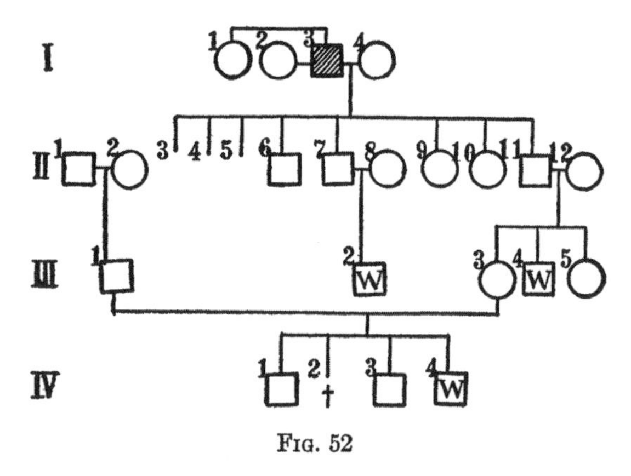

Fig. 52

roving disposition; he joined the navy and his whereabouts are unknown; his father was a lawyer and a fine character.

6. Another case of truancy (Fig. 53, III, 2) is a 7 year old boy whose home conditions are not favorable. His selfish father consorts with lewd women so that his mother has left her husband and now conducts an employment agency. She has hysterical attacks with blank periods during which she may wander. The boy is bright and able but is subject to hysterical attacks; he runs away from school and home and says he does not know why; goes for a long period without food or sleep. His father's father was erratic, a soldier, very superstitious, used to walk in a graveyard and perform incantations at Christmas time. The mother's father was also

erratic and disappeared from home about the time his mother was born. Two of his sons have hysterical fuges and one of them served a term in prison; he is now quite lost to the family. This is a remarkable history of hysteria with a slight criminalistic tendency.

7. An intelligent and esteemed physician (Fig. 54, II, 2) with training abroad as well as in this country and of a good family (his brother, II, 1, is a college professor and his father a methodist preacher) married a lady (II, 3) of good family,

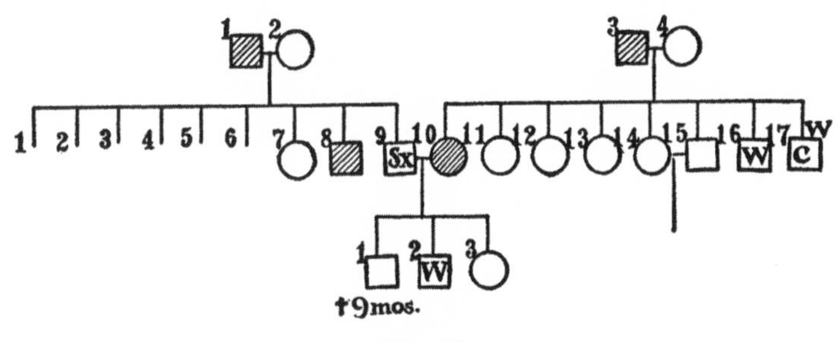

Fig. 53

with much musical talent, but subject to migraine and formerly to chorea. They have two sons born in the best of environments. The younger (III, 3) is still in the kindergarten, seems wholly normal, truth-telling and lovable; the other, (III, 2) now 13, developed normally, has had no convulsions, and has never been seriously sick and ordinarily sleeps well. He has regular, refined features and a normal alert attitude and is very industrious. He attends sunday school regularly, has excellent talent for music. At 3 years of age he walked to a near by railroad, boarded a train and was carried 12 miles before the conductor discovered him; since then he has run away very many times. From an institution for difficult boys, where he was placed, he ran away 13 times. He escapes from his home after dark and sleeps in neighboring doorways. His mother used to make Saturday a treat day. She would take a violin lesson with him and spend the afternoon in the Public Library which he much enjoyed but he would slip away from her on the way home and be gone till midnight. He is an unconscionable liar. He contracts debts, steals when he has no use for the articles stolen and has been convicted for burglary. Much money and effort have been spent on him in vain. His mother's father, (I, 3) (of whom he has never heard) was a western desperado, drank hard and was involved in a murder, but finally married a very good

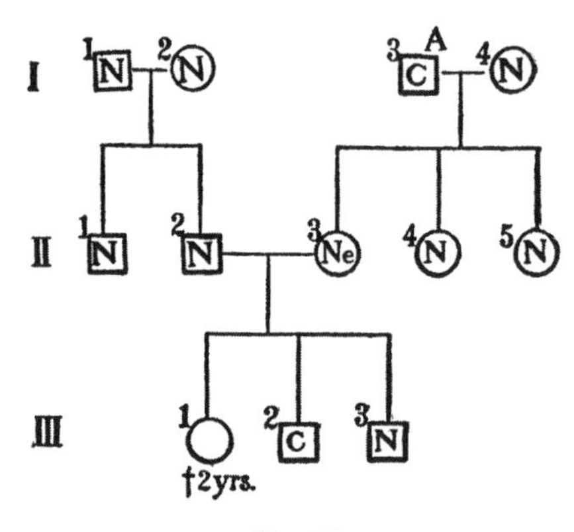

FIG. 54

woman (I, 4) and has 2 normal daughters in addition to this boy's mother.

The typical skipping of a generation, seen in these pedigrees of the wandering instinct, suggests that it is a recessive, like most neuroses—and strengthens the probability that it is due to a real mental defect.

The following case suggests the inheritance of an extremely erotic instinct also as a defect (Fig. 55).

A large, healthy man (II, 4) engaged in an engineering profession, has much ability in music and is an inventor. He drinks very little alcohol, has always been a good worker and is highly esteemed by those who employ him. But he is "crazy about women." He left his first wife and married another, was convicted of bigamy and served a term in prison; later he married a third wife without undergoing the formality of a divorce from the others and was again imprisoned for bigamy. He has had also other, even looser, relations with women. His second wife (II, 5) was a healthy young girl who comes from a long lived family. Since her husband deserted her she has had to work very hard to support their children and is much broken down in consequence. She is

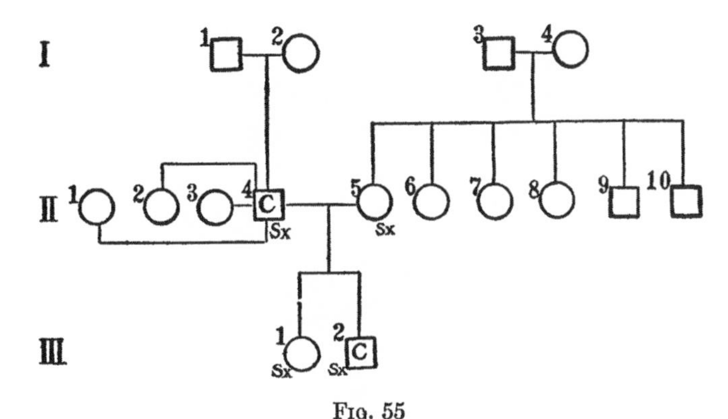

FIG. 55

not a strong character, she keeps boarders and is currently believed to be sexually immoral. Nothing is known about her parents nor those of her husband. The daughter of this pair (III, 1), is thirteen years old. She is wilful, refuses to study, runs on the streets, has stayed out all night on two occasions and has been in court as a delinquent. The son, (III, 2), eight and a half years old, has a fair physical development, but his face is unsymmetrical and his mouth open despite removal of adenoids when he was 5. His speech is thick and rough. He seems dull at times but can brighten up. He has had convulsions. Like his sister he is wilful, won't learn, and runs on the streets where he sells papers and where he has stolen many articles. He throws stones and garbage and despite his tender years he indulges in vile language, exposes his person to little girls, masturbates and is sexually misused by men. All attempts at reformation have failed,—orphan asylum, home for boys, life on a farm; from all these he runs away and returns to the life he loves.

The foregoing cases are samples of scores that have been collected and serve as fair representations of the kind of blood that goes to the making of thousands of criminals in this country. It is just as sensible to imprison a person for feeble-mindedness or insanity as it is to imprison criminals belonging to such strains. The question whether a given person is a case for the penitentiary or the hospital is not primarily a legal question but one for a physician with the aid of a student of heredity and family histories.

24. Other Nervous Diseases

a. The General Problem.—The marvellous complex of neurones (nerve cells and fibres), sustentative tissue, and blood vessels that constitute the central nervous system forms, perhaps, the most wonderful mechanism in nature. Little wonder that it should vary greatly in different individuals, or that it should become easily deranged. Such variations in structure and such derangement though ordinarily hidden from view can be inferred from the behavior of the person. For the general principle holds that every psychosis (or peculiar mental manifestation) has its neurosis (or aberrent nervous basis). Peculiar or abnormal behavior, then, is an index of peculiar or abnormal brain condition.

That heredity plays a part in nervous disease is indicated by the familiar fact of high incidence of some or other psychic disturbance in the members of a single family. We have already seen how incomplete mental development is a consequence of the absence of a definite inheritable defect in the germ plasm, such that when the factor that stimulates to complete mental development is absent from the germ plasm of both parents it will be absent from all their offspring. Varied as are the mental conditions of the persons in a family containing feeble mindedness the children do not ordinarily surpass in mental development the better developed parent.

In considering heredity of mental disease we must not forget that what is inherited is not, as in imbecility, a tendency to incomplete mental development, but rather a tendency such that a completely developed and apparently normal mentality is liable under ordinary, or still more under extraordinary, conditions to show disturbance of a temporary or permanent nature. The more intimate nature of this inherited tendency is probably varied. In some cases there is doubtless an idiosyncrasy in the neurones, in other cases there is a lack of resistance to infection or specific poisons, again the trouble may be outside the neurones in the supporting tissue or even in the blood vessels whose walls may be peculiarly liable to weaken and burst; to waste away; to thicken, occluding the lumen and shutting off nutrition to a part of the brain.

Before considering the inheritance of specific nervous diseases it may be pointed out that what is inherited is often a general nervous weakness—a neuropathic taint—showing itself now in one form of psychosis and now in another. Especially the lower types of mental defect may be carried in the higher, *i. e.*, departing least from the normal.

b. The Neuropathic Makeup.—We have seen (page 77) that imbecility, epilepsy and many forms of insanity are due merely to the absence of some factor. It remains to be considered how they behave amongst each other in heredity. A pedigree worked out by Barr (1907) gives the desired information (Fig. 56).

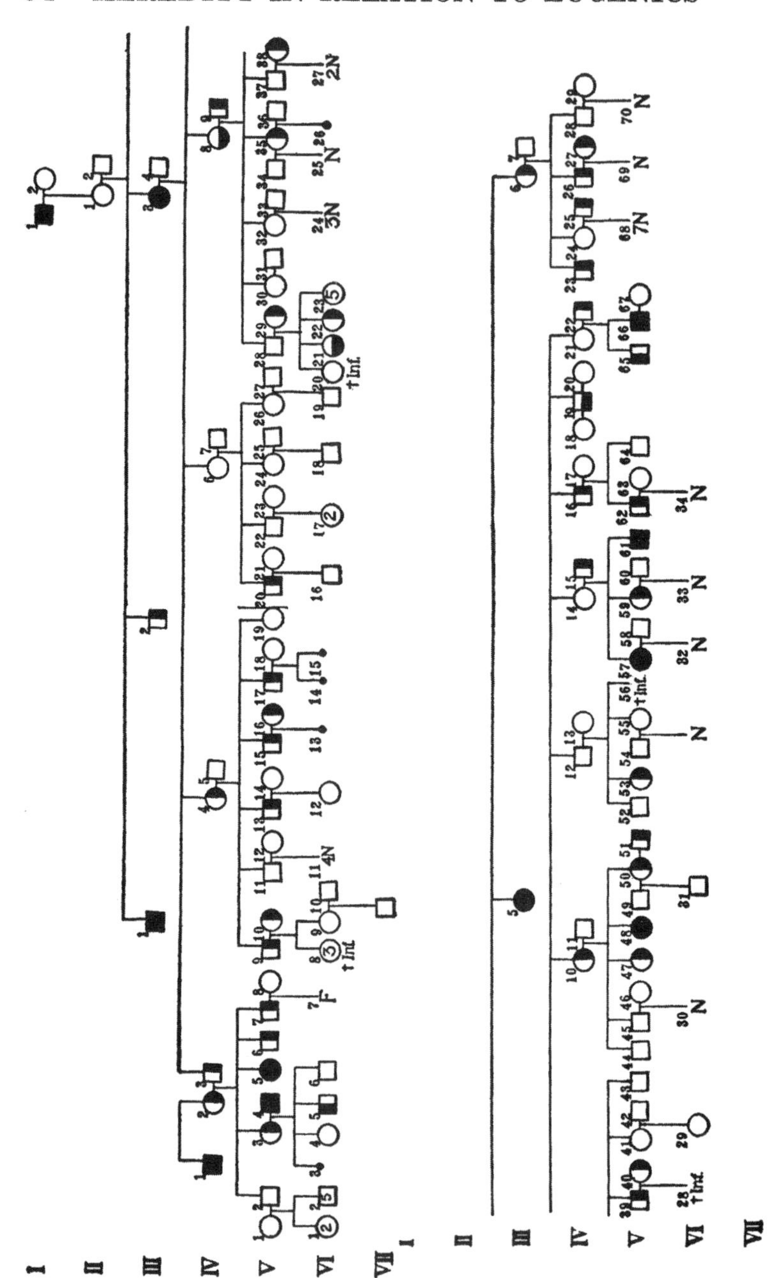

Fig. 56.—Pedigree of a family of neuropathic makeup.—I, 1, insane; I, 2, normal; II, 1, 2, normal but evidently simplex; III, 1, insane; III, 2, neurotic; III, 3, insane; III, 4, normal; III, 5, insane; III, 6, neurotic.

Gen. IV–VI. Full black symbols indicate insanity; symbols blacked on the right half represent neurotic persons; symbols blacked on the lower half indicate epileptics; symbols blacked at the left indicate feeble-minded.

III, 3, herself insane, married into good stock, had 10 children, 4 normal, 4 neurotic, and 2 epileptic. Oldest son George, IV, 3, neurotic and unprincipled, married a neurotic woman whose brother is insane; of 5 children, one, V, 5, is insane; 3 others neurotic and 1 normal. One of these neurotic daughters married a man who died insane whose father was semi-insane and whose brother was an imbecile. One child, of 3, is an imbecile and has a feeble-minded cousin. IV, 4, Rebecca, herself neurotic, married a man of good stock, had 4 neurotic and 2 normal children. One of these neurotic sons, V, 9, married a woman of loose character, a harlot, sprung from a neurotic family, and they had 3 children who died young and 1 normal. Another neurotic son, V, 15, married a harlot, also neurotic, and had one still-born child only. Other two neurotic sons married women of good stock.

Mary, IV, 6, herself normal, exceptionally beautiful and accomplished, married into a good stock. Cf her four children, only 1 was neurotic but all married into good stock, and their children were all normal. Her sister, Olivia, IV, 8, an epileptic, married neurotic stock and 2 out of 8 children only were neurotic. The oldest son, V, 28, himself normal, but peculiar, married a neurotic woman of loose character. Of 8 children, 1 was epileptic (seduced by a negro), 1 neurotic, 1 died in infancy and 5 were normal. IV, 10, neurotic, married a man of good stock, had puerperal insanity at birth of first child. Of 5 children, 2 were normal, 2 neurotic and 1 insane. IV, 12, normal, married a woman of good stock and had 9 children, 4 boys normal, 2 girls neurotic, 1 child died in infancy. IV, 14, Agnes, herself normal, married a man of neurotic stock who had himself had mania several times. Of 3 children, 2 were insane, 1 neurotic. IV, 19, an epileptic, married twice but had no issue. IV, 21, herself normal, married a hypochondriac (son of third sister's husband by first wife), and had 1 child who was imbecile and 1 who was insane.

III, 6, Jane, is neurotic and married a man of good stock. Of their children 2 are normal and 2 are neurotic with a pronounced dipsomania. Their grandchildren are all normal.

Note. On account of the length of the chart it has been cut in two in the middle and the right-hand portion has been placed on the page, below the left-hand portion.

This pedigree contains 22 significant matings (*i. e.*, that yield more than one child). The products of these matings are summarized in Table IX.

TABLE IX

PRODUCT OF VARIOUS MATINGS IN BARR'S PEDIGREE

	Mating	Offspring						
	Nos.	N	Ne	F	E	I	X	Still-births
Neurotic×neurotic	4	1	3			1		
Neurotic×neurotic	16	1					3	
Neurotic×epileptic	7	4	2			2		
Neurotic×insane	15	2		1				1
Insane×normal	2	4	4		2			
	10		1			2		
Neurotic×normal	3	2	2					
	5	2	4					
	8	2	2			1		
	11	1	1					
	12			1		1		
	13	7						
	22	2						
Neurotic×unknown	18							2
	20	1	1		1			
Normal×normal	1		2			3		
	6	3	1				1	
	9	2	1					
	14, 17, 19, 21	16						

E, epileptic; F, feeble-minded; I, insane; N, normal; Ne, neurotic; X, unknown.

In Table IX there is no marriage of two insane persons. Where a nervous person marries a neuropath, of 11 known offspring 6 are normal and 5 neuropathic; when two neurotic marry, 2 out of 6 children are normal and 1 insane; when an insane and a normal marry, of 13 children 4 are normal and 2 insane; when a neurotic and a normal marry, of 28 children 16 are normal, 9 nervous, 1 feeble-minded and 2 insane. Even some normal parents (of this strain) have insane or epileptic children. One sees what a variety of gametic conditions may be carried by a "nervous" or even a "normal" person, just as blue eyes may be carried by brown eyed parents, or light brown hair by dark haired parents. A "nervous" person is thus frequently simplex in the factor that makes for mental strength and is apt to carry defective germ cells (Figs. 57–59).

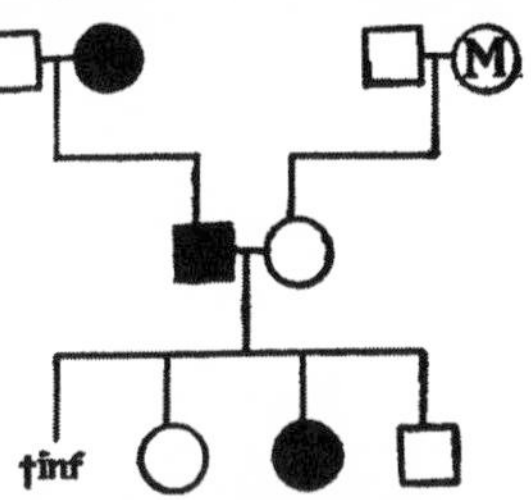

Fig. 57.—Pedigree of "nervous trouble." I, 2, was typically affected and I, 4 suffered from migraine. II, 1, had the same nervous trouble. Of three grandchildren who survive, 1 already shows at 6 years, a tendency toward nervous weakness. F. R; Hug. 1.

c. Cerebral Hemorrhage.—However numerous the causes that weaken the walls of the cerebral arteries or raise abnormally the pressure upon them, there can be little doubt that hereditary predisposition plays an important part. (Figs. 60 and 61). Cerebral hemorrhage is commonly found in the parentage or grandparentage of the mentally

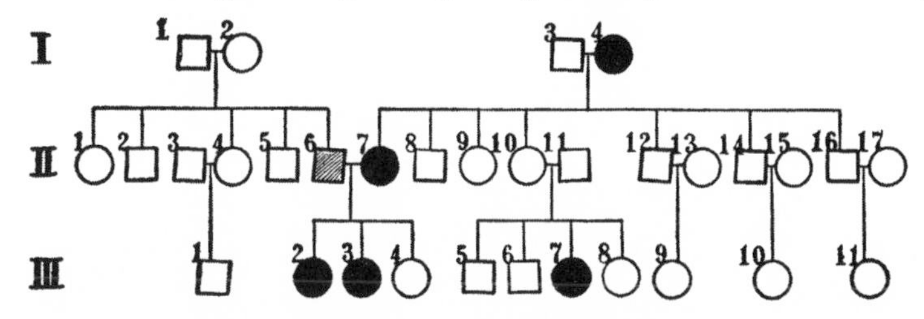

Fig. 58.—Inheritance of nervousness and brilliancy. I, 4, is subject to headaches and nervousness. Her daughter, II, 7, is similarly affected. She married a man, II, 6, who has had temporary attacks of paralysis. One of their children, II, 2, has nervous prostration and one, III, 3, is subject to headaches and nervousness. F. R.; Cla. 3.

weak as well as brilliant. (Fig. 61). See also arteriosclerosis, page 162.

d. Cerebral Palsy of Infancy.—This disease, of obscure origin, affects infants within a few years of birth; it leads

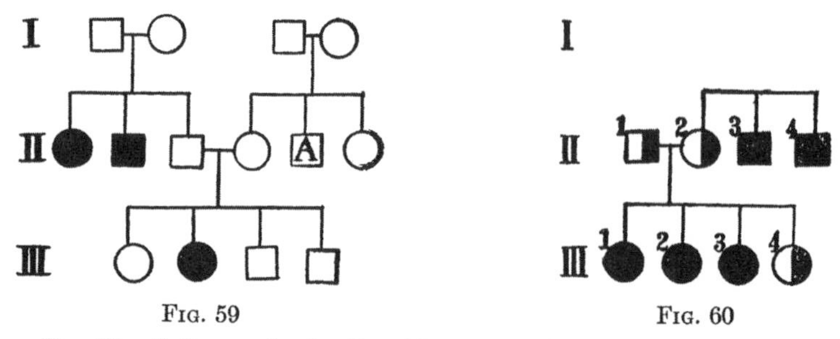

FIG. 59 FIG. 60

FIG. 59.—Pedigree of a family with nervous disease. I, 3, was a heavy drinker; I, 4, died of apoplexy after suffering from paresis. The father was normal, but he had a brother, II, 1, who was eccentric and committed suicide, and a sister, II, 2, who was a good linguist but deteriorated mentally. The mother, II, 4, is normal but she had a brother who while a civil engineer and excellent draftsman was alcoholic, and a sister who was a good musician. One child, III, 2, is suffering at 23 from dementia precox. F. R.; Coi. 1.

FIG. 60.—Pedigree of a family with high incidence of cerebral apoplexy. The father and mother, I, 1 and 2, both have apparently a tendency toward cerebral congestion. I, 2, had recently had an attack which was relieved by nasal hemorrhage. Two of the mother's brothers, I, 3 and 4, died after a brief attack of apoplexy. Three of the daughters have died of the same disease at 32, 30 and 46 years respectively; the remaining suffers from cerebral congestion. HARRINGTON, 1885.

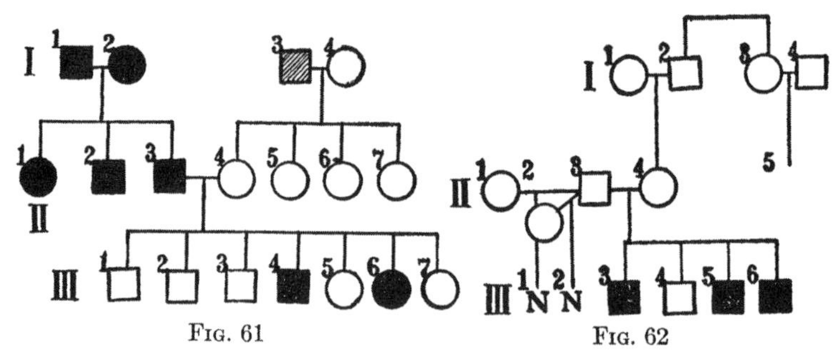

FIG. 61 FIG. 62

FIG. 61.—Pedigree of a family with "nerve weakness." The father's father, I, 1, had a "nervous weakness," his wife died at 28 of encephalitis, the mother's father, I, 3, was subject to apoplexy and died of a stroke at 71. The father, II, 3, and all of his fraternity had encephalitis—the father three times—and one died of it, while the others were left with a nervous weakness. The children were not vigorous. III, 1, had always a low vitality and died at 8 years; III, 3, had a low vitality and died at 14 of "congestion of the lungs"; III, 4, was feeble-minded; III, 5, a laborer, suffered much from "bowel trouble"; III, 6, has a nervous weakness; and III, 7, engaged in housework and, with III, 2, is the strongest of the family.

FIG. 62.—Pedigree of a family with cerebral diplegia. The father in the central mating, II, 3, has been three times married. By two of the marriages

to general paralysis of one or both sides and, in later development, is associated with feeble-mindedness. Pedigrees are given by Dercum (1897) Fig. 62, Pelizaeus (1885) Fig. 63, Freud (1893) and others.

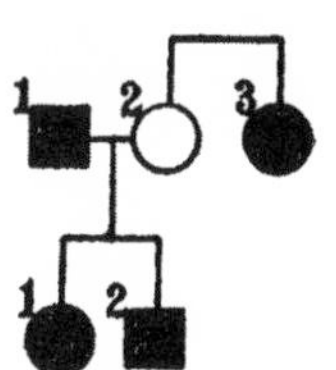

FIG. 63.—Illustrates the pedigree of a man that now has cerebral diplegia who married a woman who had a sister similarly affected. Both children are affected. PELIZAEUS, 1885.

Since the tendency is carried by normal persons and since (as in Freud's case) it is apt to occur with consanguineous marriage it is probably due to a specific defect. To avoid the reproduction of the disease, marriage with unrelated blood is essential.

e. Multiple or Disseminated Sclerosis.—This is a diffuse degenerative disease of the spinal cord. It leads to tremors in the arms and trunk, disturbance of speech and eventual paralysis. It is usually not regarded as hereditary but an interesting pedigree showing its appearance in 3 generations has been investigated by Merzbacher (1909), Fig. 64.

As the pedigree table shows, the disease is transmitted through unaffected females. The eugenic conclusion is, consequently, that even unaffected females who have affected brothers should not have children.

f. Hereditary Ataxy (Friedrich's disease).—This disease causes a slowly but surely progressive loss of directed movements, first of the legs and then of the arms; speech becomes elusive and indistinct; scoliosis (curvature of the spine) may appear and the feet become drawn up. These symptoms accompany a degeneration in the upper part of the spinal cord.

he had only normal children, but by the third (to a normal woman who had a first cousin, II, 5, with cerebral diplegia) he had 4 sons of whom 3 were affected with this disease. The eldest, III, 3, was normal until 16 months old, then had general convulsions, after which spastic symptoms gradually appeared, becoming pronounced later. Now he can walk only a few steps and is quite idiotic. The third son was normal until 2 years old, but is now deteriorating after an attack of measles and the youngest, only 2 years old, has just become diplegic and epileptic. DERCUM, 1897.

Some extensive pedigrees of ataxy have been published. One of the most extensive is by Mott (1905). It is reproduced in Fig. 65.

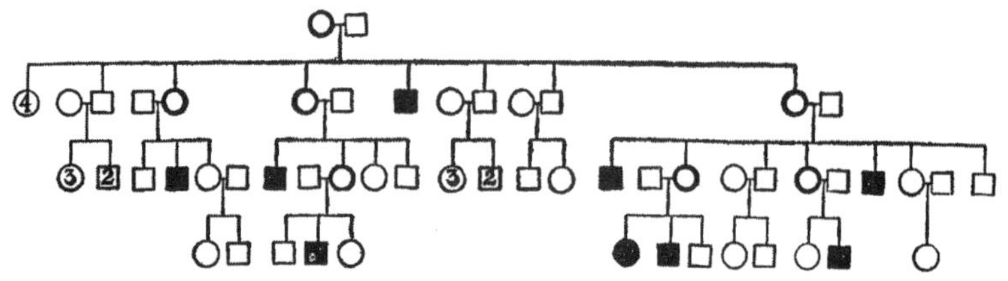

FIG. 64.—Part of EICHOLD-FLEMING-STOSSEL-HERZER pedigree showing multiple sclerosis (black symbols). One notes the skipping of a generation (indicating a recessive trait). The trouble is usually carried by unaffected females (heavy circles) and appears in their sons. Interesting because same family was independently noted by two neurologists. PELIZAEUS, 1885; MERZBACHER, 1909.

Since, as the pedigrees show, normals may have affected offspring the disease is probably dependent, as in insanity, on the lack of something necessary for normal development. The disease seems to be in no way sex-limited (Fig. 65).

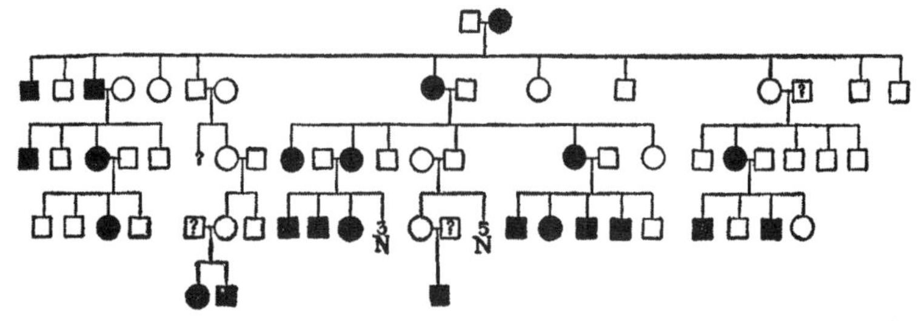

FIG. 65.—Pedigree of a family with hereditary ataxy (black symbols). Consorts not in direct line mostly unknown. Note that affected persons have (for the most part) one affected parent; the trouble is due to the presence of some positive character. MOTT, 1905.

The eugenic teaching is that affected persons and also normals of the affected fraternities should marry only outside the strain. Whether *all* cases of atactic offspring of one normal parent are derived from consanguineous marriage is still uncertain and warrants hesitation in advising the marriage of any atactic person.

g. Ménière's Disease is apparently due to a disturbance in the auditory nerve or its centre. It is accompanied by dizziness and roaring in the ear, often so severe as to force the patient to fall to the ground. Simon (1903) describes a family with these symptoms, consisting of an affected father, son and two daughters. The onset of the attacks varied from the 25th to the 50th year.

h. Chorea (St. Vitus's dance) is a disease of the cere-

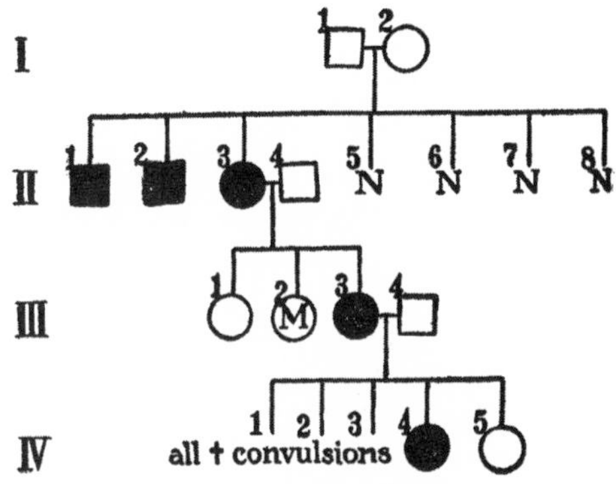

FIG. 66.—Pedigree of chorea (black symbols). II, 1, became affected with chorea at 8 years before his death; II, 2, has suffered many years; 4 other brothers and sisters are healthy. II, 3, became sick at 35 and suffered until her death at 46; she also had a marked loss of memory and died in a hospital. III, 1, is healthy; III, 2, suffers from severe sick headaches. III, 3, has chorea. IV, 4, is 11 years old and has been afflicted with chorea and epileptic fits for past 2 years. Her sister is still healthy at 10 years. JOLLY, 1891.

bral hemispheres characterized by involuntary, irregular movements of the limbs or other parts of the body. It commonly occurs in families with neuropathic make-up. Ordinarily the disease appears in the children and ends in recovery; occasionally it appears only later in life and runs various courses, sometimes ending in death through exhaustion. This disease is commonly sharply separated from Huntington's chorea, but transitional conditions occur. A case cited by Jolly is shown in Fig. 66. In this case nothing is known about the first generation; the second comprises 4 normals and 3 affected persons, 2 males and 1 female. II, 1 became affected with chorea "8 years before his death"; II, 2 "has been affected for many years"; II, 3 became ill with chorea at 35 and suffered until her death at 46. These look like cases of Huntington's chorea. III, 2 suffers from migraine; III, 3 has chorea, IV, 1–3 died at birth of convulsions; IV, 4 at 9 years began to show choreiform movements. These have continued for two years until the present time. This girl also has epilepsy; but her chorea has appeared at the age for St. Vitus's dance.

i. Huntington's Chorea.—This is said to be a "rare" disease in Europe, but not so in the United States. It is characterized by appearing typically first in middle life and progressing with ever increasing disorder of movements until dementia and death occur. It affects both sexes about equally. Two pedigrees are given in Figures 67 and 68.

The method of the inheritance of this disease was recognized by its original describer, Dr. George Huntington. He states that those exempt from it cannot transmit it. An examination of the extensive pedigrees shows only one exception to his rule and this a doubtful case. Huntington's chorea is, consequently, a typical dominant trait, the normal condition is recessive; or, the disease is due to some positive factor. The eugenic lesson is that persons with this dire disease *should not have children.* But the members of normal branches derived from the affected strain are immune from the disease.

This disease forms a most striking illustration of the principle that many of the rarer diseases of this country can be traced back to a few foci, possibly even to a single focus; certainly in this case many of the older families with Huntington's chorea trace back to the New Haven Colony and its dependencies and subsequent offshoots. The subject of foci of origin of traits will be discussed more fully later (page 181)

j. Hysteria.—This term is applied to a variety of symptoms that indicate a functional disturbance of the psychic centres usually combined with a derangement of the lower

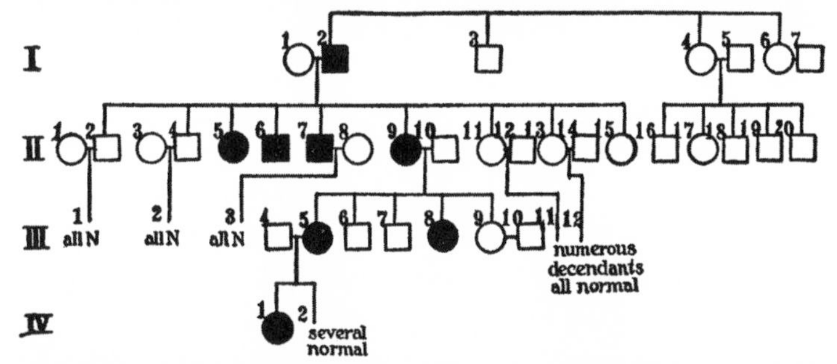

Fig. 67.—Pedigree of a family showing Huntington's chorea. Affected persons (indicated by black symbols) are always derived from affected parents. From original data furnished by Dr. S. E. Jelliffe; Smi-family.

cerebral or spinal centres. The psychical symptoms approach mania on the one hand and show a more or less complete loss of the moral sense on the other, so that many

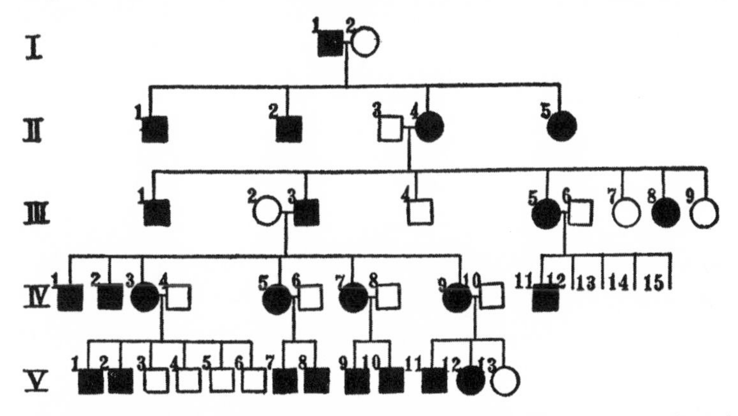

Fig. 68.—Pedigree of a family with Huntington's chorea. All affected persons (black symbols) have at least one affected parent. Hamilton, 1908, p. 453.

cases of larceny, assault, and sexual immorality are consequent upon this disease. The emotions usually are disturbed. The motor symptoms are frequently profound.

Thus paralysis, or spasmodic contractions, or even convulsions not unlike, if not identical with, those of epilepsy, make their appearance.

The greatest social importance of hysteria lies in its relation to crime and responsibility. A large proportion of "criminals" doubtless are in need of hospital care. The family history of the offender will give the best possible clue to his probable mental condition and, where a "neuropathic blood" is evident, the patient should be segregated, not to punish him but to care for him at the expense of that "society" which still permits his kind to breed unrestricted; and to prevent, or at least to limit, the further spread of his tainted germ plasm.

In studies made on 175 families containing epileptics which the author has had the privilege of making with the coöperation of Dr. David F. Weeks hysteria was frequently found associated with chorea, migraine and a "neurotic" condition in the parentage of epileptics and in the offspring of an epileptic or insane parent married to a normal. It acts like a condition induced by a simplex determiner such that the patient produces some defective germ cells.

25. Rheumatism

Rheumatism, as is well known, is often associated with chorea. An example of such association is given in Figure 69.

A second instructive case is that cited by Cheadle (1900). A man who had subacute arthritis and muscular rheumatism and whose sister died at 8 years of heart disease following acute rheumatism and chorea married a woman who had suffered from acute rheumatism, heart disease and chorea and had had a nephew affected with rheumatic fever and heart disease and a niece with subacute rheumatism. The child of this pair at 9 years of age had chorea in a most severe form, repeated attacks of inflammation of the heart and pains in joints with formation of nodules beneath the skin. Finally the girl died a victim to extreme, uncontrollable rheumatism and chorea.

The exact laws of inheritance in these cases are not clear and eugenic instruction cannot be drawn from them.

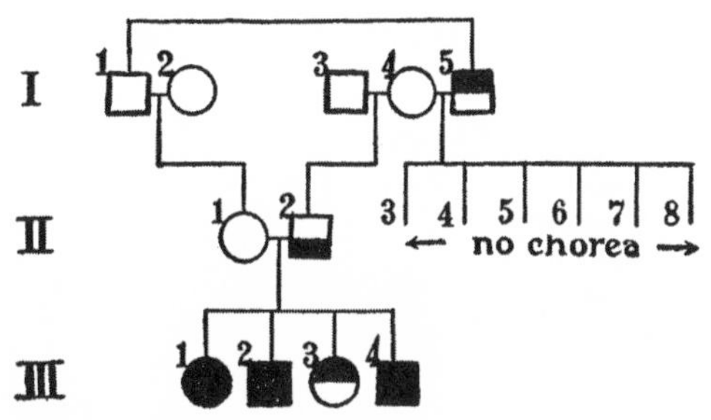

Fig. 69.—Pedigree of family showing chorea and rheumatism. I, choreic at 15 years; still has slight twitchings; II, 2, is not choreic but is subject to migraine and has had several attacks of rheumatism. He has had 2 daughters and 2 sons. III, 1, is 18 years old and since her eighth year has had chronic and severe chorea; at 12 she had an attack of rheumatism and since then attacks of rheumatism and chorea have alternated. Her elder brother, 16 years of age, was attacked a year before by chorea which lasted 2 months; recently has had another attack preceded by rheumatic pains. The third child, III, 3, now 13 years old, has had no rhematism but was first attacked by chorea at 12 and has had other attacks since. The youngest, III, 4, now 11, had a first attack of chorea at 8 years, lasting 2 months; a second attack at 10 and a third recently; in his eighth year he had articular rheumatism. Apert, 1907, p. 235.

26. Speech-defects

While the minor speech defects of stammering, stuttering, lolling, lisping and poltering correspond to no yet recognized abnormality of the central nervous system or organs of articulation, nevertheless, aside from imitation, they clearly have an hereditary basis and while the slighter grades may be cured by practice the more profound disturbances remain a permanent affliction. Especially are these defects found in children of a neuropathic inheritance and, in such, yield the strongest evidence of inheritance.

The exact method of inheritance of stuttering will not

become known until more extensive pedigrees of stuttering families have been obtained. Two pedigrees have been obtained (Figs. 70, 71).

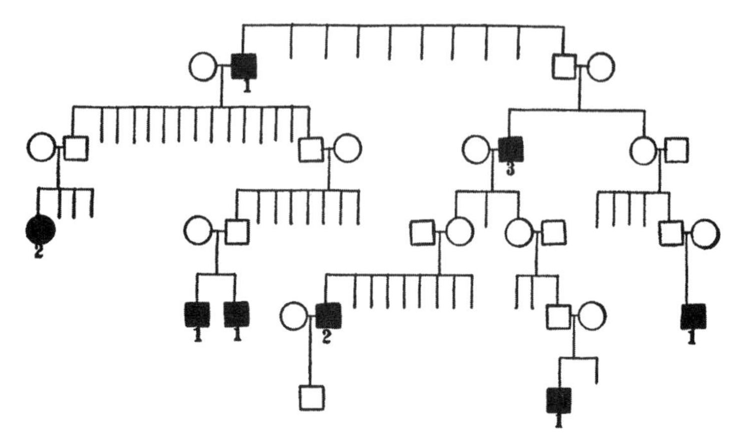

Fig. 70.—Pedigree of a family that contains stutterers (black symbols); 1, stutterer; 2, impediment in speech; 3, impediment, if excited. F. R.; Bar. 4.

Stuttering is seen to affect both sexes. It can hardly be a dominant trait because it is found so often in children of unaffected parents. It *might* be due to the absence of

Fig. 71.—Pedigree of a part of a family of stutterers (black symbols).
Fig. 72.—Pedigree fragment of poltering family. Affected individuals in black. Berkan.

some factor if consanguineous marriages were common in these pedigrees.

The trick of repeating short words and syllables is sometimes called poltering. A case of it occurs in three generations and is given by Berkan (Fig. 72). The peculiarity is found in each of three generations; it may of course be assisted by imitation.

Lolling is speech in which the articulatory mechanism is not used with precision, as in young children. There is some evidence that this defect may be a family one. Thus Moyer (1893) records a family in the first generation of which there were a normal sister and three brothers; one who was quite normal in speech, one who did not learn to speak until 6 years old, and one who lolled his life long. The latter had 6 children, all normal save one who lolled. The other affected brother had 12 children of whom, however, 5 died in infancy, leaving 7. Of his four daughters one had defective utterance, while all three boys were defective in speech, although after puberty the defect gradually disappeared. One of these boys has 3 sons, all normal. The case illustrates segregation but hardly suffices to demonstrate the law of inheritance of the peculiarity.

27. Defects of the Eye

Apart from albinism, the effects of which are most strongly felt in the increased sensitiveness of the retina to strong light, the chief optical defects whose inheritance has been studied are as follows; (a) absence of or defect in the iris and displacement of the pupil; (b) reduction in size of the whole eyeball to complete absence; (c) atrophy of optic nerve; (d) cataract; (e) dislocation of the lens; (f) degeneracy of the cornea; (g) glaucoma or excessive production of fluids of the eye; (h) megalophthalmus, or big eye; (i) nystagmus or "swimming eye;" (k) paralysis or imperfect development of muscles of the eye and lids; (1) pigmentary degeneration of the retina (retinitis pigmentosa); (m) night blindness (hemeralopia); (n) color blindness; (o) astigmatism; (p) myopia.

a. Anomalies of Iris.—Coloboma is a defect in the development of the optic cup such that it fails to close completely and leaves an open suture running from the pupil to the optic nerve. The commonest external evidence of

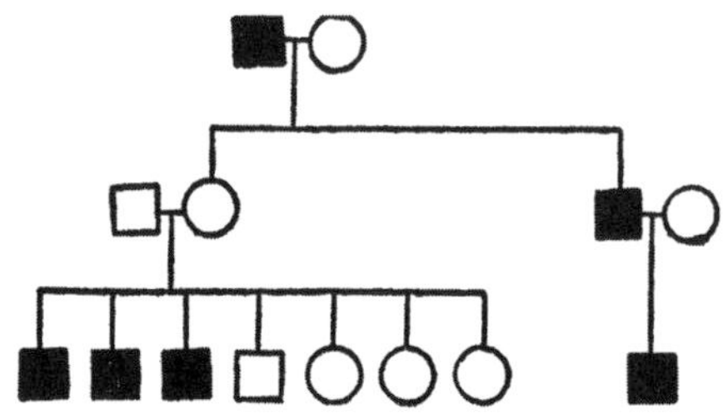

FIG. 73.—A pedigree of a family affected with coloboma. Black symbols stand for affected persons; all are males. A normal female in the second generation transmits the defect to about half of her children, but her sons alone show the defect. STREETFIELD, 1858.

this defect is the incomplete iris; but the lens, retina, choroid coat, etc., may be involved. The cause of the defect is conceded to be an hereditary defect in the developmental impulse (Von Hippel, 1909).

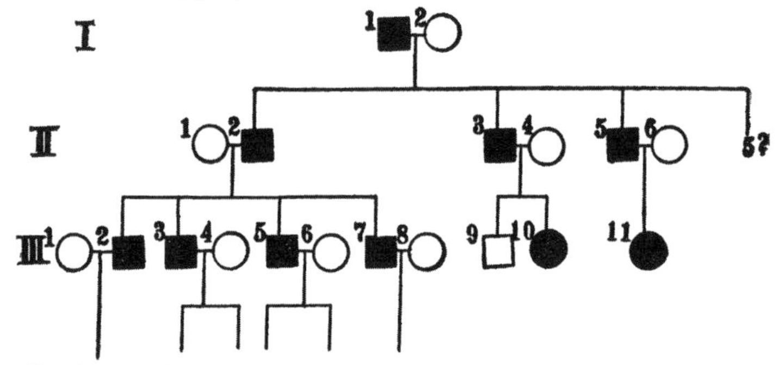

FIG. 74.—Pedigree of a family that shows absence of iridae (black symbols). Here, too, only males show the defect, except for III, 10 and 11. Hypothesis, in this case, requires that II, 4 and II, 6, shall be related to their consorts and carry germ cells with the inhibiting factor. GUTBIER, 1834.

The method of inheritance is shown by the pedigrees (Figs. 73, 74, 75). These lead to the conclusions that the defect is a positive character and is due to an inhibitor of development; the affected male is either simplex or duplex in this inhibitor; the affected female is typically duplex, rarely simplex; unaffected males are always nulliplex, and unaffected females are either nulliplex or simplex.

The eugenic conclusion is: No female with the coloboma defect should have children since *all* sons will be defective in the structure of the pupil. For males with the defect the danger in marriage is also great, for either all or half of the

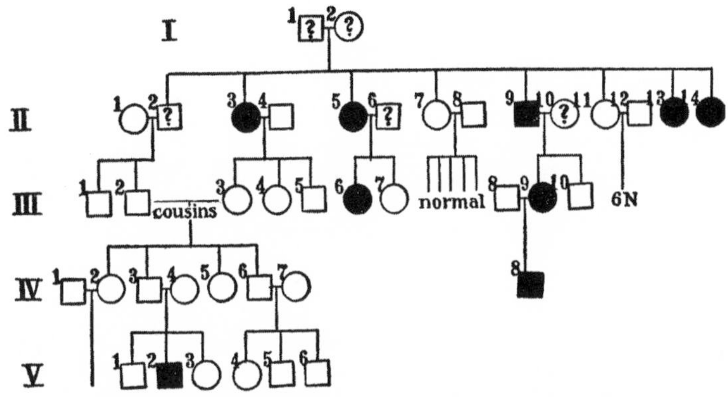

FIG. 75.—This is the pedigree of a family (Payne) with coloboma of the iris. I, 1, and 2 are not definitely known; at least 1 of their sons and 4 daughters are affected. As for the rest, two normal parents have normal offspring. The apparent exception, V, 2, may not be such as the mother, IV, 2, is wholly unknown. The number of affected females in this pedigree is extraordinary. DEBECK, 1886.

sons of such a father, although married to a woman from a normal strain, will be defective, but the daughters will not be defective in this respect unless the wife belongs to a strain with this defect. Two normal persons may marry with impunity except that if the woman belongs to the abnormal strain it may be that half of her sons will be affected.

b. Reduction in size of the Eyeball.—All grades in the size of the eyeball down to complete disappearance are known, but usually only the extremely reduced condition has been studied. Such a condition seems to be due to an inhibitor so that, when present in a marked degree, all offspring shall have it. Both sexes seem to be equally affected.

It is not particularly apt to occur in consanguineous marriages. An illustrative case is given by Martin, 1888 (Fig. 76).

The two sexes are equally affected. A person with the defect in a marked degree will have at least half of the children similarly defective.

It is not, at the moment, possible to say that, when both parents are unaffected the children will all be normal, but there is a strong presumption that such will be the case.

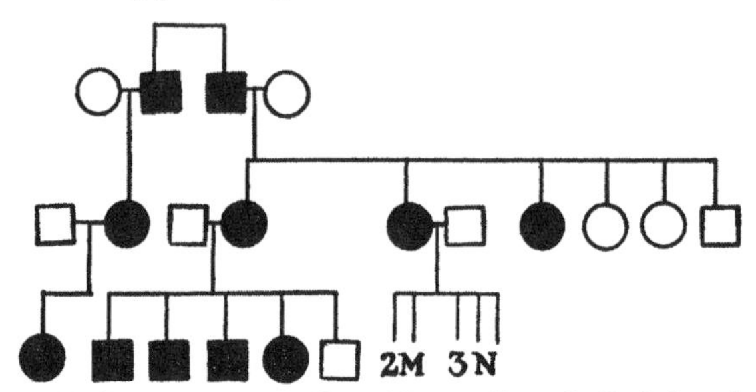

FIG. 76.—Pedigree of a family with small eyeball (microphthalmus). Every affected person (black symbol) that has married has affected offspring. Actually, there are 11 affected progeny to 7 normal; but as frequently happens in practitioner's records, some normal children are probably not recorded. MARTIN, 1888.

c. Atrophy of the Optic Nerve.—This disease usually begins "at about the 20th year with a rather sudden disturbance of the central sight of both eyes while the peripheral parts of the field of vision remain normal." "The course of the disease is generally the same in the same family, so that the prognosis depends in the main upon the degree of malignancy which the malady exhibits in that particular family" (Senator-Kaminer, 1904).

The method of inheritance in this case resembles that of coloboma (except that even duplex females rarely exhibit the trait) and is shown in the ideal scheme of Figure 77 in which the heavy ring means without somatic defect but with defective germ cells.

The eugenic rule is: a normal son of an abnormal male may marry quite outside the family with impunity, but a normal daughter may transmit the defect to her sons. But such a woman may marry with impunity if all of her brothers are without defect and there are more than two of them. A defective male should abstain from having children, for some of his sons, at least, will probably be defective.

d. Cataract.—This is an opacity of the lens which may result from abnormal conditions originating in other parts of the eye or body or they may seemingly originate inside the lens itself, in which case their heredity is marked. Prob-

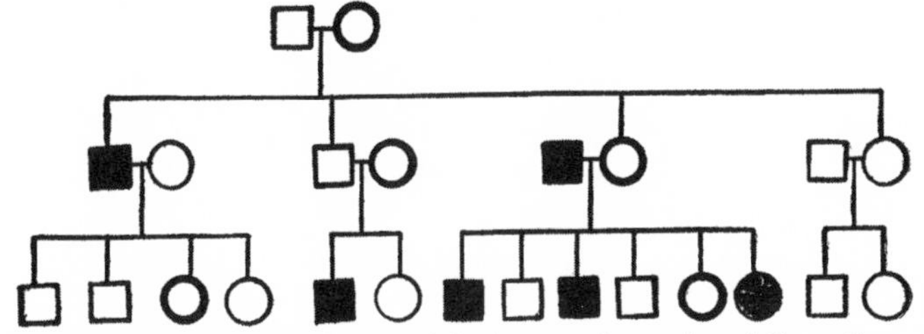

Fig. 77.—Ideal scheme showing inheritance of atrophy of the optic nerve. The solid black squares indicate affected males; the heavy rings represent non-affected females with defective germ cells.

ably more pedigrees of cataract have been published than of any other eye defect. Loeb (1909) refers to 304 families of which accounts have been printed. Of the 1012 children in these pedigrees, 589 were affected, or 58 per cent.[1]

The usual method of inheritance is that of a positive character. Affected individuals have either half or all of their offspring affected, while two unaffected parents will probably not have defective offpsring. However, as cataract usually appears late in life it is not always possible to predict whether the parent will become affected or not (Fig. 78).

The eugenic rule is this:—If either parent has cataract at least half of the offspring will have it also. If a person belongs to a strain that has cataract but is free from it, advice must depend on the nature of the cataract. If in

[1] The report of the medical officer (education) to the London County Council, 1909, contains 9 additional cases.

the family strain cataract appears early, before the age of the person who contemplates marriage, then such marriage may be advised; but if in the given family the cataract occurs late in life it is not possible to predict as to the immunity of the parent, but in that case also, since the potential defect will not greatly interfere with the effectiveness of the children, fertile marriage may not be gainsaid.

e. Displaced Lens (ectopia lentis).—This malposition of the lens always causes distorted vision. Fortunately it is not so common as cataract, for Loeb found only 42 families

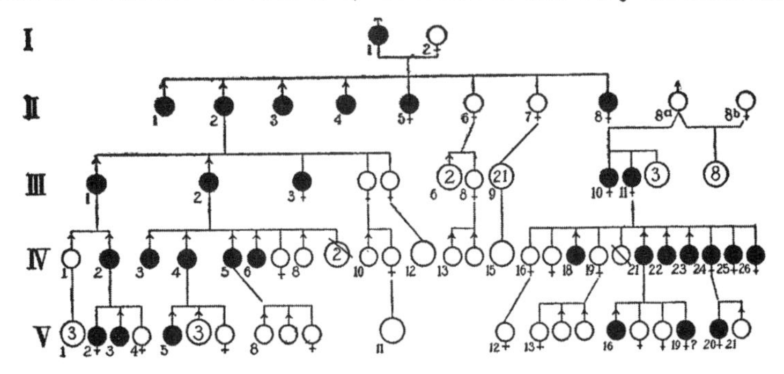

FIG. 78.—Pedigree of "coralliform" cataract. Affected persons represented by black symbols; ♂, male; ♀, female; numbers in circles indicate number of individuals. From NETTLESHIP, 1910.

described, with 150 children, of whom 70 per cent were affected. The details of the condition and the degree of injury to sight vary from strain to strain (Fig. 79).

In this case, also, it appears that the defect is due to some positive factor and that when present in either parent it will be present in about half the offspring; but if present in neither parent it will be absent from all descendants.

The eugenic teaching is clear; persons with displaced lens should have no children; but normal persons of the same strain will not reproduce it in their offspring.

f. Degeneracy of the Cornea.—While several causes of corneal opacity are known that seem not to be hereditary,

18 cases of hereditary degeneration of the cornea are recorded. So far as the studies that have been made go they indicate that persons with such hereditary corneal opacity should not have children but that normal members of such a strain will have normal offspring.

g. **Glaucoma.**—This is a swelling of the eyeball due to excess fluid in the chambers of the eye. It appears to depend upon the presence of something that prevents the escape of the fluids of the eyeball. In the study of the inheritance of this disease we meet with the difficulty that, like cancer and many forms of cataract, its outset is late in

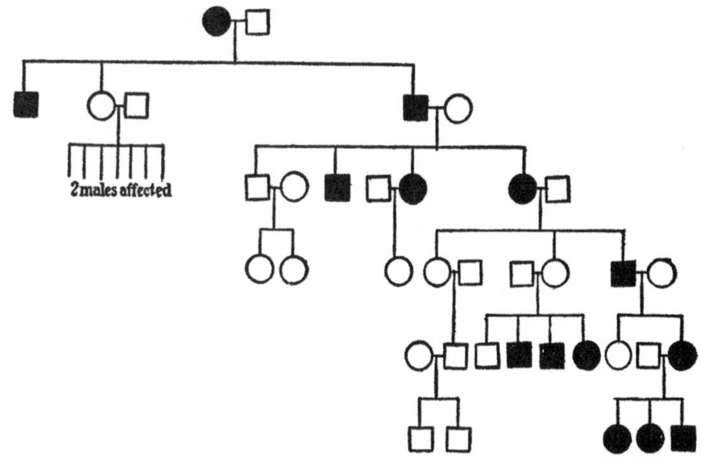

FIG. 79.—Pedigree of a family with dislocation of lens, resulting in imperfect vision, vertigo, flashes of light, etc. The amount of displacement varies in the different individuals. In the third generation 2 individuals are affected in one eye only but in all other cases both eyes are affected. LEWIS, 1904.

life—so that many persons with potential glaucoma die before realizing it. However, the age at onset is variable, in some families high and others low; but in the children the onset is frequently earlier than in the parents; thus, in one family the father shows the disease at 70, his daughters at 45, and 40; in another case father is attacked at 49 and his sons at 18 and 16; again, a father has glaucoma at 60, his 4 children at from 55 to 40; and a mother is affected in one eye at 60 and the other eye at 81, while her 3 children are affected at 60. In one family strain, Von Graefe noticed an unusually long

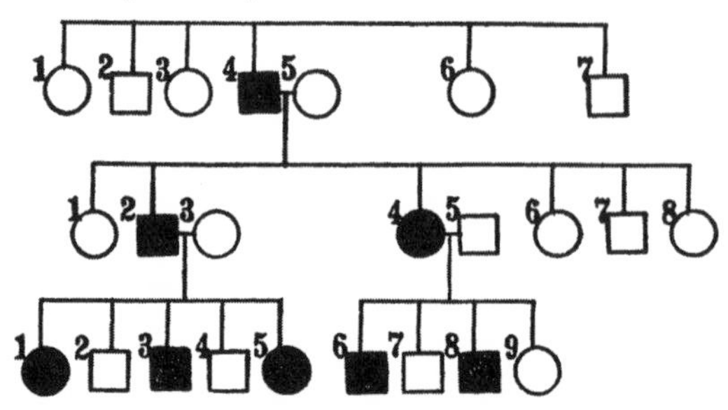

Fig. 80.—Pedigree of family with glaucoma, showing simple dominance of the trait. In I, 4, the disease appeared at 40 years of age; in II, 2, at 28; in II, 4, at 25; in generation III, at 28 to 17 years—an extraordinarily early age. Howe, 1887.

prodromal stage (10 to 15 yrs.), before the fully developed attack. This is one of the special family strains.

Glaucoma is said to have various inciting causes. The type that follows a characteristic inflammation shows the

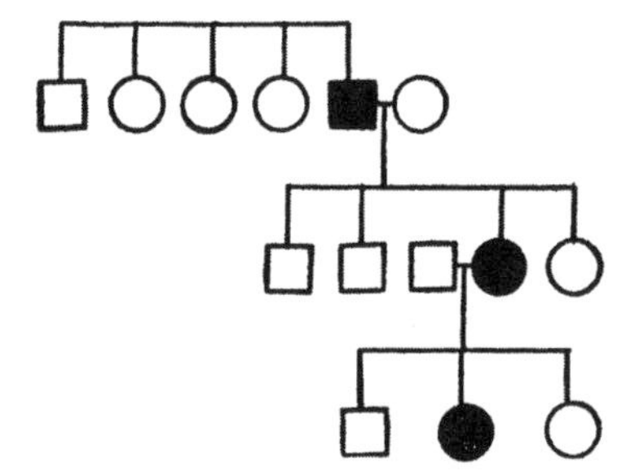

Fig. 81.—Pedigree of family with glaucoma, percentage of incidence of disease small, owing perhaps to early deaths (?). In the first generation the disease began at 71 years, in the second at 40; in the third at between 25 and 30 years. Nettleship.

best evidence of heredity. A pedigree or two will illustrate the method of its inheritance (Figs. 80, 81).

The eugenic teaching is rendered more difficult by the fact that glaucoma usually first appears toward the end of the reproductive period. But certainly affected persons should avoid having children, while non-affected may marry if the disease first appeared in the grandparents at 50 or after. If it appeared earlier it would seem to be prudent for the normal persons to delay reproduction until within ten years of the time that the defect appeared in *their* parents. Then if no trace of the disease has occurred they may have children with impunity.

h. Megalophthalmus or protruding eye. A rather rare disease of whose inheritance there can be no doubt, although the exact method of that inheritance is uncertain. Persons with a well marked case had best avoid reproduction.

i. Nystagmus, or "swimming eyes." This is due to spasmodic contractions of the eye muscles and may or may not be associated with other defects of the eye. The disorders with which it is most apt to be associated are: strabismus, retinitis pigmentosa, coloboma, albinism, microphthalmus and cataract.

In some of the pedigrees that have been published (Clarke's, 1903), nystagmus, like optic nerve atrophy, is not expressed in the (simplex) females [1] but is expressed in all males capable of transmitting it. When it is unexpressed in the males of the strain, it will probably not (in non-consanguineous marriages) appear in the offspring. But marriages of even non-affected females (unless from large families of non-affected brothers) and of all affected males are pretty certain to yield offspring with nystagmus.

k. Paralysis or imperfect development of the muscles of eye and lids.—This includes ptosis, or drop of the upper eyelid; epicanthus, a fold of skin passing from nose to eyebrow over the inner corner of the eye; blepharophimosis, or smallness of opening of eyelids; ophthalmoplegia, or paralysis of eye muscles; strabismus or squinting. Every one of these peculiarities shows clear evidence of heredity.

[1] In other families nystagmus appears also in the females.

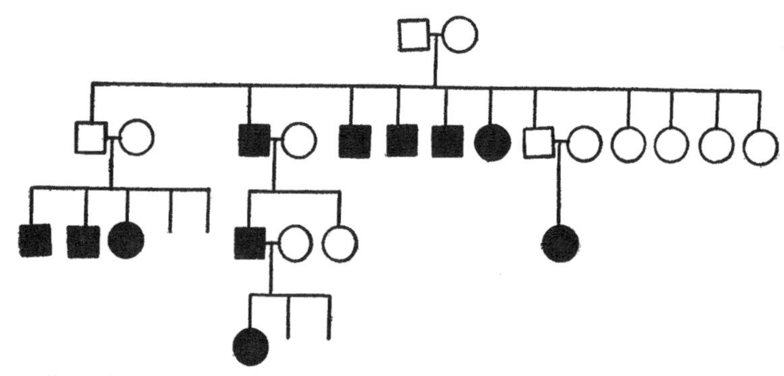

FIG. 82.—Pedigree of a family, every affected member of which (black symbols) has drooping eyelids, a fold over the inner corner of the eye, and narrow eye opening. VIGNES, 1889.

One family pedigree is reproduced in Fig. 82. This is remarkable because every affected person showed the same combination of characters, namely, drop of upper eyelid, epicanthus, and ophthalmoplegia.

In Cutler's case (Fig. 83) the parents are first cousins; all affected persons have strabismus. Expectation in this group of cases is that an affected person will have affected offspring but that two normal parents will rarely have offspring with the defect, even though one belongs to the defective strain.

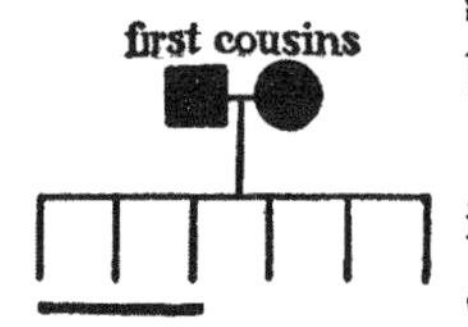

FIG. 83.—Pedigree of a family in which the parents are first cousins and both have strabismus (squint). Three of their 6 children are similarly affected. CUTLER.

1. **Pigmentary degeneration of the retina** (retinitis pigmentosa).—This degenerative process is accompanied by an atrophy of the optic nerve and leads to eventual blindness. It is frequently associated with consanguineous marriage, 27 per cent of the marriages which yield it being (according to Feer's list, 1907, p. 14) consanguineous. The method of inheritance is well illustrated by Fig. 84 which is a portion of a chart prepared by Nettleship. This figure illustrates the general law of this disease; namely, that two normal

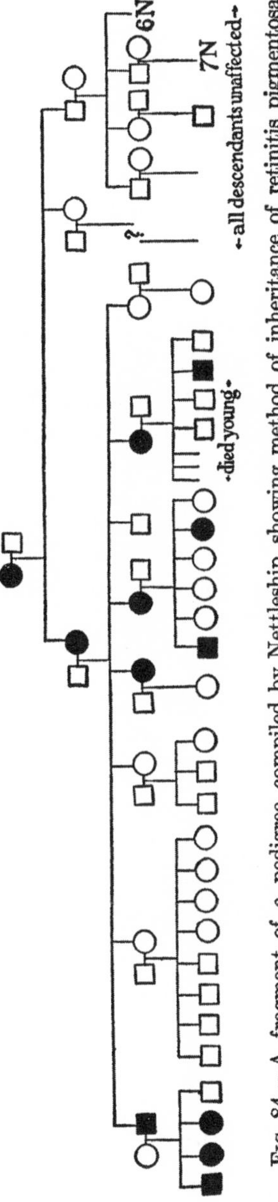

FIG. 84.—A fragment of a pedigree compiled by Nettleship showing method of inheritance of retinitis pigmentosa. Black, affected individuals. Note that two normal parents have only unaffected offspring. NETTLESHIP, 1910, p. 13.

parents produce no abnormal children. The condition that makes for retinitis is something added to the normal condition.

The extent of the degeneration varies with the family. In a pedigree recorded by Leber (Fig. 85) the characteristic, throughout the family, was an increasing dimness of vision accompanied by night blindness; but later the degeneration was stayed.

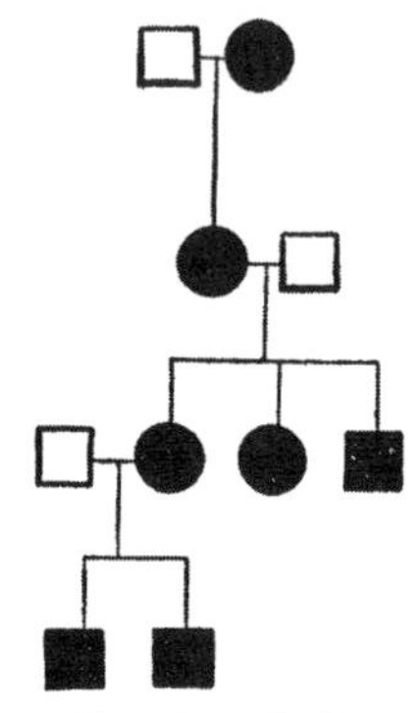

FIG. 85.—Pedigree of retinitis pigmentosa in a family in which the disease becomes checked before blindness becomes complete. LEBER, 1871.

The eugenic instruction is clear. An affected man or woman should not marry even into stock without taint of retinitis. Above all, in retinitis stock, cousins, especially if affected should by no means marry.

m. Night Blindness (hemeralopia). — This disease is accompanied by no loss of perception of form, but at sunset the affected persons must cease working. Artificial light helps little unless very intense. The lamps of the street are of no assistance in guiding these people at night. Eventually, in most strains, the affected persons become totally blind often with a retinitis. This disease is probably due to a defect in the brain and not as has been suggested merely to lack of the visual purple of the retina (Bordley, 1908).

Through the researches of Cunier (1838) and Nettleship (1907) we have a pedigree of a night blind strain that is the most extensive that has yet been compiled for any disease. It includes 2,116 persons. A part of it is reproduced in Fig. 86. Fig. 87 is a pedigree of an American (colored) family furnished by Dr. Bordley.

The disease is due to a positive factor. The normals lack this factor. Usually, however, the factor must be duplex

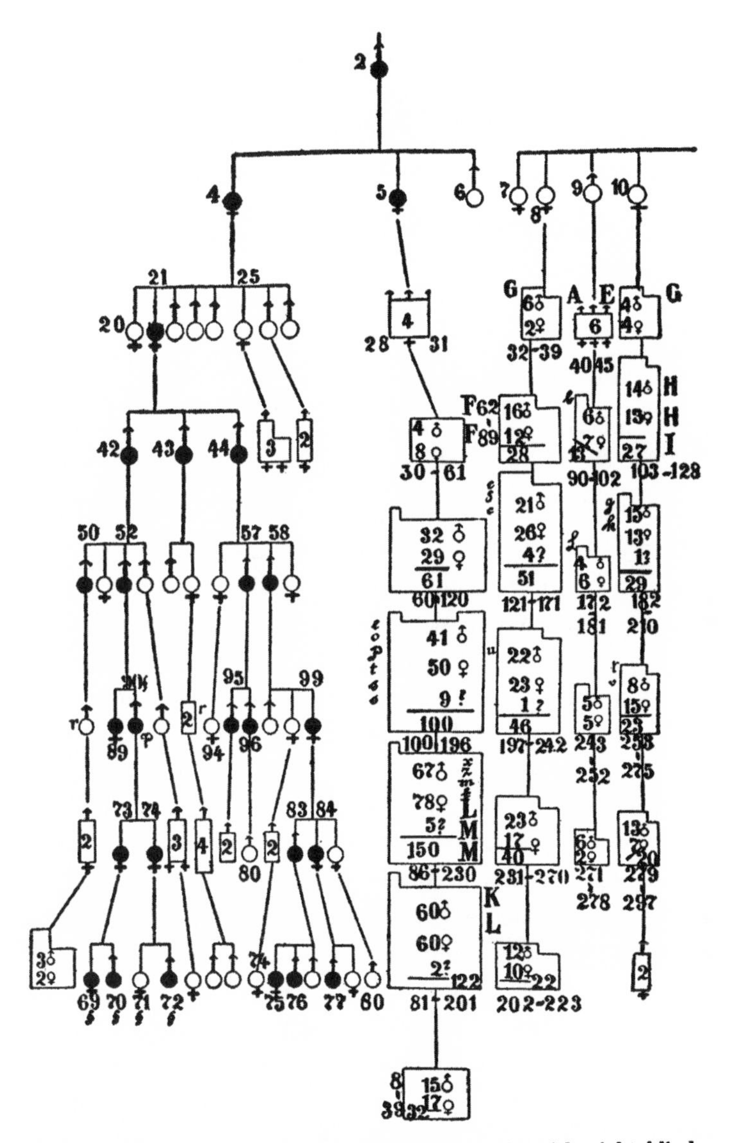

Fig. 86.—Pedigree of chart of an European strain with night blindness (black symbols). The rectangles indicate numerous normal individuals. Two normal parents have only normal children. Nettleship, 1907, from Grüber and Rudin, 1911.

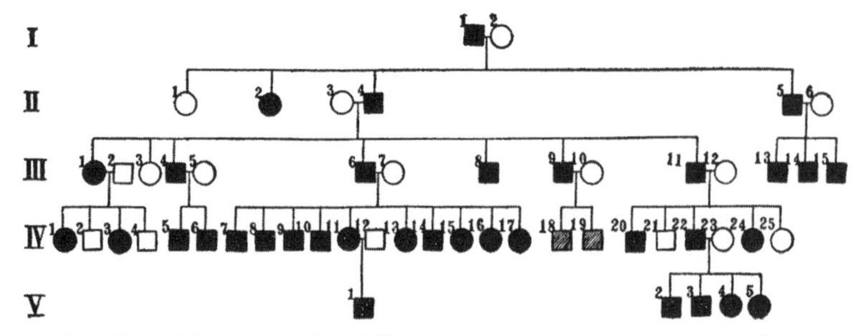

Fig. 87.—Pedigree of night blindness in a negro family, many of whom were personally examined by Dr. Bordley. IV, 18, 19, are doubtful. All solid block symbols stand for affected persons; clear symbols unaffected. The blindness is progressive and ends in death within 16 months after blindness becomes complete. All affected persons have an affected parent. Night blindness is a positive trait. Bordley, 1908.

in females in order to develop; but in both Nettleship's and Bordley's families even simplex females have night blindness. Ordinarily, consequently, while night blind people should not reproduce, normal males from such stock may do so with impunity, but normal females may have children only when all their brothers (more than two) are without the defect; for normal females, in most night blind families, *may* carry the disease.

n. Color Blindness.—The inability to distinguish certain colors, notably red and green, is not a rare condition but much less common in women than men (in Europe, 4 per cent males, 0.5 per cent females). The method of inheritance of the condition is much the same as that of atrophy of the optic nerve and night blindness; namely, that color blind males do not have color blind sons but that females free from color blindness may have sons with it (Fig. 88).

The eugenic conclusion is that while color blind males will have no color blind sons and, typically, no color blind offspring of either sex yet their daughters, married to men of normal stock, will have color blind sons.

To the ordinary rule there are various exceptions. Daugh-

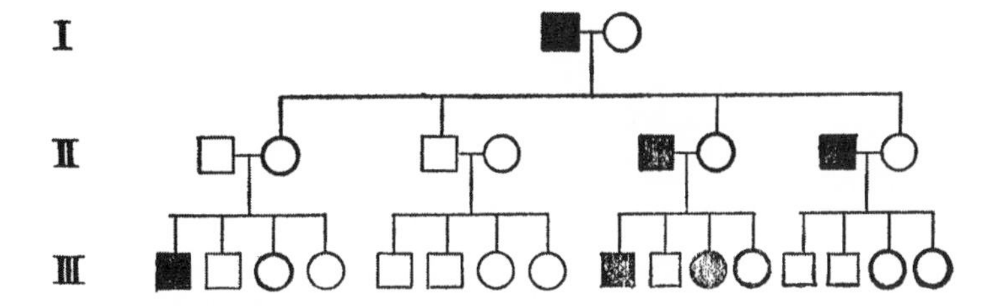

Fig. 88.—Ideal scheme, showing method of inheritance of color blindness. Typically it appears in sons only of simplex females, represented by a heavy ring. The third mating in second generation is illustrated in Fig. 89, II, 6.

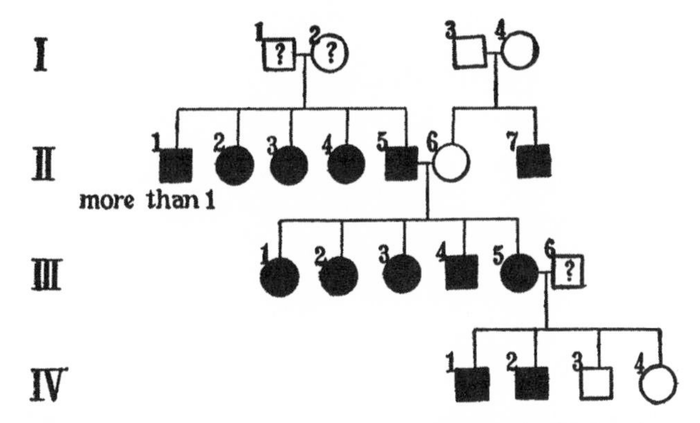

Fig. 89.—A remarkable and exceptional pedigree of color blindness. The fraternity, II, 1–5 (which comprises the grandfather, his brothers, and his 3 sisters), were said all to be color blind. The grandmother, II, 6, had the normal color sense but had an affected brother. The entire fraternity, III, 1–5, including 4 females, has impaired color perception. Details are given about III, 5, as follows: She is about 50 years old, a physician's wife, and a test shows complete confusion of dark green, dark red and brown. While lighter tints are better distinguished, rose and blue are confounded. The sons show exactly the same conditions. Reber, 1895.

ters may inherit color blindness from fathers. At least such is the history given by Reber (1895), Fig. 89; an exceptional history that is not entirely without precedent. In the case of these exceptional families a color blind parent may have color blind offspring of either sex.

o. Myopia.—That the shape of the eyeball is largely

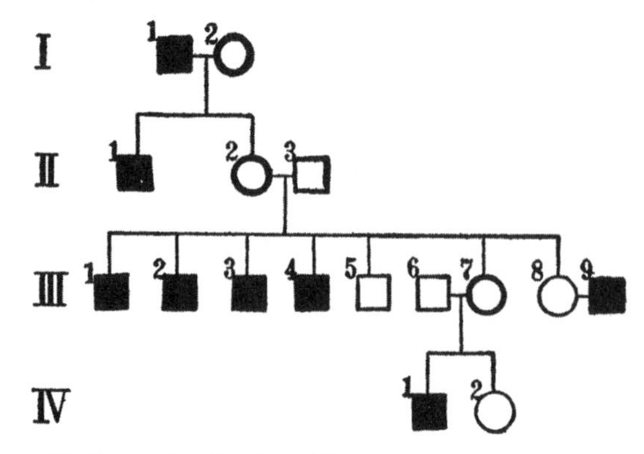

Fig. 90.—Pedigree of a family with myopia. In the first generation the man had myopia and strabismus while his wife was normal. Their son, II, 1, had myopia and died unmarried. His normal sister married a normal man and had 7 children. III, 1 and 2, had both myopia and strabismus; the eyesight of III, 3 and 4, was defective but in what way is unknown. A normal sister, III, 7, had a son with defective sight—probably myopia. From Oswald, 1911. Note that males only are affected and are derived only from 2 normal parents. Simplex mothers indicated by heavy circles.

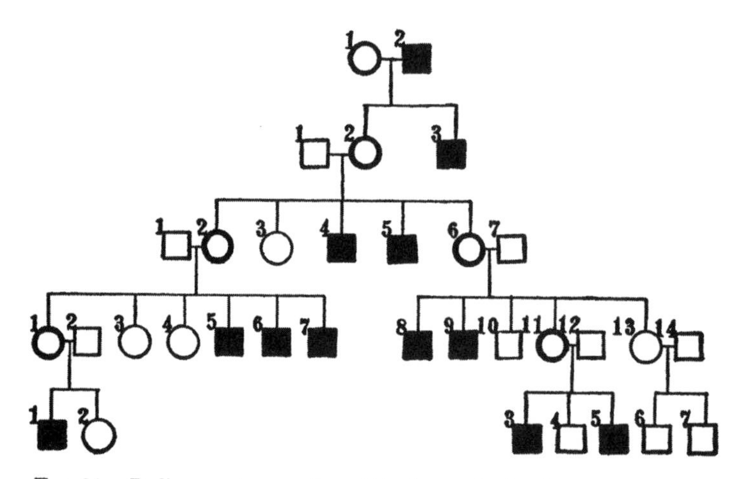

Fig. 91.—Pedigree of myopia. Members of the 3 youngest generations were personally examined. Nearly all males of the family are myopic, and none of the females, but myopia is transmitted through the female line. Myopia is about the same in all cases, 10 or 12 *D*, with some astigmatism. From Worth. The defect shows in males only and these are always descendants of normal females. Their simplex mothers are represented by heavy circles.

controlled by heredity has been shown by Hertel (1903), as a result of measuring the refraction in children and their parents.

That myopia, or near sightedness, is inheritable has long been known. A typical case has been recorded by Oswald (1911), Fig. 90, and a second pedigree is given by Worth (Fig. 91). In both pedigrees inheritance is sex-limited as in color blindness. A normal female has some, at least, of her

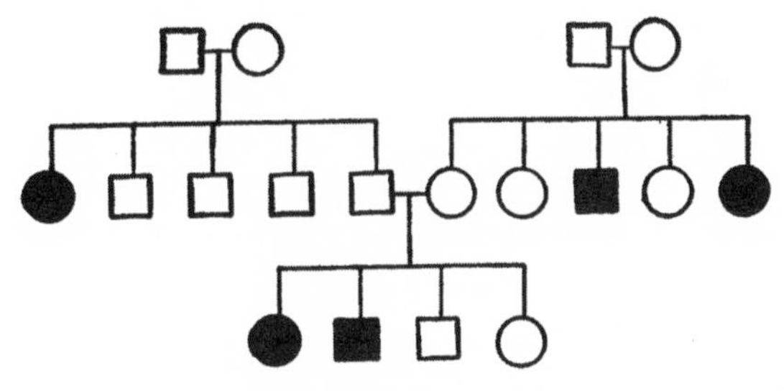

Fig. 92.—Pedigree of astigmatism, affected persons represented by black symbols. F. R.

sons myopic, but all daughters are normal. In such a family, then, normal daughters in a myopic fraternity may expect myopic sons.

p. **Astigmatism.**—This condition of improper curvature of the lens belongs to the list of family traits. A correspondent submits the pedigree of his family shown in Fig. 92.

From this pedigree it appears that, in this family, astigmatism is a recessive trait, since normal persons may transmit it and since it is equally apt to appear in either sex. It would be desirable, other things being equal, for a person belonging to an affected strain to seek a partner from a strain that has normal eyes.

28. Ear Defects

The ear is the most complicated of the sense organs and though its important elements are deeply hidden in the head yet the lining of the middle ear is continuous with the mucous membrane of the throat—in some respects the most vulnerable portion of the human body. Hence it is subject to the weaknesses of that membrane. On account of its very complexity it is especially liable to exhibit *deformations or deficiencies.*[1] In view of the great variety of changes any one of which may result in deafness it is clear that *deafness* can hardly be a unit defect. Consequently it will not be inherited as a simple character.

The facts justify the *a priori* conclusions. Deafness of certain sorts is clearly hereditary but it is not possible to predict certainly the outcome of a particular mating. Nevertheless something can be done; and it will be worth while to learn what is known of the actual incidence of deafness in the offspring of deaf parents.

Inheritable deafness is of three general types. (a) That due to defects or changes before birth or shortly after, giving rise to *deaf mutism;* (b) *otosclerosis*, or hardness of hearing, with usually progressive symptoms; (c) *catarrhal weakness* of the mucous membranes, rendering them liable to infection with inflammation and suppuration.

a. Deaf Mutism.—This kind of deafness is characterized by its early appearance in life, before speech has been acquired. It is the less likely, consequently, to be due to disease and, as a matter of fact, it is that form which shows clearest evidence of pure inheritance. So clear is the evidence of inheritance of congenital deafness that some coun-

[1] *Politzer* (1807) gives among others the following anatomical causes of congenital deafness: impaired development or absence of middle ear, defects and rachitic deformities of the labyrinthine windows; narrowing of the recess of the round window to a cleft with connective tissue; atresia of the same; atrophy of the cochlear nerve and spinal ganglion in the first turn of the cochlea; abnormalities of the membrance of the otoliths, organ of Corti and ductus cochlearis; faulty development of the sensory epithelium; defects of the crista and sulcus spiralis; lack of development of the labyrinth and of the auditory nerve; malformations of the central nervous system. In addition there are numerous changes in structure due to inflammations.

tries have forbidden the marriage of persons of this class. Yet the inheritance of deaf mutism has been disputed and, indeed, without careful consideration of the separate family histories the method of inheritance seems truly obscure.

The most extensive data on the marriage of deaf are those collected by Fay (1898). He finds that, when both parents are congenitally deaf (Figs. 93, 94), of the 335 matings 25

FIG. 93 FIG. 94

FIG. 93.—Pedigree of deaf mutism. Parents both deaf; the father at 3 years; the mother before birth. The first two children died shortly after birth; the other two are deaf mutes—one born so; the other following a slight blow on the head. SAINT HILAIRE, 1900, p. 31.

FIG. 94.—Pedigree of deaf mutism. Father mother, and 3 children, all deaf mutes from birth. SAINT HILAIRE, 1900, p. 31.

per cent yield some deaf offspring; and of the total of 779 offspring 26 per cent are deaf. It is clear that such marriages are, in the long run, dangerous. That all children of such marriages are not deaf is doubtless due to the fact that the parents are not deaf in the same way and that one parent brings into the combination what the other lacks. The contrast between the result of marriages of two *congenitally* deaf parents and two who are *adventitiously* deaf is shown by the fact that the latter yield only 2.3 per cent deaf children.

If, on the other hand, the partners belong to the same deaf mute strain, i. e., are related, the percentage of marriages yielding some deaf mute offspring rises to 45, and the proportion of deaf offspring to 30 per cent (Fig. 95). But that is not the whole story, for the closer the relationship of the parents the larger the proportion of deaf children as the following table shows:—

	Per cent deaf offspring
Partners "cousins," degree unreported	19.4
" first or second cousins	34.6
" nephew and aunt (1 family)	75.0

The interpretation of this fact would seem to be that the nearer the relationship the greater the chance that both parents lack the same element and so all of their children

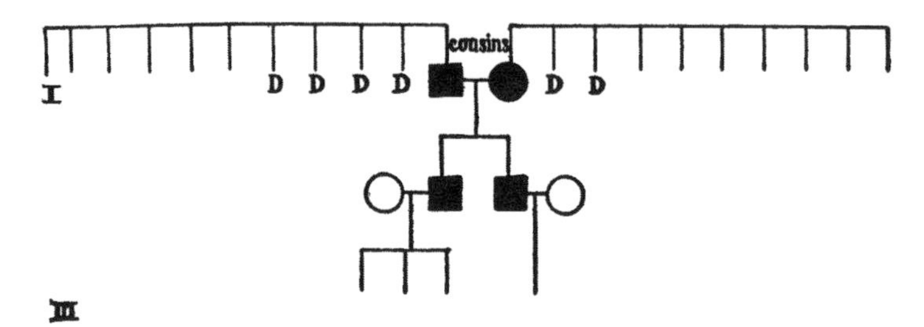

FIG. 95.—Pedigree of deaf mutes. Two deaf mute cousins each belonging to fraternities having several deaf mutes marry one another. Both of their children (II) are deaf. Each child marries a hearing wife and of 4 children all hear. FAY, 1898, No. 2621.

tend to lack it. In Figs. 96 to 100 are given some pedigrees of deaf mute families. They show that, under certain circumstances, probably identity of defect in parents, the children will all be similarly defective.

The studies of Bell (1906) based on the census returns of a large proportion of the deaf population of the United States show the importance of consanguineous marriages in favoring the production of deaf mute offspring. He finds (p. 17) "of the 2,527 deaf whose parents were cousins, 632, or 25 per cent, are congenitally deaf, of whom 350, or 55.4 also have deaf relatives of the classes specified; while among the 53,980 whose parents were not so related the number of congenitally deaf is 3,666 or but 6.8 per cent, of whom only 1,023 or 27.9 per cent have deaf relatives."

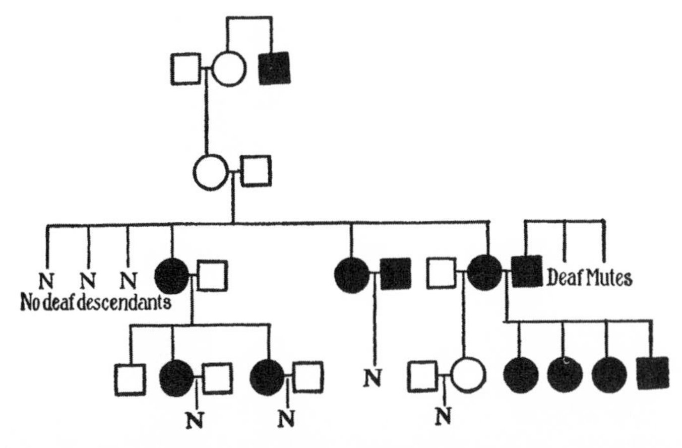

Fig. 96.—Three sisters (Gen. III), deaf mute from birth, had several perfectly normal brothers and sisters. Their mother's uncle had been a congenital deaf mute. The first sister married a hearing man and had 3 children, 1 hearing son and 2 mute daughters, who married hearing men and had only hearing children. The second sister was educated and married an educated mute but died soon after the birth of her normal child. The third sister married, first a hearing man and had a normal daughter whose children were in turn normal. But she married for a second husband a deaf mute belonging to a fraternity with 2 other deaf mutes and all 4 children who survived infancy were deaf mutes. Report, N. Y. School for Deaf and Dumb, 1853, p. 96.

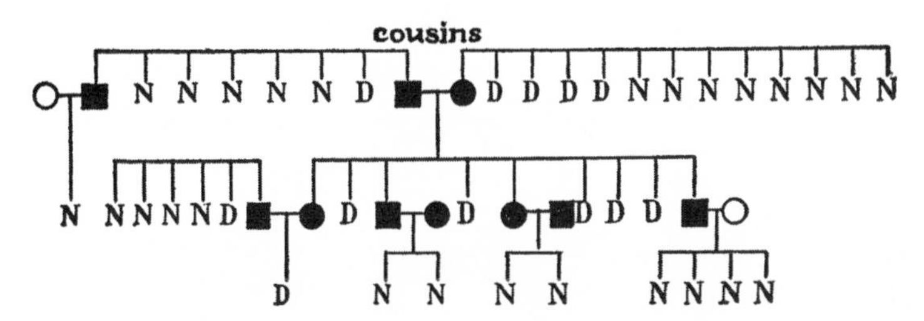

Fig. 97.—Pedigree of deaf mutes—black symbols or *D*. Note the fraternity of deaf mutes derived from the central mating of cousins. Most of those who outmarried, even though their consorts were deaf, had hearing children. Fay, 1898, No. 810.

In view of the foregoing data the first eugenic recommendation clearly is that two deaf mutes should not have children, especially if they come from the same long-settled community or are known to be blood relatives.

If one partner be congenitally deaf and the other have no ear defect and knows of none in his family the chances for deaf offspring are small. In 72 such marriages considered

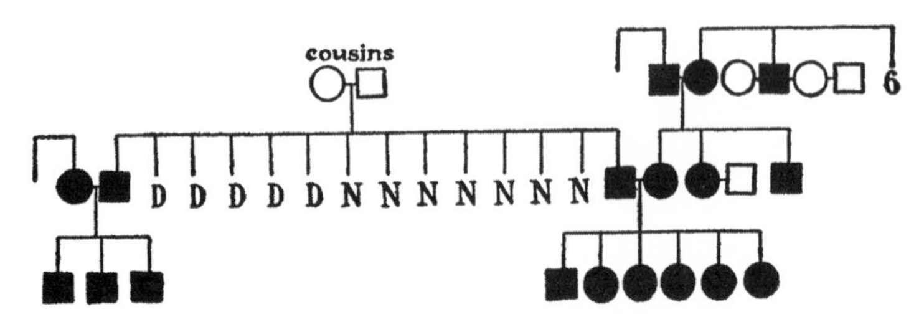

Fig. 98.—Pedigree of deaf mutism. In the first generation 2 hearing cousins marry. They have 14 children of whom 7 are dead. Two of these marry deaf wives belonging to fraternities with other cases of deafness. Of 9 children, altogether, all are deaf. Fay, 1898, No. 7.

by Fay only 5 resulted in deaf offspring. It is quite likely that in some even of these five matings the normal parent had unknown deaf relatives.

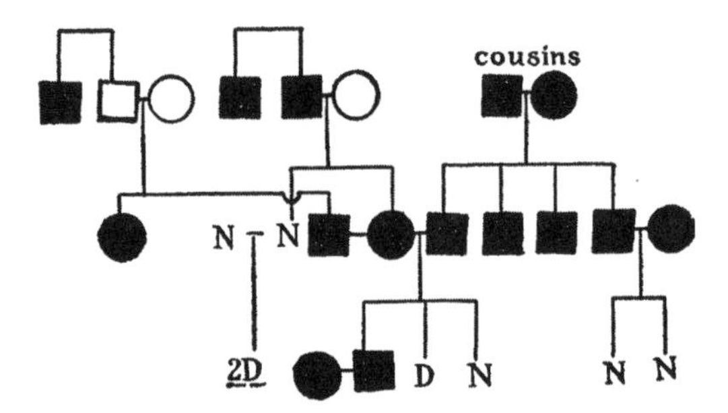

Fig. 99.—Pedigree of deaf mutism. Two deaf mutes, first cousins, marry and have 4 children, all deaf mutes. One of these marries a wife whose father, an uncle and two nephews or nieces were deaf mutes, and two out of three children were deaf mutes. Another child of the original pair married a deaf mute and had two hearing children. Fay, 1898. Nos. 3292, 2260, 442, 3290, 3291, 3234.

But if the hearing partner have deaf relatives then the proportion of resulting fraternities containing deaf mutes increases to 35 per cent.

Even though both partners hear, if they belong to the same strain with a tendency to deafness the liability to deaf offspring is so high as to warrant warning strongly against such a marriage (Fig. 99).

Finally if one or both partners are adventitiously deaf and have no deaf relatives then there is no eugenic obstacle to marriage, for such marriages result in a negligible proportion of deaf offspring—in Fay's statistics only 2 out of 552.

b. Otosclerosis.—This disease consists of a progressive rigidity of the mucous coat of the tympanic membrane;

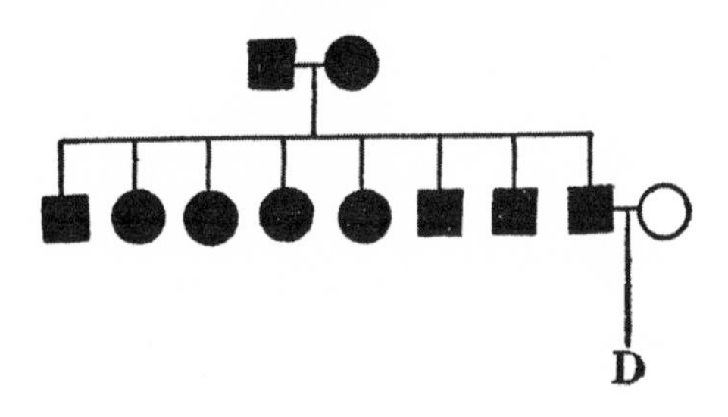

Fig. 100.—Pedigree of "fistula auris congenita." Both of the original pair were affected with a congenital aural fistula, with a fistulous canal anterior and close to the ear; all persons represented by black symbols had a similar fistula. Hartman, p. 56.

usually associated with adhesions in the inner ear and alterations of the windows (fenestra). It shows itself in an ever increasing difficulty in hearing conversation.

The inheritance of otosclerosis is a familiar fact. Most persons know families many of whose members become "hard of hearing" as they grow older. The deafness is frequently attributed to climatic causes and this belief is increased by the presence of many cases in the same locality. But it will be found on inquiry that the affected persons are relatives and that their unrelated neighbors are not affected by the same climate. This makes it clear that a severe climate merely brings out the latent weakness of the mucous lining of the ear. Some examples of strains showing otosclerosis are given in Figures 101–104.

An examination of the available pedigrees indicates that otosclerosis is due to a defect—perhaps to the absence of a resistance to infection and inflammation of the lining membrane of the inner ear. Like other defects it is relatively common in the progeny of cousin marriages.

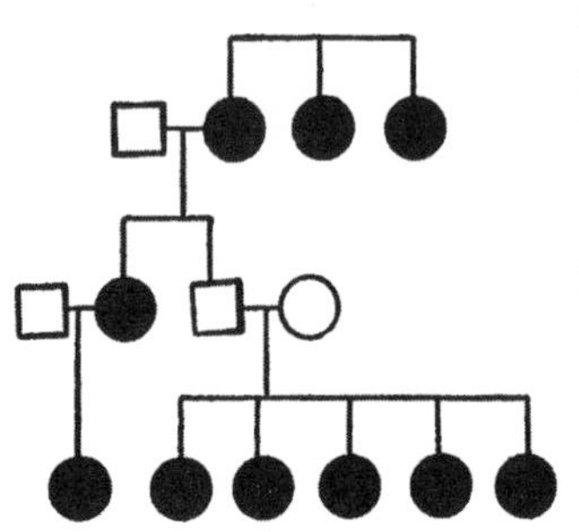

FIG. 101.—Pedigree of otosclerosis. In this pedigree all affected individuals, so far as known, are females. LUCÆ, 1907.

The eugenic indications then are, two persons with a tendency towards otosclerosis should refrain from marrying, as probably all of their children will he hard of hearing. But a person with otosclerosis and an unaffected person of an untainted strain may marry with impunity as their children will probably all have strong hearing.

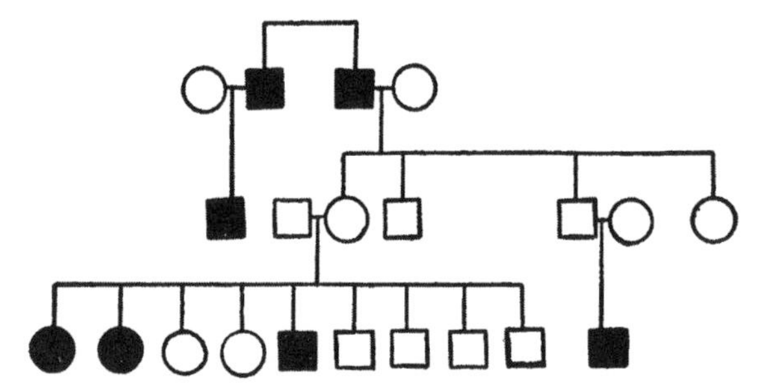

FIG. 102.—Pedigree of a family with otosclerosis. Two deaf brothers marry; one has a single son, who is deaf; the other has four unaffected children. Of these latter two marry consorts who are, so far as known, normal. From one pair three out of nine children are affected; from the other only one child is known and he is hard of hearing. HAMMERSCHLAG, 1906.

c. Catarrhal affections.—That a weakness of the mucous membranes permitting catarrh is hereditary, we shall see in speaking of the weakness of mucous membranes in general, and it cannot be doubted that such a weakness plays a rôle in deafness. Thus Bell (1906) has shown that, in the census returns, over 55 per cent of the deaf children in the country come from parents who became deaf in adult life and he states that this "confirms the conclusion reached upon other grounds that heredity sometimes plays a part in the production of catarrh of the middle ear—the chief cause of deafness occurring in middle life."

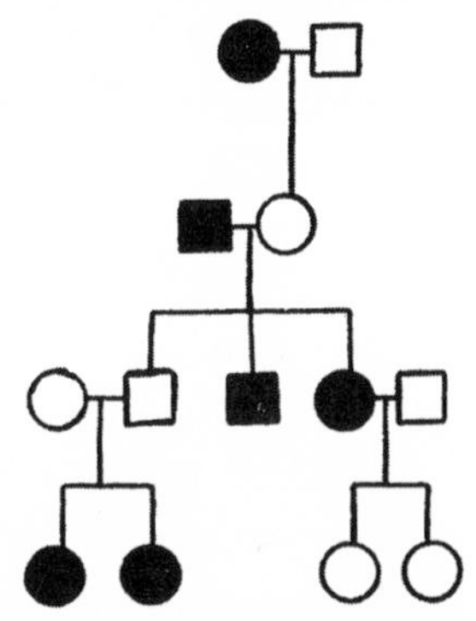

FIG. 103.—Pedigree of otosclerosis. Affected persons (black symbols) for the most part, but by no means always, have an affected parent. LUCÆ, 1907.

29. SKIN DISEASES

The skin is an admirable organ for the protection of the delicate internal parts not only from desiccation but also from the entrance of the numerous parasites that thrive on mammal-

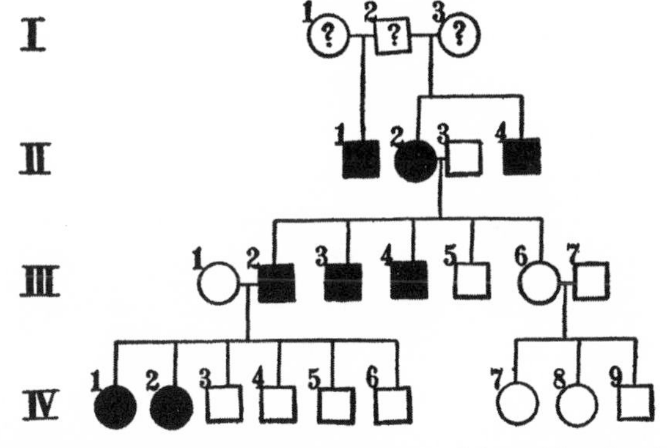

FIG. 104—Pedigree of otosclerosis. The condition of hearing in the first generation is unknown and some of the children in the fourth generation have not reached the age of incidence; thus, IV, 4–6. are 22 to 18 years old and IV, 7–9, are 20 to 14 years.

ian blood and tissues. Nevertheless, its exposed position renders it liable to attack by the various germs that are ubiquitous. Abrasions and the openings of the sebaceous glands and the hair follicles offer vulnerable points. The main reliance of the organism must be its internal means of defense. The efficiency of specific means of resistance is undoubtedly an inherited quality. We find families characterized by low resistance to specific germs of particular diseases.

Thus liability to boils and eczema appears as a family trait in the Dow-1 family. One of the parents is subject to boils and the other to eczema. Of five children three are subject to eczema and one to boils. It seems probable that we are here dealing with a lack of resistance to infection through the skin in both parents, leading to a non-resistance in all of the children. A few cases of inheritance of more specific types of skin diseases are cited below.

a. Congenital Traumatic Pemphigus (epidermolysis bullosa).—The children are born with a liability to form fluid filled vesicles after the smallest physical provocation. The excessive vulnerability shows itself in the first month of life and is said to diminish from 40 to 50 years of age and to cease altogether in old age. It is strongly hereditary, often through several generations (5 in Bonajuti's case); it shows also a prevalence in particular families and is rather more frequent in males than females. The slightest injury, blow, pressure, friction or scratching is followed by the formation of a bulla. The bullae are often full of blood and of large size, 5 centimeters or more across and their shape may be irregular instead of round or oval depending upon the nature of the injury. Fingers and nails are often deformed or altogether destroyed. The pathology of the disease is obscure; it seems to be influenced by arsenic (Radcliffe-Crocker, 1903, p. 293).

The case described by Bonajuti is given in Fig. 105. Of an affected parent about half the offspring are affected. Two normal parents usually produce only normal offspring. In case the single known parent is normal and has affected offspring it is presumed that the unknown spouse was affected. On the whole, epidermolysis seems to be due to the presence of a distinct factor, absence of which results in normality.

The eugenic teaching is then that two normals belonging to such a family as that of Fig. 105 may marry with impunity

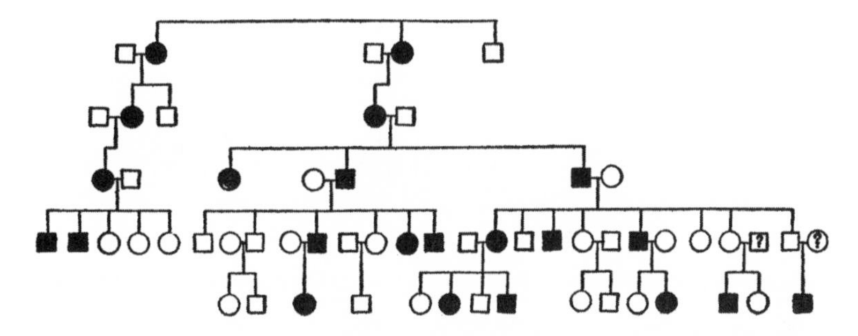

Fig. 105.—Pedigree of a family showing epidermolysis bullosa, behaving like a dominant trait—appearing in each generation. Only in two instances, at the right of the chart, does a case arise from a parent not known to have the trait. Gossage, after Bonajuti.

but that in the case of parents who have, or had in childhood, epidermolysis probably at least half of the children will be similarly affected.

b. Psoriasis (itch).—The question of the inheritability of this disease has been much discussed. Some declare it is due to infection, others deny it. Various experiments have been tried. Schamberg (1908) performed auto-inoculation in 23 cases and got a positive result in only 3. Inoculation into normal human subjects—usually the experimenter's own body—have produced the disease in only one case (that of Dr. Destot). On the other hand in about a third of the cases observed by various physicians psoriasis was recognized as a family disease. The most reasonable explanation is that the disease is due to a parasite to which most persons are immune; and that lack of immunity is an inheritable trait.

Besides skin diseases due to infection there are other abnormal conditions consisting of irregularities or exaggera-

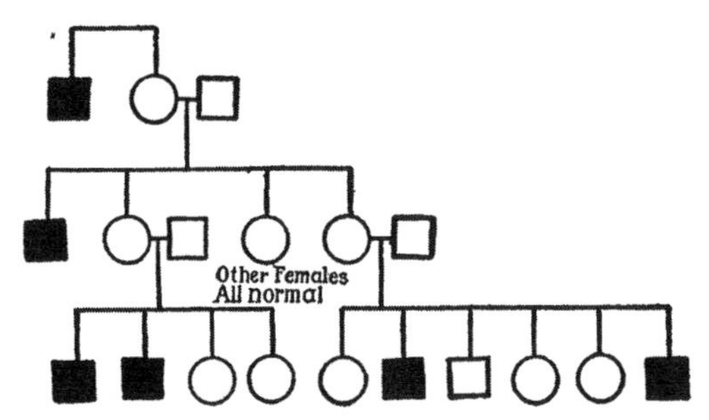

FIG. 106.—Pedigree of ichthyosis. All affected persons are from non-affected females. BRAMWELL, 1903, p. 77.

tions of the process of rendering the outer layer of the skin horny. The liability to these diseases is usually recognized to be hereditary.

c. Ichthyosis or **xerosis** (xeroderma).—This is a dryness of the skin in which plates are formed like the scales of a fish. The disease is remarkable because, apparently, it is sometimes limited in heredity by sex and sometimes not, —in different families. At least, in two of the pedigrees (Figs. 106, 108) males only are affected and inheritance is through a normal female. But in other cases (Figs. 107, 109) the females seem to be affected equally with the males and the peculiar skin condition is transmitted either by normal or by affected females. Ichthyosis is especially apt to be found in families in which consanguineous marriages occur and this fact, together with the pedigrees,

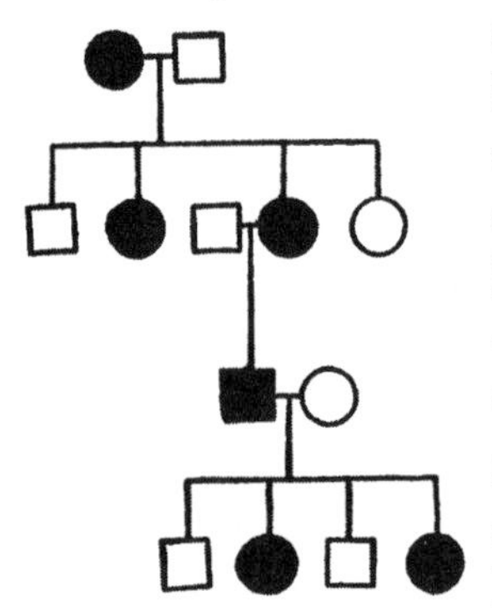

FIG. 107.—Pedigree of ichthyosis, behaving like a positive trait. BRAMWELL, 1903.

suggests that it is due to the absence of some factor that controls the process of cornification of the skin. On this hypothesis a normal person who belongs to an affected family

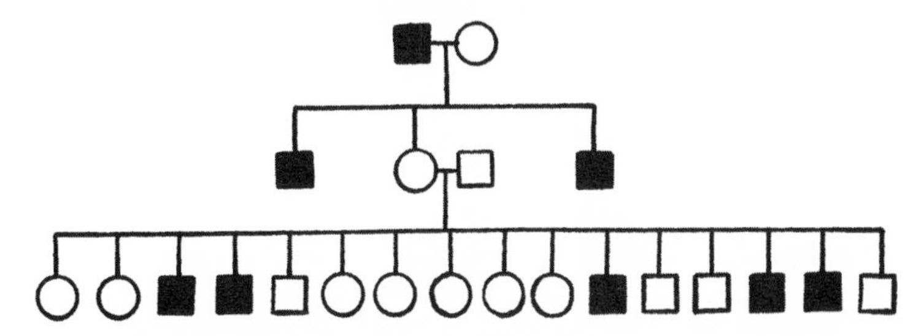

FIG. 108.—Pedigree of a family with ichthyosis. Note that only males are affected. BOND, 1905.

may marry into a normal family with impunity, but cousin marriages are to be avoided.

d. **Thickening of the outer layer of the skin** is a disease that is closely related to the foregoing. In the generalized

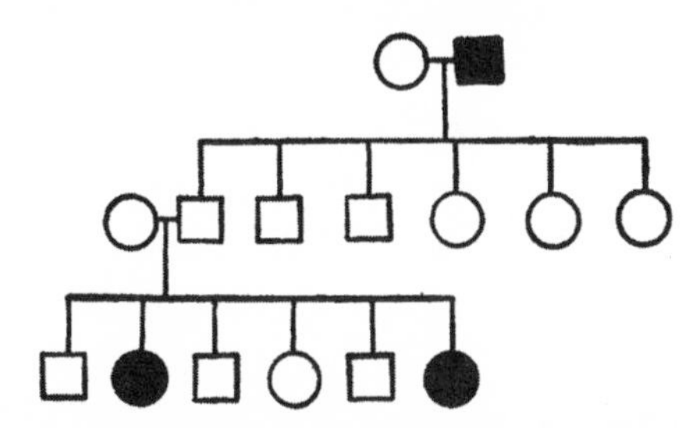

FIG. 109.—Pedigree of a family showing general ichthyosis, giving evidence that it is a positive trait. GOSSAGE, 1907, p. 342.

forms (called hyperkeratosis) infection has been alleged as a cause; but if infection plays a part it seems to be effective only where there is a susceptibility. Evidence for contagion is said to be given by the case where the only two affected children were those who, alone, were nursed by their mother, an affected woman. But, on the other hand, the fact that the mother had the disease proves her susceptibility.

Finally, the peculiar thickening of the palm of the hand and the plantar surface of the foot known as *Tylosis* seems to follow the same rule as keratosis of which it is only a special case. Both males and females are affected and two normal parents, even of an affected family, rarely transmit the defect (Figs. 110, 111).

The records of 45 families with this abnormality have been studied by Gossage. In the 39 that can be used, it appears that males and females are equally affected (166 to 140) and transmit equally. As affected persons always mate with normals, affected offspring are always simplex and expectation is that half of their offspring shall be abnormal. In 28 families 222 children are abnormal and 184 normal. Only one exception appears to the rule that two normal parents have only normal children.

30. Epidermal Organs

Heredity in these organs may be considered under the four heads of glands, hair, nails and teeth. The inclusion of teeth is justified since their true epidermal origin is now recognized; they are equivalent to the scales of fishes, but, in the higher animals, including man, they are confined to the mouth and jaws. On account of the close interrelationship of these four types of organs a modification of one may mean a change in all, and so it is not possible in discussing one of them always to avoid a consideration of another.

a. The Skin Glands are principally the sebaceous and sweat glands, associated functionally with the hair and morphologically with the milk glands. The latter are usually reduced to two in man but cases of supernumerary mammae are not exceedingly rare. This condition is doubtless hereditary for Leichtenstern (1878) refers to the case of a woman with three mammae on the chest who bore a daughter who in turn also had three mammae (though the additional

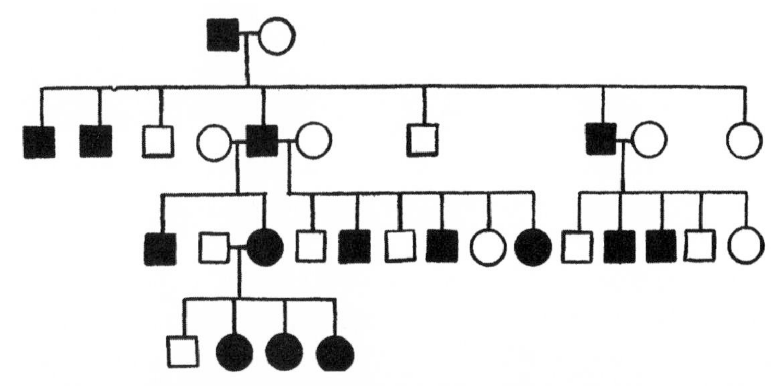

FIG. 110.—Pedigree of a family with tylosis (black symbols). Note that all affected persons have at least one parent affected—showing that tylosis is due to a positive determiner. UNNA, 1883.

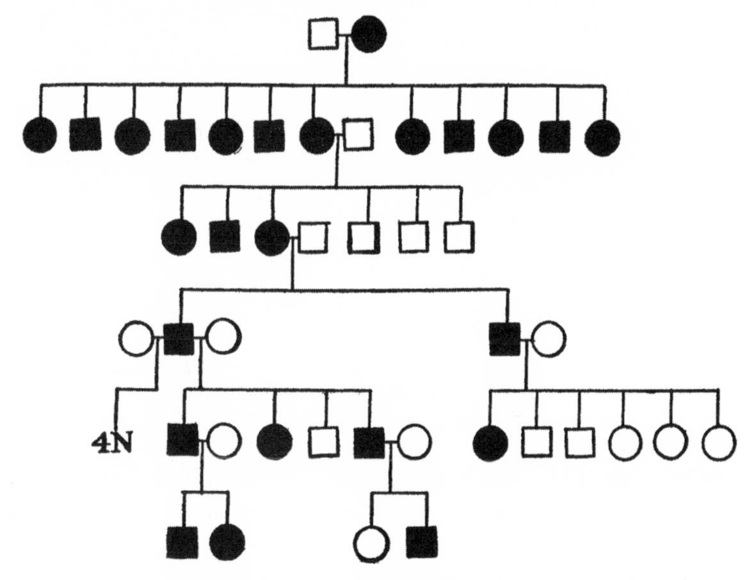

FIG. 111.—Pedigree of a family with tylosis palmae plantaris (black symbols)—proof of its positive nature. 4*N*, four normals. GOSSAGE, after RIZZOLI, 1907.

one was on the thigh), and Iwai (1904) cites many cases of a mother and five to one of several children who possessed supernumerary pectoral nipples.

b. Hair.—Peculiarities of hair, apart from pigmentation, are not infrequent as family traits. Thus a family with curled, woolly hair is described by Gossage, the curly condition being clearly dominant over its absence. Hair may be entirely absent even from birth. Such a case is described by Molenes (1890). There was brought to him a girl of 4 years who was hairless from birth until 19 months old. She had a brother who was bald at six and the mother lost her hair at 19. Another case, described in the Medi-chirurgical Transactions, is that of a boy of three who was nearly bald. His sisters had normal hair but his mother had complete alopecia areata from the age of six.

A third case is that described by White who knew a family that came from France to Canada. One grandfather was nearly hairless and the nails were faulty; the parents were normal; but in the next generation of 6 sons and 2 daughters one daughter was almost hairless and the nails abnormal in her and in two sons. This daughter married (presumably a normal man) and had a son who at 19 retains on his scalp the nearly invisible downy coat with which he was born. His only sister has a thick, downy scalp-covering quite different from normal hair. One of the uncles of these children has a son of 9 and a daughter of 4; the latter was born entirely without hair or nails. The data are not very full but the fact that normals carry the trait indicates that it may be accompanied by a definite defect in the germ plasm. Baer describes a family of ten children of two normal parents of which one was born hairless and has continued so while three were born with heavy hair but lost it; in two cases at 14 days and in one at 9 months.

The *form of the hair* may show family peculiarities. Thus, in some cases, it is thickened at intervals resembling a string of beads—hence called "monilithrix." A pedigree of a family of this sort has been recorded by Anderson (Fig. 112). Unaffected parents apparently yield only normals and abnormal parents are usually simplex, so that about half of the offspring have the new character.

The facts of inheritance of curliness have been considered on page 35.

Hair-coat Color.—Ordinarily the hair of the scalp is of uniform color but in man, no less than in horses, a piebald condition is possible. This shows itself in locks of white hair in the midst of a prevailing brown or red. This spotted condition is due to a definite positive factor, even as in the coat of mice, and two parents who lack spotted hair-coat will have only uniform-coated children. This is illustrated

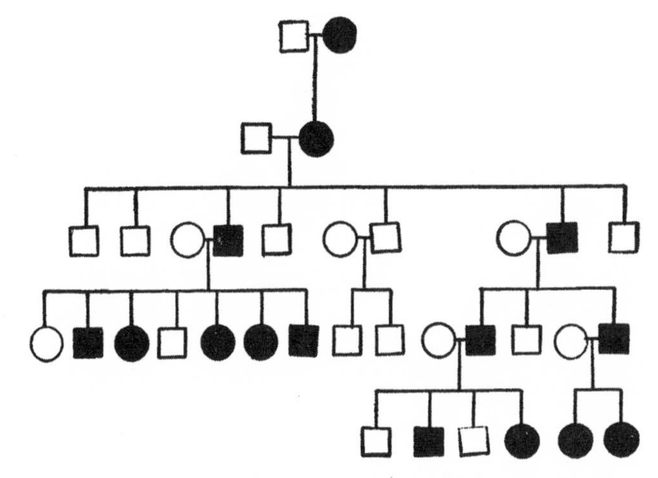

FIG. 112.—Inheritance of monilithrix—a positive character. Black symbols represent affected individuals. ANDERSON.

in the pedigree (Fig. 113) from Gossage. The hair-coat also varies in thickness and that this quality runs in families can hardly be doubted (Fig. 114).

c. **Nails.**—Hereditary nail defects are almost always associated with hair defects, as in the cases of hair peculiarities already described. One family pedigree must suffice for nail and hair defect (Fig. 115).

d. **Teeth.**—As is well known each half of either jaw has typically 2 incisors, 1 canine, 2 bicuspids and 3 molars. To this formula there are, however, exceptions and these exceptional conditions may run in families. Thus McQuillen records a family in which father, son and grandson lacked the lateral incisors of the upper jaw, a second son had them exceedingly dwarfed and some of *his* children had them so stunted that they were unsightly. The absence of the last

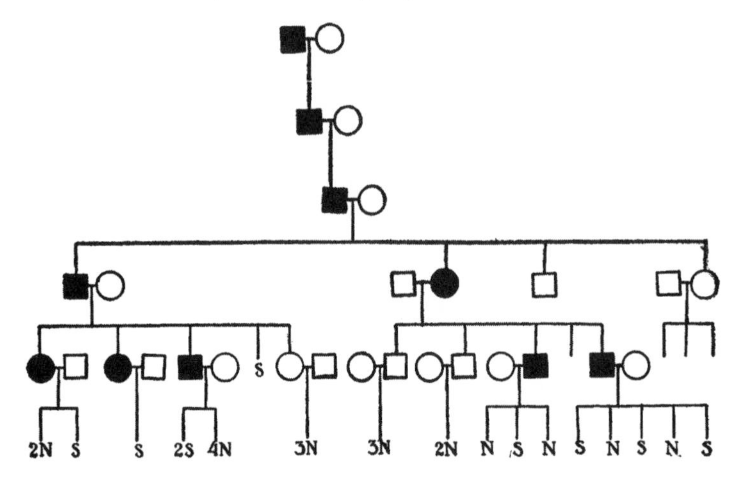

FIG. 113.—Pedigree chart, showing inheritance of spottedness in human hair covering—"congenital lock of white hair." Affected persons in black symbols. *S*, spot in hair-coat, sex unknown. GOSSAGE, after RIZZOLI.

molar is perhaps the commonest variation but no good evidence of its extended occurrence in families is at hand.

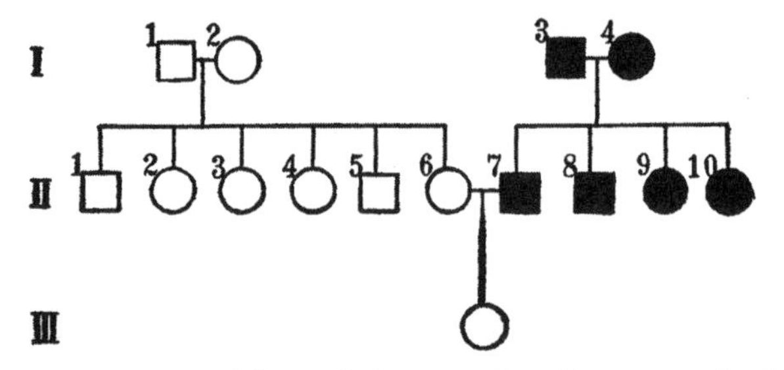

FIG. 114.—Pedigree of heavy hair-coat. I, 3, heavy growth of hair on head and face; I, 4, heavy growth of hair on head; II, 7, 8, heavy growth of hair on head and face; II, 9, 10, heavy growth of hair on head. F. R.; Tin. 1.

Entire absence of teeth is occasionally found as a family trait—there are said to be several such families in America but they have not yet been studied in detail. Guilford

(1883) records the case of a woman who never had teeth nor hair. Her sister was normal but her son was edentulous, and hairless. The sister (by an undescribed consort) had 18 children who grew up. Of these, one is edentulous while some of the others have failed to erupt all of their teeth.

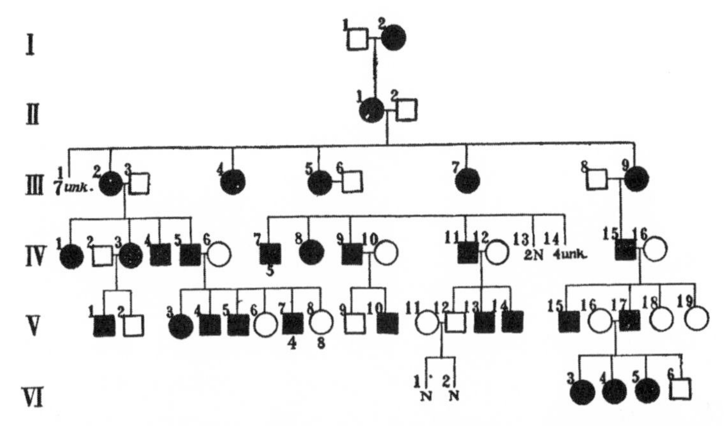

FIG. 115.—Pedigree of a family with peculiarities of hair and nails. I, 2, wife of PIROUT, poorly nourished nails and hair; II, 1 wife of QUIMBEL, *born* Rouen, 1775, poorly nourished nails and hair; III, 2, mar. DELAF, bald with bad nails; III, 4, bald, bad nails; III, 5, DELAU, bald, bad nails; III, 7, bald, bad nails; III, 9, bald, bad nails; IV, 1, bad nails; IV, 3, bald and bad nails; IV, 4, chestnut hair, bad nails; IV, 5, bald and bad nails; IV, 7, stands for 5 boys who were bald and had bad nails; IV, 8, a girl who is bald and has bad nails; IV, 9, rachitic in childhood, bad hair and nails; IV, 11, bad nails and hair; IV, 15, bad nails and hair; V, 1, had bad nails and hair, he died insane but his brother was normal. Of the children of IV, 5, 6, three had bad nails and hair, four (V, 7) were bald as well and nine others were normal. Of the children of IV, 11, 12, two had bad nails and hair. Of the children of IV, 15, 16, two had bad nails and hair and there were three granddaughters similarly affected. NICOLLÉ et HALIPRÉ, 1895.

The edentulous son married a normal (?) woman and had eight children. One, 14 years of age, who was examined, had many teeth undeveloped; another, at 16 years of age, had only 14 teeth when 28 were to be expected. Further data are necessary to determine whether or not imperfect development of the dental arcade is due to a genuine defect in the germ plasm.

Abnormalities in excess number of teeth are also found. Tomes refers to the occurrence of "well defined additional lingual cusps in the upper molar" in both "father and his

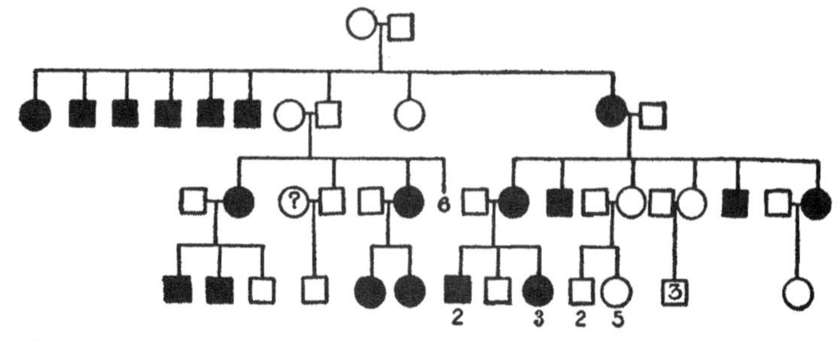

FIG. 116.—Pedigree of family with faulty enamel of the teeth—"brown teeth." Numbers below, or inside of, symbols indicate the number of individuals of the sex and condition of teeth. With one possible exception affected persons have at least one affected parent. SPOKES, 1889.

children." An American family with whom the writer has corresponded has a double set of permanent teeth as a family trait.

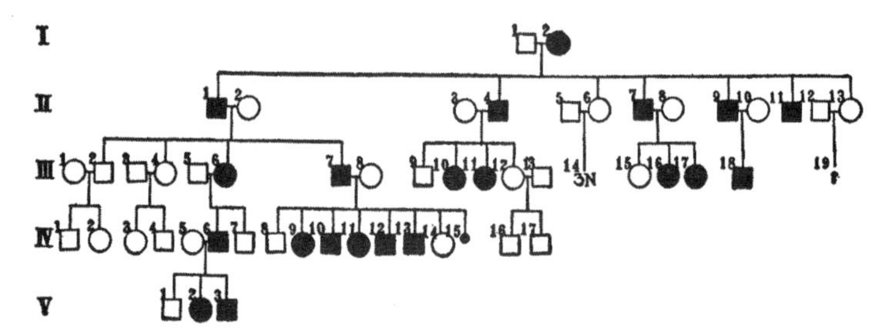

FIG. 117.—Pedigree of hypoplasia of enamel in THROWER-WALSINGHAM-CHESSUM family of Ware, England. I, 2, original parents of strain; II, 1, at the age of 84 two stunted teeth in the upper jaw; III, 6, two stunted upper teeth; III, 7, at 51 years has the fourth upper right and fifth lower teeth broken down; IV, 6, some teeth never erupted; some broken down; IV, 9, at 30 some teeth small, some never erupted. This dental peculiarity appears only in the offspring of an affected parent, consequently it is a positive trait. TURNER, 1907.

More complete are the studies made on families with *faulty enamel* of the teeth. In Fig. 116 is given the case of "brown teeth" due to faulty enamel. In Fig. 117 is given

B

A C

FIG. 118.—A case of reappearance of peculiarities in the features of three generations; namely, upturned nose and receding lower jaw. *A*, the grandfather; *B*, his daughter; *C*, his granddaughter. V. H. JACKSON, Orthodontia, 1904.

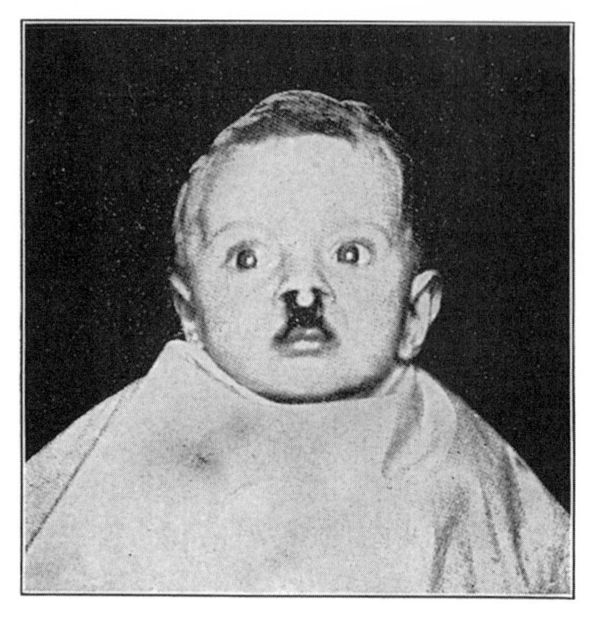

Fig. 119.—Case of harelip at one year of age. R. W. Murray, "Harelip and Cleft Palate," 1902.

a second case of insufficient enamel together with failure of some teeth to erupt. In these cases the abnormal condition seems to be due to some additional factor, inhibiting, as it were, the normal development of the enamel.

There is a close relation between the form of the jaw and peculiarities of dentition. That the form of the jaw is inheritable is nicely shown in figure 118.

e. Harelip and Cleft Palate.—These are intimately associated deformities, due to a more or less complete failure of the foundations of the upper jaw, which are paired, to grow completely to the middle line of the roof of the mouth. If the failure to close is in front harelip results, if behind cleft palate or merely cleft uvula. Occasionally both cleft palate and harelip may be present (Fig. 119).

A number of fairly extended pedigrees have been published (Rischbieth, 1909) yet they are not as critical as one would like (Figs. 120, 121), particularly, the consorts are

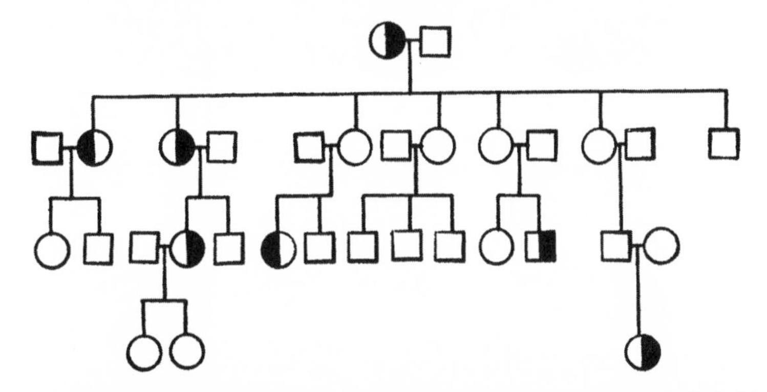

FIG. 120.—Pedigree of a family with harelip (right half of symbol dark) and cleft palate (left half dark). Frequently the affected persons descend from affected parents. APERT, 1907, after SCHMITZ.

rarely given. One can say, however, that the defect seems not to be sex-limited. So often are some of the children

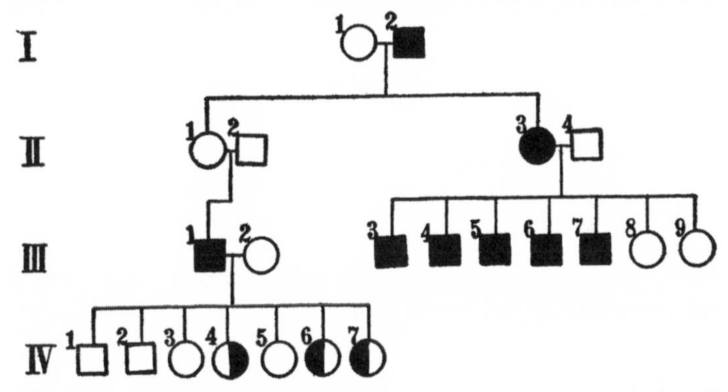

FIG. 121.—Pedigree of harelip (solid black symbol) and cleft palate (half black symbol). The type of defect is not constant. I, 2, simple fissure; II, 3, bilateral fissure; III, 1, palatine fissure; III, 3–7, lip fissures; IV, 4, harelip with cleft palate; IV, 6, 7, palate cleft without harelip. This particular pedigree is interesting because of an alternation of the affected sex in successive generations. SCHMITZ, 1904.

of one affected parent defective that the first impression is that the trait is dominant. But, if so, two normals should not have affected offspring—but this is just what is alleged commonly to happen. These cases, however, deserve careful study. Frequently when both parents of the defective child are normal one of them will belong to a fraternity with the defect; occasionally, however, one must go back to the second ancestral generation to find an affected relative. No eugenic instruction is, as yet possible. Correspondence from affected persons, or their relatives, who will volunteer to coöperate in studying the method of inheritance of this trait is solicited.

31. Cancer and Tumor

The question of "inheritance of cancer" has been much discussed and nothing but difference of opinion has resulted. This is largely due to the bad formulation of the problem. In the first place, if, as seems probable, the stimulus to cancer growth is an inoculable something—germ or ferment—it does not follow that the consequence of stimulus is not determined by an inheritable factor. It is known that certain strains or families of mice are uninoculable while others will acquire cancer upon inoculation. The question is, are there human strains that are easily and others with difficulty inoculable? The whole question is complicated by the fact that cancer is a disease of middle or later life. Thus in the census for 1900 we find that the heavy incidence of deaths from cancer occurs between 40 and 80 years (84.4%). The detailed distribution is shown in Table X. Here we see that the death rate of cancer (as compared with deaths from all causes) reaches its highest point at between 50 and 60 years, but that absolutely more deaths occur from that disease between 60 and 70 years. On account of this heavy mortality late in life many who are inoculable never reveal the fact, owing to their death before the cancer age. If cancer is communicable,

TABLE X

DISTRIBUTION OF DEATHS FROM CANCER IN AGE GROUPS

At death period	40–49	50–59	60–69	70–80
Per cent of all deaths from cancer	17.1	24.4	25.8	17.1
Proportion of cancer deaths to all deaths at that age period	8.3	11.2	10.1	7.0

like typhoid fever, still not all who are non-resistant will die from cancer because some will not become inoculated. The answer to the question of the "heredity of cancer" is not to be sought in mass statistics—in the correlation of

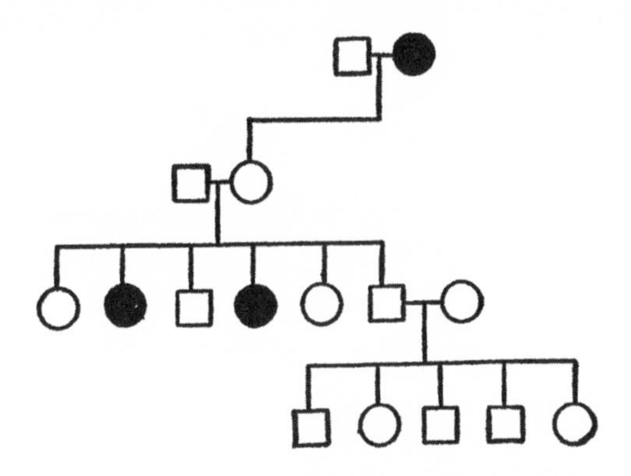

FIG. 122.—Pedigree of cancer. In the first generation cancer is admitted. In the second it is not known to have occurred, but the father died at 71 of a somewhat mysterious disease. In the third generation were two cases of cancer (one "bone cancer"). The fourth generation contains persons who are still young.

deaths from cancer between parents and children, but only by a careful analysis and comparison of individual families. One then sees in many families no deaths from cancer among 10 to 20 persons dying at cancer age, while in other families there will be 2 or 3 or even 4 deaths from cancer among those dying at the cancer age (Fig. 122). Thus in a pedigree that lies before me, half of those who have died at 35 years or over have died of cancer or tumor or have been operated on for cancer (4 cases in all) and two others have been operated on by a cancer surgeon, but details were not furnished. Two others in the family are suspected of having died of the disease. Now such families as these are by no means rare and this is the basis for the conclusion that there is a family liability to cancer.

Moreover, there is a specificity of the disease in each particular family. In one family non-resistance shows itself in the females in cancer of the breast, in another, in cancer of the uterus, in another in cancer of the intestine. Silcox (1892, Fig. 123) gives a fragment of a pedigree showing that a father, four daughters and a granddaughter all probably have sarcoma of the eyeball; and Broca records the case of a woman and three daughters who, at about the same age, possessed fibrous formation on the breast. Considering the few pedigrees of cancer families extant and the large number of organs subject to cancer these cases of cancer in the same organ strengthen materially the view of specific inheritability.

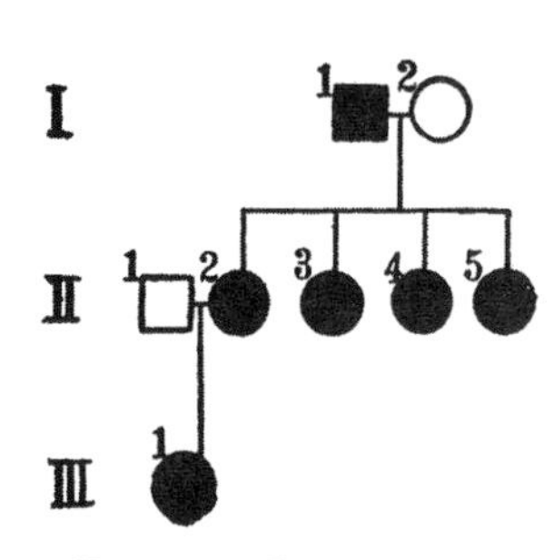

FIG. 123.—Fragment of a pedigree showing a specific inheritance of sarcoma of the eyeball. All persons indicated by black symbols are similarly affected. SILCOX, 1892.

That certain "benign" tumors are hereditary is indicated by various records in the literature. Thus Atkinson cites the case of a man whose body was covered with countless tumors varying in size from that of a canary seed to that of a pullet's egg. His sister and their father were similarly affected. The disease is not a common one in this form and this fact gives its high incidence in this family the greater weight as evidence that internal conditions have at least molded the form taken by the disease.

32. Diseases of the Muscular System

Since most muscular response is controlled by the nervous system it is frequently difficult to determine whether a peculiarity of muscular response is due chiefly to the one organ or the other. The classification of these diseases is therefore somewhat arbitrary.

a. Thomsen's Disease is a rather rare one in most localities. It is characterized by lack of tone and prompt re-

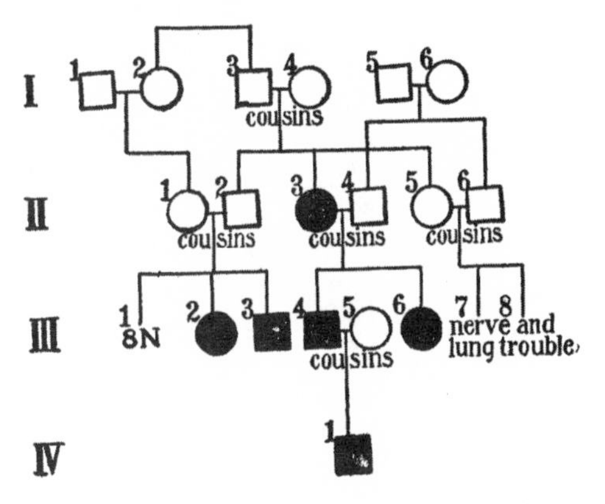

Fig. 124.—Pedigree of Thomsen's disease. Appears in cousin marriages even from unaffected parents; hence due to a defect. Bernhardt, 1885.

sponsiveness in the voluntary muscles. A striking pedigree has been recorded by Thomsen (Fig. 124). It shows a remarkable reappearance of the disease in the offspring of cousin marriages and this indicates that the disease is due to some sort of a defect whose nature has yet to be elucidated. The clear eugenic advice is *outmarriage.*

b. Certain Muscular Atrophies appear to be secondary to diseases of the nervous system while others seem to originate in the muscles themselves, without corresponding defects in the nervous centers. In a family described by Herringham (1885) sometimes all appendages, sometimes the arms only,

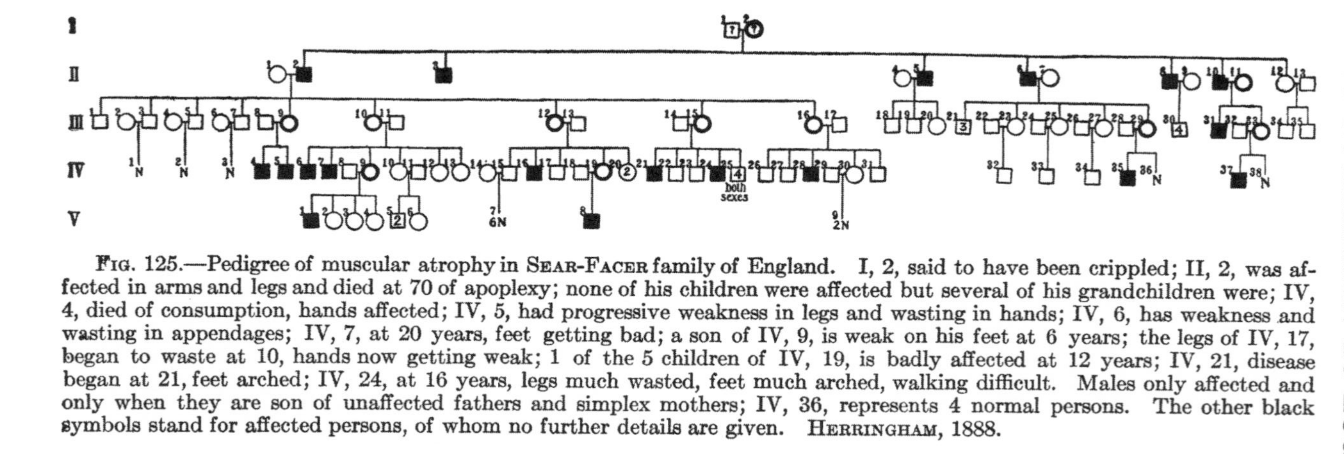

Fig. 125.—Pedigree of muscular atrophy in Sear-Facer family of England. I, 2, said to have been crippled; II, 2, was affected in arms and legs and died at 70 of apoplexy; none of his children were affected but several of his grandchildren were; IV, 4, died of consumption, hands affected; IV, 5, had progressive weakness in legs and wasting in hands; IV, 6, has weakness and wasting in appendages; IV, 7, at 20 years, feet getting bad; a son of IV, 9, is weak on his feet at 6 years; the legs of IV, 17, began to waste at 10, hands now getting weak; 1 of the 5 children of IV, 19, is badly affected at 12 years; IV, 21, disease began at 21, feet arched; IV, 24, at 16 years, legs much wasted, feet much arched, walking difficult. Males only affected and only when they are son of unaffected fathers and simplex mothers; IV, 36, represents 4 normal persons. The other black symbols stand for affected persons, of whom no further details are given. Herringham, 1888.

underwent a slow atrophy starting as early even as the twelfth year. The method of inheritance in this family is striking. Only males are affected and they, as well as the unaffected females, may transmit the defect; but unaffected males have no affected children. Femaleness in this family is incompatible with atrophy. (Fig. 125).

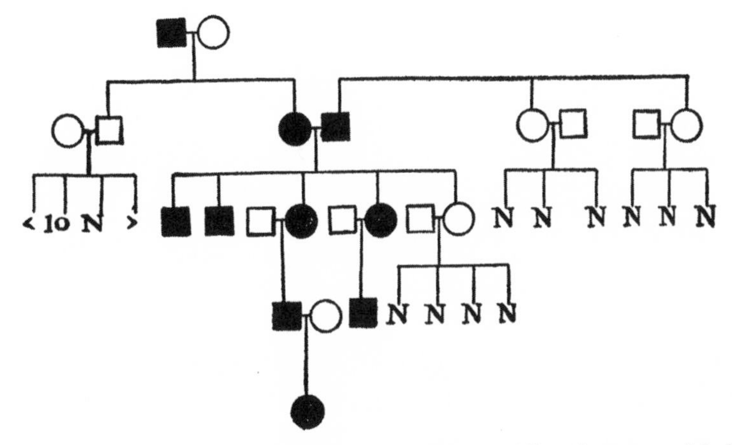

Fig. 126.—Pedigree of a family of tremblers. Affected persons (black symbols) are derived from at least 1 affected parent, and 2 normal parents have only normal offspring. Trembling is thus due to the presence of a special character. From Debore and Renault, 1891.

c. A family of *tremblers* has been recorded by Debore and Renault. In this family all normals produce only normal offspring while two affected parents may have a normal child. The pedigree deserves no great stress since details are lacking (Fig. 126).

d. **Hernia.**—Man's erect position is accompanied by physical dangers from which his quadruped ancestors were free, for in man the weight of the viscera has largely to be borne by the pelvis and lower abdominal wall. The erect position has subjected the muscles of the inguinal region to a peculiarly rigorous test. They often fail and an inguinal hernia is the result. Such hernias usually are consequent to a strain but the strain merely reveals, and does not cause, the weakness.

That such weakness or liability to hernia is inherited admits of little doubt. Just how, there is hardly sufficient data

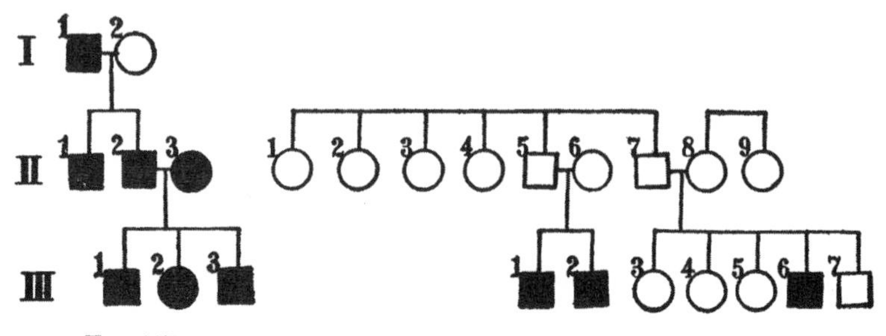

FIG. 127 FIG. 128

FIG. 127.—Pedigree of inguinal hernia. Probably only affected persons (black) are shown. All males have a right handed scrotal hernia and both affected females have a femoral hernia. COUCH, 1895.

FIG. 128.—Pedigree of inguinal hernia (black symbols). F. R.; Rei. 3.

to determine with certainty. It is probable that a weakness from both sides of the house will yield only weak offspring. This is indicated in Figs. 127, 128; all males have a right handed scrotal hernia and both affected females have a femoral hernia.

33. DISEASES OF THE BLOOD

These are generally classified into two groups; the anemic and the hemorrhagic; in both, the evidence of an inheritable tendency is clear.

a. Of the Anemic Diseases, *chlorosis* is the commonest, is found almost exclusively in females, and occurs frequently enough in many or all of the females of one family to render it probable that eventually it will be found to accompany a distinct inheritable weakness.[1] A careful study of pedigrees is highly desirable.

[1] Potain (Article, Anemia, Dict. encycl. des sci. med.) says "The children of a chlorotic woman are often all chlorotic—and in certain cases even the male children do not escape."

b. Progressive pernicious anemia.—This is a relatively rare disease which has been little studied from the standpoint of heredity. A case described by Bramwell (1876) is suggestive (Fig. 129).

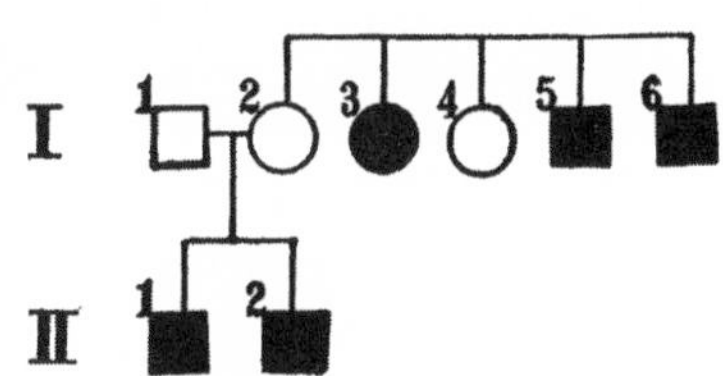

FIG. 129.—Pedigree of a family with progressive pernicious anemia. The mother, I, 2, died of cardiac weakness and chronic diarrhea; it is uncertain in how far a tendency to anemia was responsible for the result. I, 4, died of a heart trouble which was not further diagnosed. The other three members of the fraternity died of anemia. Both children, II, 1, 2, were affected with progressive anemia. BRAMWELL.

c. Nosebleed (epistaxis) — This representative of the hemorrhagic diseases of the blood may be a family disease, characterized by its frequency and severity and occasionally by its fatalness. In some of the fraternities from an affected parent all, in others about half, of the children are affected. An example is the family described by Babington (1865). Unfortunately no facts are given about consorts (Fig. 130). In this case most of the persons were violently affected. The fact that no cases are recorded from normal persons in so far raises the suspicion that the disease is due to the presence of a positive trait, which should tend to make persons having a violent form of the trait hesitate about having children.

d. Telangiectasis.—Nosebleed is often associated with red spots in the skin from which bleeding may occur. This condition is called telangiectasis; its behavior is well illustrated in Figs. 131, 132. Like epistaxis it seems to be a dominant trait, so that normal children who outmarry will probably have no affected offspring.

e. Hemophilia.—This remarkable condition is characterized by a proneness to hemorrhage and by difficulty in blood-clotting, so that a hemorrhage once started is stopped with difficulty. Families with this peculiarity (fortunately not very frequent) are known as "bleeders." In such fam-

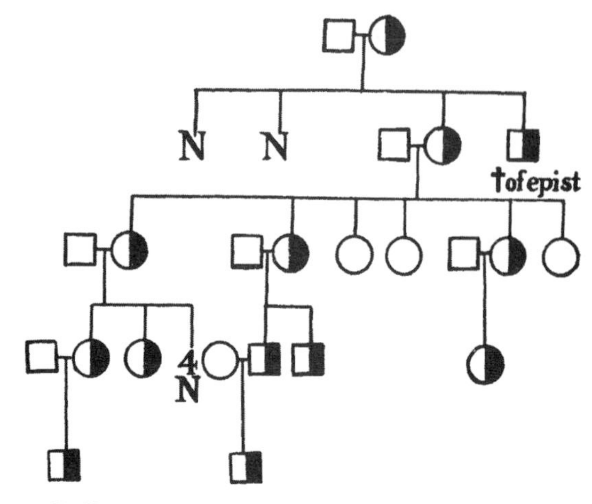

FIG. 130.—Pedigree of a family showing epistaxis or nosebleed. Affected persons indicated by half shaded symbols. All affected persons arise from an affected ancestor. *N*, normal. Consorts unknown. BABINGTON, 1865.

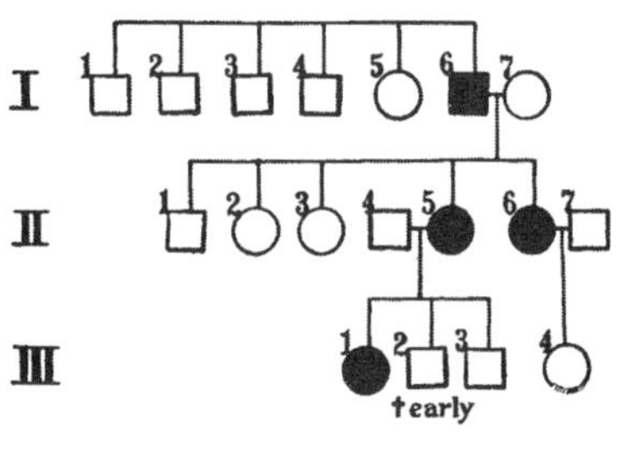

FIG. 131.—Pedigree of family showing multiple telangiectasis. Affected persons (solid black) from affected parent only. I, 6, had "spots" on face, subject to vomiting and to nosebleed, from which latter he died. II, 5, spots appeared at between 38 and 48 years, epistaxis increased and led to her death. Her daughter, III, 1, is gaining telangiectasis but the younger son at 20 years shows no sign of trouble; II, 6, has red spots that first appeared in her 27th year and are extending.

ilies there are more than fifty times as many affected males as females. In general as age advances, the severity of the hemorrhages diminishes and finally they cease altogether.

As in other diseases so in hemophilia special variants appear in particular families. Thus among some of the de-

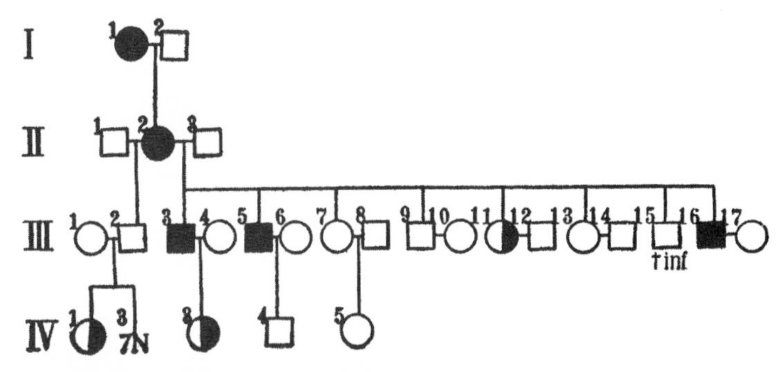

FIG. 132.—Pedigree of multiple telangiectasis. I, 1, is an English woman who was subject to epistaxis (nosebleed) and had red spots on her face; her daughter, II, 2, 60 years old, has a number of bright red angiomata distributed over face, ears, lips, tongue, mucous membrane of mouth, and inner surface of all 4 eyelids. During last 6 years has had recurrent attacks of epistaxis. By her first husband she had a son and 8 grandchildren of whom 1 suffers from epistaxis. By her second husband she had 8 children of whom III, 3, has had epistaxis since 8 years and 2 small "spider naevi" on left cheek and has a child of 11 who suffers from epistaxis; III, 5, has nosebleed and 3 small spots on cheek; his son is normal as yet; III, 11, has epistaxis; III, 16, has slight attacks of epistaxis but no spots visible. WEBER, 1907.

scendants of the early settlers of Sullivan Co., Pennsylvania, occur "nine-day bleeders." "After the wound is received, instead of healing a sort of core, of very dark color, composed mostly of coagulated blood forms in the wound, which in about nine days opens, and the blood begins to flow as if from a freshly severed artery. It usually continues to bleed about two weeks, or until the patient is thoroughly exhausted, when the "core" falls out and the wound heals. Binding up the wound does no good. The only death known to have occurred through bleeding is supposed to have been caused by binding the wound lightly to stay the flow of blood."

That hemophilia has an hereditary basis is generally conceded and the conclusion would not be weakened were a specific hemophilia germ some day demonstrated. The particular method of inheritance is well illustrated by Fig. 133 of the Sullivan County strain. The males alone are af-

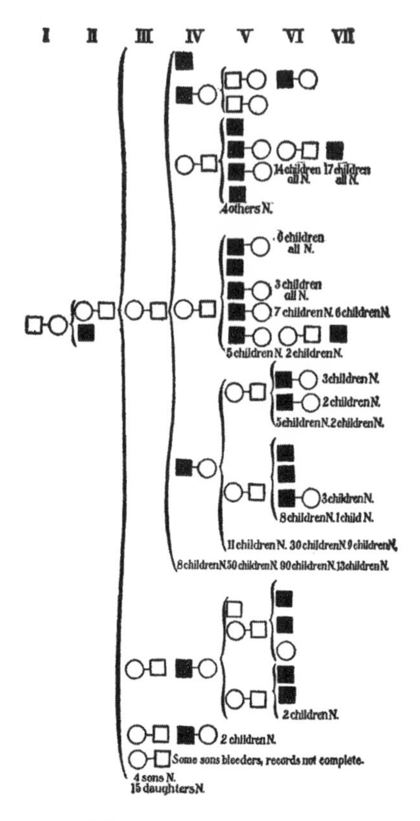

FIG. 133.—Pedigree of the Sullivan Co., Pa., bleeders. Roman numerals at top of columns indicate generations. Of the two symbols connected by a horizontal line that at the left is the direct descendant, that at the right the consort; the bracket includes their children. Only males are bleeders, and bleeding children are derived always from non-bleeding females of the family. PARDOE, 1904.

fected. No male of the family, whether affected or not has affected offspring so long as he marries outside of the family. Hence, all "bleeding" children are derived from the females of the family.

Fig. 134 gives the pedigree of the family Mampel from Kirchheim near Heidelberg (Lossen, 1905), and Fig. 135 is the pedigree of a family that settled in Carroll Co., Maryland,

and has since spread over the country. It is remarkable because it contains records of female bleeders, whose occurrence has been doubted by Bulloch (1911).

The eugenic teaching that holds for practically all families is clear. Sisters of bleeders should not have children. Males if not actual bleeders may, so far as this trait goes, marry and reproduce with impunity—their germ plasm is free of taint of hemophilia.

Hemophilia is a particularly difficult disease to control in descent because it is disseminated by normal females. On this account it is liable to produce a community of bleeders as it formerly did at Tenna, Canton Graubunden, Switzerland. Even normal females from the old world families of bleeders may well be prevented from landing in America.

f. Splenic Anemia with enlargement of the Spleen.—This condition, usually recognized as hereditary, not infrequently appears in the offspring of two unaffected parents. In such a family reported by Bovaird (1900) 2 children out of a fraternity of 10 were affected. In a family reported by Brill there were affected 3 out of 6 (Fig. 136). In both families together there were, then, 5 out of 16. In another family, when one parent is affected, of 15 children of whom details are known, 5 were certainly affected, two doubtful and 8 were normal. Of the two matings involved one is consanguineous (Wilson, 1869, Fig. 137). Though the data are still meager the result favors the view that the liability to splenic anemia is due to the absence of some factor that usually gives strength. A person having or fearing such a defect should marry into a normal strain. It may be added that Gossage (1908, p. 321) suggests that splenic anemia is due to the presence of some dominant factor so the matter must be regarded as still unsettled.

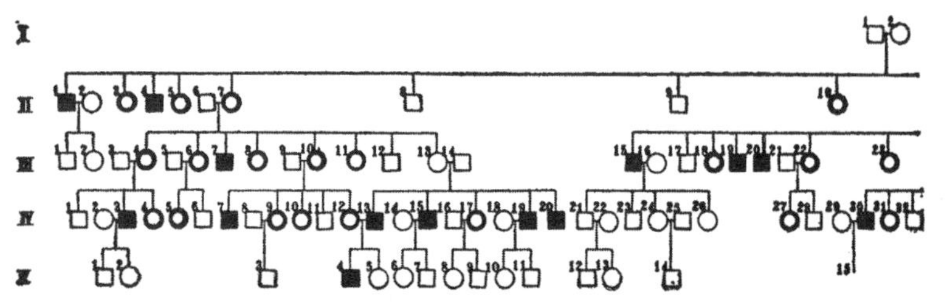

FIG. 134.—Pedigree of hemophilia in the MAMPEL family, originally of Kirchheim near Heidelberg, Germany. Black symbols indicate bleeders; it is seen that they are males only, but they, in turn, have no bleeding sons.

34. Disease of the Thyroid Gland

This may lead to a variety of effects, cretinism, goitre, myxedema, exophthalmic goitre, etc. Many of these show evidence of an inheritance of the liability to thyroid degeneracy.

a. Cretinism.—This is characterized by arrest of growth, by large pendulous abdomen, poor teeth, coarse, scanty scalp hair, mongolian face, feebly developed genitalia, and marked impairment of intelligence. The thyroid gland is often absent and a goitre frequently present. The distribution of the disease is interesting. It appears chiefly in mountainous countries where close intermarriage is more likely to occur than on the plains. Thus it abounds in Switzerland and is said to occur in some parts of Scotland. It is a cause of deportation when it occurs in immigrants to this country. That it is hereditary admits of no doubt. Aosta, at the southern base of Mount St. Bernard, was once a great breeding place of cretins, since their marriage there was permitted. For some years they have been segregated and kept from marrying and now, we are told, they are nearly all gone (Jordan, 1910).

b. Goitre.—That goitre frequently occurs repeatedly in families is well known; but in how far this is due to common sources of infection is still disputed. Buschan states that

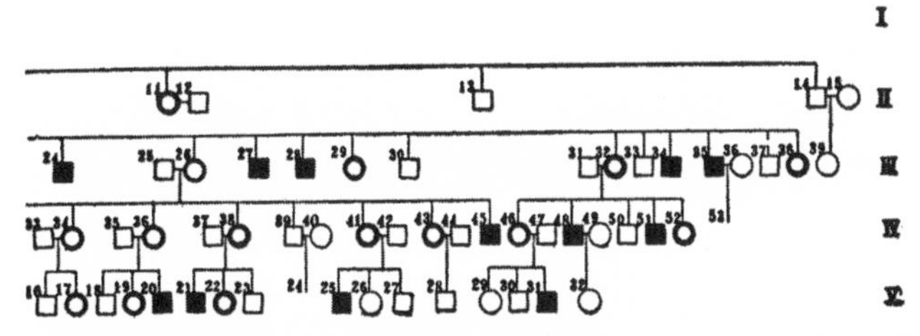

Heavy ringed circles are normal females who transmit the trait. Lossen, 1905. The details of Lossen's paper are translated in the "Treasury of Human Inheritance," Parts V and VI, pp. 267–271.

family histories of goitrous patients usually show a neuropathic ancestry. A pedigree from Buschan is given in Fig. 138.

c. **Exophthalmic Goitre.**—This peculiar condition is characterized by an enlargement of the thyroid gland, protrusion of the eyeballs, and extreme nervousness. It more commonly affects women than men. Although, in the country as a whole, it is not common yet it is more prevalent in some districts than in others, doubtless owing to the interrelationship of the members of the district with heavy incidence of the disease.

The disease is common in females; yet it is not inherited strictly in sex-limited fashion. It is, however, clearly inherited; as certainly as epilepsy, with which it is not infrequently associated. Not many family pedigrees seem, however, to have been studied (Fig. 139).

35. Diseases of the Vascular System

This system consists, in the narrow sense, of the heart, arteries and veins. Less is known about heredity of defects and diseases of such an internal system because it is so inaccessible to observation and study in the living person. Nevertheless we shall see that "blood tells" in respect to the traits of this set of organs.

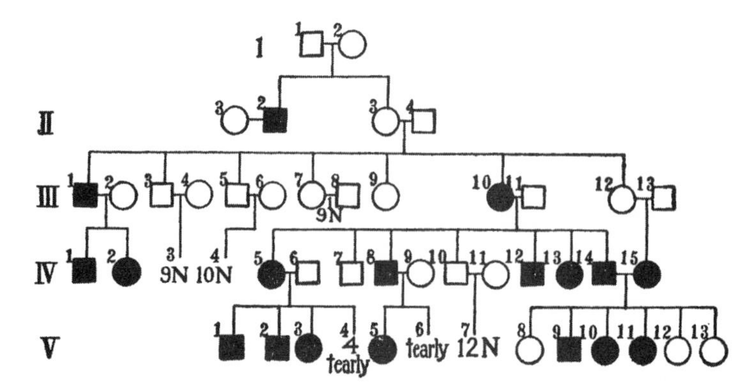

FIG. 135.—Pedigree of a family of "bleeders"—the K. family, located in and about Carroll Co., Maryland. Their son, II, 2, was a bleeder but died without issue. The eldest son, III, 1, of the daughter was a bleeder from 18 up to 45 years, "often bled till he fainted." He had 2 unaffected brothers and 3 normal sisters but 1 sister, III, 10, was "a bleeder until 40." He had a son, IV, 1, who was a very bad bleeder from 18 until toward middle life and a daughter, IV, 2, who often "bled until she fainted" and eventually died of dysentery. All 19 children of the 2 normal brothers were normal and 9 children of the normal sister, III, 7. The affected sister, III, 10, had 3 sons and 2 daughters who were affected. IV, 5, is stated to be "a bleeder" and had by an unaffected husband 2 bleeding sons and 1 bleeding daughter besides 4 others who died of scarlatina. Her brother, IV, 8, had a daughter, V, 5, who was a bleeder until 15, and then died of a hemorrhage of the lungs consequent upon tuberculosis. There were other children all of whom died young of scarlatina. The normal brother, IV, 10, had 12 normal children. The next 2 had no offspring. The youngest son, IV, 14, began to bleed while an infant, grew worse until he was 25 and has since improved. He married a cousin who is also a bleeder and they have 6 children. Three of the daughters have not bled as yet. V, 9, has been a bleeder since he was 8 months old and bleeds until he faints; V, 10, has been a bleeder since she was 8 months old and V, 11, bleeds occasionally but not very severely. Original data, contributed by Dr. J. H. STICK.

a. Heart.—That congenital heart defects are hereditary has long been known and the striking evidence for it has been brought together by Vierordt (1901). His summary deserves translating entire: "Friedberg mentions 3 sons suffering from cyanosis (due to imperfect structure of heart) from one father, 2 from his first, 1 from his second marriage; likewise Foot records 3 cases in one family; Haillet reports on 4 children with open foetal canals (in the heart)

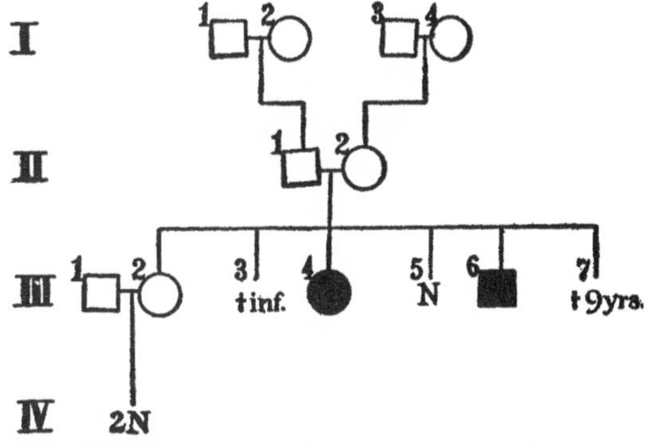

FIG. 136.—Pedigree of a family with splenic anemia. I, 1, died at 73 of gall stones; I, 2, died at 94 from a fall; I, 3, died at 72 of pneumonia; I, 4 died at 38 from childbirth; II, 1, died of pneumonia and II, 2, is in perfect health at 62 years. In the third generation all are well except that III, 3, died in infancy of diarrhea; III, 4, was well until an enlargement of the spleen occurred, which has continued; III, 6, 30 years old, suffers a continued enlargement of the spleen; and III, 7, died at 9 years of an enlargement of the spleen. BRILL, 1901.

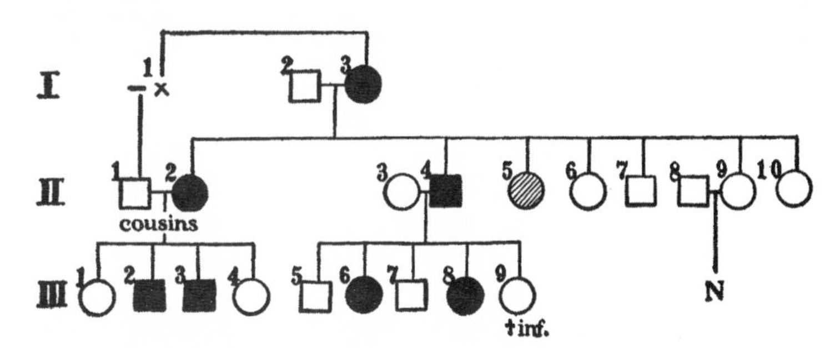

FIG. 137.—Pedigree of splenic anemia. A. P., I, 2, has a form of nervous deafness but otherwise healthy until attacked by diabetis mellitus. His wife gained sallow complexion and enlarged spleen at 33 years. Of their children one, II, 2, had enlarged spleen, at 7; she married a cousin and had 2 boys with projecting spleen. A son, II, 4, is subject to epistaxis and fainting spells; since 35 years old his spleen has been enlarged; he has 2 affected girls; II, 5, became deaf at 4; she is becoming sallow, but the spleen is not palpable. II, 6, is slightly deaf. WILSON, 1869.

from one marriage; Strehler of a rachitic woman who bore 5 cyanotic children, 3 boys and 2 girls; the father (who later died of phthisis) has by a second wife a normal daughter. In Kelly's case of transposition the mother had borne 11 children of whom one died at 5 months from congenital heart disease. In the case of Schmaltz, that of a seven year old boy, the father and father's mother had heart defect. The patient of Potocki who, 29 years old, died of brain abscess and had a pulmonary stenosis with closed septum and defect of the interauricular septum, descended from a mother with a congenital heart disease. Rezek observed 8 cases of heart disease in 4 generations of one family, including 2 congenital defects; the mother probably having got her heart disease from the grandmother (Fig. 140). Two sisters afflicted with ichthyosis congenita, descended, according to Leuch's report, from a mother who suffered from a defect of the bicuspid valve; the oldest

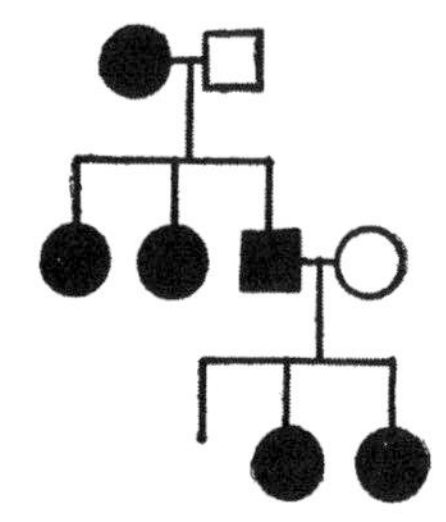

FIG. 138.—Pedigree of goitre. Affected persons come from at least one affected parent. BUSCHAN.

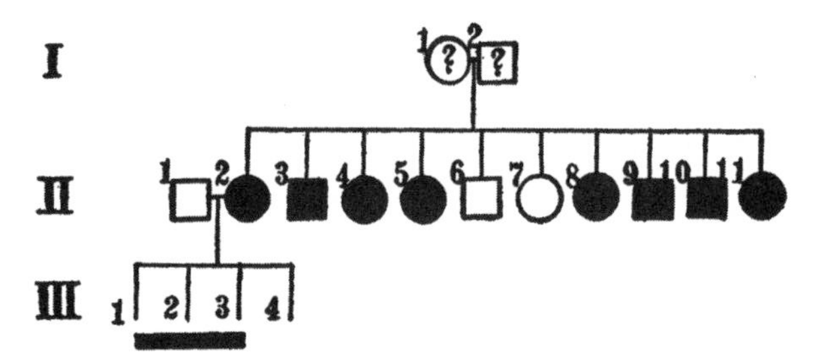

FIG. 139.—Pedigree of a family showing heavy incidence of exophthalmic goitre. III, 1, 2, 3, also affected; sex unknown.

child, the son, had also congenital heart disease. . . . Eger found in 12 cases of congenital heart disease, three times lues patris as well as consanguinity of the parents." To these cases it would be possible to add almost indefinitely. "Heart disease" is very common, but it does not fall upon individuals at random, but prevailingly upon strains with an inherent liability or weakness (Figs. 140–143).

b. Arteriosclerosis.—While degeneration of the wall of the arteries is ascribed to numerous inciting causes there can be no doubt that the cerebral hemorrhages, even of old age,

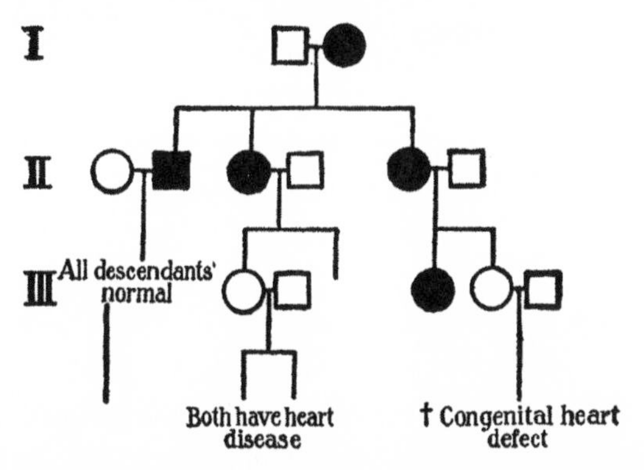

FIG 140.—Pedigree of heart disease. I, 2, probably had heart disease, II, 2, 3, and 5 had heart disease. The descendants of II, 1, 2, are normal for two generations. Those of II, 3, 4, are healthy but 1 of them has 2 children with heart disease. II, 5, has a daughter and a grandson who died of congenital heart defect. REZEK from VIERORDT, 1901.

are dependent in large part upon an inherited strength or resistance. Cases of arteriosclerosis have been reported in infants and here heredity must play an important rôle.

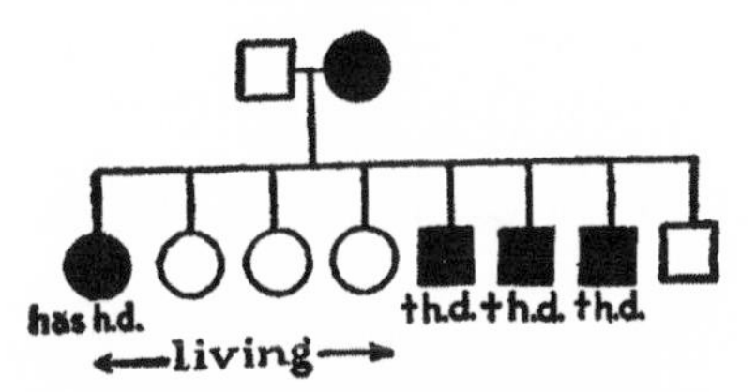

FIG. 141.—Pedigree of "heart disease."

36. DISEASES OF THE RESPIRATORY SYSTEM

The respiratory organs, including the passages to it that are lined by mucous membranes, are the weakest part of our body. This is probably because our remote ancestors, at the beginning of the vertebrate series, were aquatic animals and we land animals have not yet become fully adjusted to life in the air. The dry, dusty and often germ laden air is a difficulty with which our mucous membranes can hardly grapple; little wonder that they, and the whole body, so often succumb.

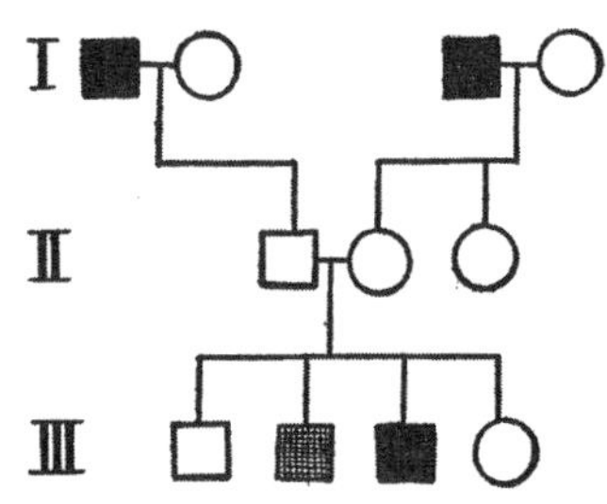

Fig. 142.—Pedigree of heart trouble. The father's father, I, 1, died of anguina pectoris at 69 years; and the mother's father, I, 3, died of ossification of the valves of the heart at 59. Father and mother are living and said to be well. Of their children, III, 3, died of heart disease at 9 months and another, III, 2, had temporary heart trouble. F. R.; All. 1.

Of the diseases of the lungs the most fatal is tuberculosis. We know that it is induced by a germ and that if there is no germ there will be no tuberculosis of the lungs. The first impulse of the modern sanitarian is to eliminate the germ. But this is a supra-herculean task; for germs of tuberculosis are found in all cities and in the country amongst most domesticated animals. The germs are ubiquitous; how then shall any escape? Why do only 10 per cent die from the attacks of this parasite?

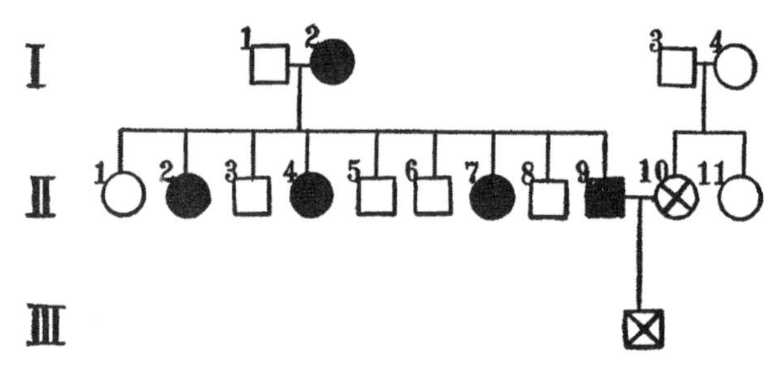

Fig. 143.—Pedigree of family with heart disease and migraine, I, 2, died of heart disease at 72 years; II, 2, 4, 7, died of "heart disease;" II, 9, died of "heart failure" at 59 years, hardworking physician; III, 1, suffers from migraine; her mother is a semi-invalid from migraine. F. R.; Bra. 1.

The answer is given by autopsies and the experiences of many physicians. Autopsies show that nearly all mature persons have the germs of tuberculosis in their lungs, but, for most part encysted and, perhaps, even completely destroyed.

Those who die of tuberculosis are those whose bodies have not been able successfully to combat the germs—their bodies have lost in the battle. Family physicians know cases where under bad conditions, overwork, depression of mind and body their patient will begin to decline and, then, under more favorable conditions begin to build up again. The battle wages now in favor of the one side, now of the other. The result depends quite as much on internal resistance as virulence of the germ.

That families vary in their internal resistance is well known. Dr. Coolidge of the Lakeville Sanitarium, Massachusetts, tells me that he classifies his patients on the basis of their resistance as measured by their response to good treatment in the first few days; and he states that the old New England families now show a relatively high resistance to tuberculosis as compared with recent immigrants.

The Family Histories that have been placed in my hands show the same thing. Though one in ten die of tuberculosis it was not difficult to pick out ten families in each of which about ten persons had died of whom not one had died of tuberculosis. On the other hand there are families with an incidence of consumption of 75 or 80 per cent. That this is not merely communication of the disease in the families with high death rate follows, of course, when we grant that practically all grown persons are infected anyway. It seems perfectly plain that death from tuberculosis is the resultant of infection added to natural and acquired non-resistance. It is, then, highly undesirable that two persons with weak resistance should marry, lest their children all carry this weakness.

Pneumonia.—Since the germ of pneumonia is a normal resident of our throats, the disease is not due merely to infection; but to a weakening of a natural or acquired resistance. Our Family Records show again and again the heavy incidence of pneumonia in certain families causing the death even of infants (Fig. 144).

Likewise a general weakness of the mucous membranes,

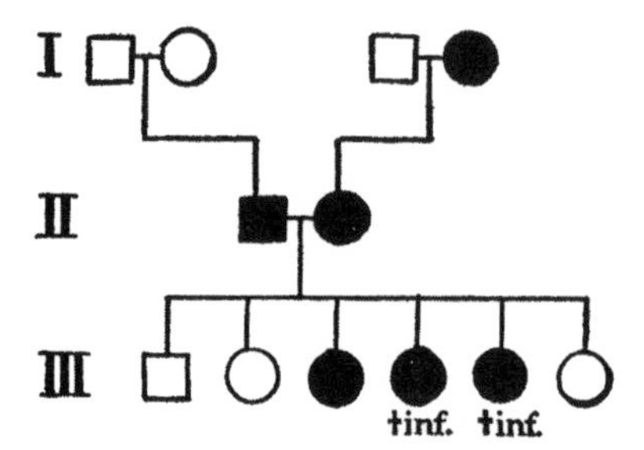

FIG. 144.—Pedigree of a family with tendency toward lung disease. I, 4, died of pneumonia at 82 years. II, 1, had an attack of pneumonia which terminated in tuberculosis from which he died at 43 years. His wife, II, 2, died at 62 years of tuberculosis. Of their 6 children 3 are still living; the others all died of pneumonia, 2 in early childhood. F. R.; Mor. 1.

leading to catarrh, adenoids, tonsilitis, deafness, bronchitis, etc., seems clearly to run in families. Such a case is illustrated in Fig. 145.

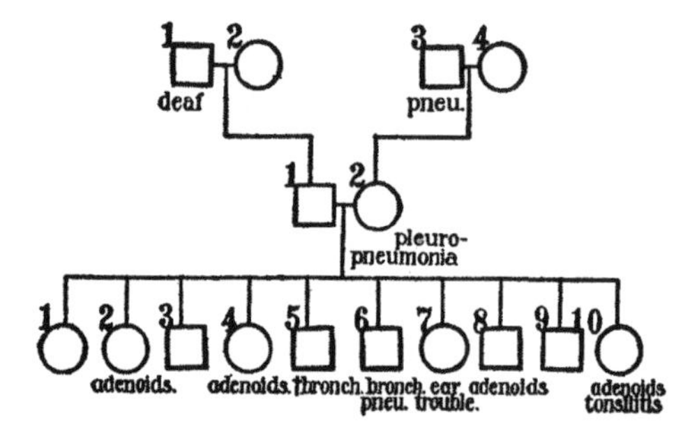

FIG. 145.—Showing "inheritance" of throat and ear weakness in a family. F. R.; New. 1.

37. DISEASES OF THE ALIMENTARY SYSTEM

The diseases of the alimentary tract are so largely due to bad habits in eating, exercising and attending to the demands of nature that most physicians consider a possible hereditary basis relatively unimportant. It is, to be sure, recognized that the "nervous temperament" may be largely responsible for disordered digestion by disturbing the ordinary secretory functions. So, likewise, it is probable that there are family characteristics which favor peculiarities of the liver resulting in its abnormal functioning. Especially jaundice and gout may have hereditary basis. An example of family pedigrees with high incidence of dyspepsia and more specific alimentary troubles is given in Fig. 146.

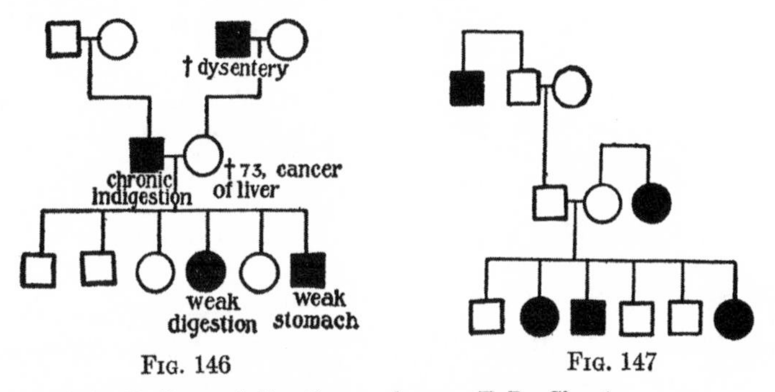

FIG. 146 FIG. 147

FIG. 146.—Pedigree of digestive weakness. F. R.; She. 1.

FIG. 147.—Pedigree of diabetes mellitus (black symbols). In this case the parents of affected offspring are not themselves affected; the trait is due to the absence of something that is present in normal persons. BRAMWELL, 1908, p. 265.

a. Diabetes Insipidus.[1]—This term has been applied to the symptoms of passing large amounts of greatly diluted urine. The affected persons have to drink much water to meet the rapid drainage through the kidneys. Numerous families are known that show this peculiarity in several close blood relatives. The typical condition is that two unaffected parents, even of diabetic strains, will have only normal children; diabetic offspring have at least one diabetic parent. This would indicate that diabetes is due to a positive factor (Fig. 148). Nettleship (1910) points out that age of incidence tends to diminish in successive generations.

[1] The hereditary behavior of diabetes mellitus or "sugar in urine" has been less studied. (Fig. 147).

The eugenic teaching is that persons with diabetes insipidus will probably have some diseased children, but unaffected persons, even of diabetic origin, will probably have only normal children.

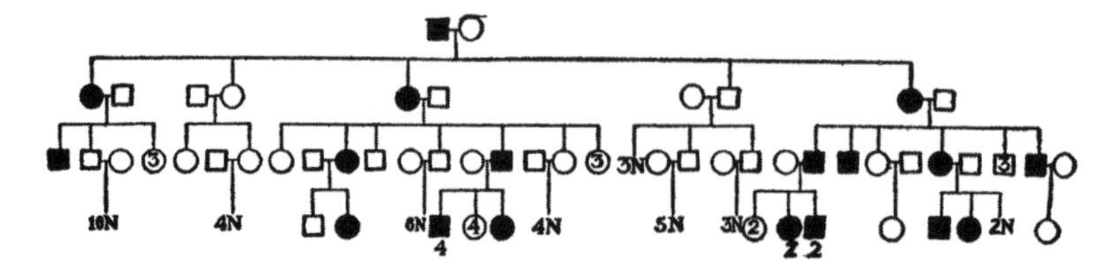

FIG. 148.—Pedigree of a family with diabetes insipidus. Affected persons (black symbols) are derived only from affected parents—thus diabetes is a positive trait. GOSSAGE, 1907.

38. DISEASES OF EXCRETION

Since the urine is the main stream carrying waste products of metabolism from the body it gives the best evidence of disorders of metabolism, hence much attention has been directed toward its study. Some of its peculiarities are known to be family traits.

a. Alkaptonuria. —This condition is marked by the constant excretion of homogentisic acid which darkens upon oxydation so that the urine darkens after passage; it is not injurious to the individual and has no special eugenic interest except as it illustrates the law of heredity. The transmission of this trait has been studied by Garrod (1902). The disease is a rare one and, apparently, occurs only in the offspring of two persons belonging to alkaptonuric strains. This condition is most easily met in cousin marriages and, as a matter of fact of the 17 alkaptonuric fraternities studied 8 were offspring of first cousins. When neither parent of an alkaptonuric fraternity is alkaptonuric about 1 in 4 of the children have the peculiarity. It appears then that alkaptonuria is due to the absence of a condition found in other (normal or ordinary) persons; and it is lost in the product of marriage of an alkaptonuric and a normal person.

b. Cystinuria and Cystin Infiltration are both family diseases though so rare that the method of inheritance has not been precisely determined.

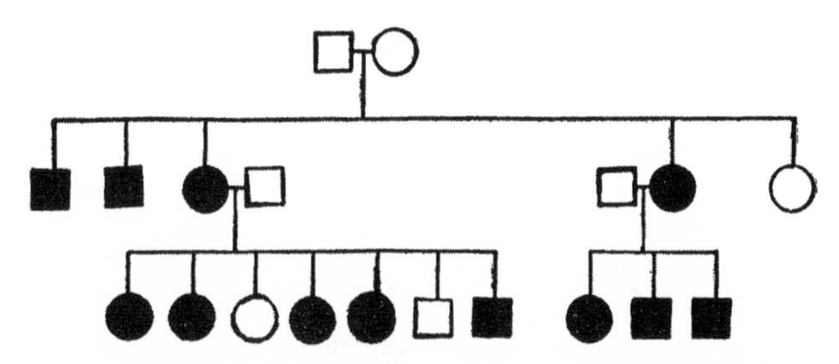

Fig. 149.—Pedigree of a family showing hematuria (red urine). Affected persons (black symbols) are descended from an affected parent, evidence that hematuria is a positive trait. Guthrie.

c. Hematuria, or red urine, may also be a family characteristic as the pedigree chart worked out by Guthrie shows (Fig. 149).

d. Urinary Calculi.—This is frequently hereditary. A pedigree recorded by Cluble (1872) illustrates this fact, though it does not give sufficient data to determine the law of inheritance. He says:—"During the last four or five years I have cut three of his sons [i. e., of the Lowestoft fisherman] at the respective ages of 2, 3, and 8. Two of the stones were lithic acid, one apparently lithate of ammonia. The father and mother of the lads always have lithic acid sediment, often gravel, deposited from urine. Their grandfather passed one stone, their grandmother seven. A great uncle was cut for stone. There are six uncles and four aunts who suffer from fits of gravel or from gravelly or sedimentary lithic acid deposits; and a cousin, an uncle's child, gets rid of urinary calculi."

e. Gout.—The hereditary tendency to gout is generally

recognized—a pedigree recorded by Garrod illustrates the fact. A man who has very severe gout is married to a woman who when 70 years old began to suffer from it. They had 7 children; all have suffered from gout, 5 have died from gout and its various complications; the other two are still living.

39. Reproductive Organs

a. Cryptorchism, or retention and atrophy of testicles. This condition, a semi-"hermaphroditic" one, is characterized by the fact that the normal descent of the testis into the scrotum fails to occur. A pedigree of a family exhibiting this condition is given, in Fig. 150. In the third generation one boy out of four is normal. This trait is probably inherited just like hypospadias.

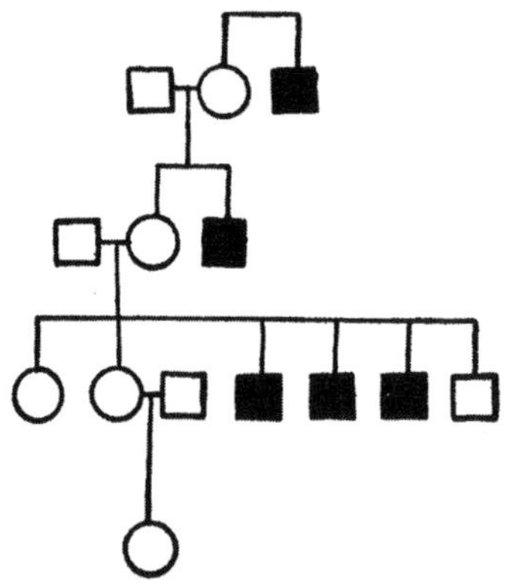

Fig. 150.—Pedigree of cryptorchism. Affected persons represented by black symbols. On account of the sterility of the males all affected persons are derived from sisters of affected persons. All affected persons are natural eunuchs. Bronardel, p. 169.

b. Hypospadias.—Like the last this is evidence of an imperfect development of the external secondary sex characters and possibly indicates an imperfect stimulus to sex dimorphism. The defect is characterized by the more or less complete failure of the male genital papilla to close along the median raphe up to the apex of the glans. An affected man may have by a wife who belongs to a normal strain some or all of his sons affected. His normal daughters may have abnormal sons even when the father belongs to a normal strain. It seems that there is an inhibitor to complete sex-differentiation in the males. Usually males who show no trace of the inhibitor when married into a normal strain have normal sons. But occasionally apparently normal fathers in whom the "inhibitor" is inactive may have abnormal sons (Fig. 151.) The eugenical conclusion is that females belonging to hermaphroditic (hypospadic or cryptorchitic) strains, if married, will probably have at least half of their sons defective, particularly if they have defective brothers; but normal males of such strains may marry females from unaffected strains with impunity.

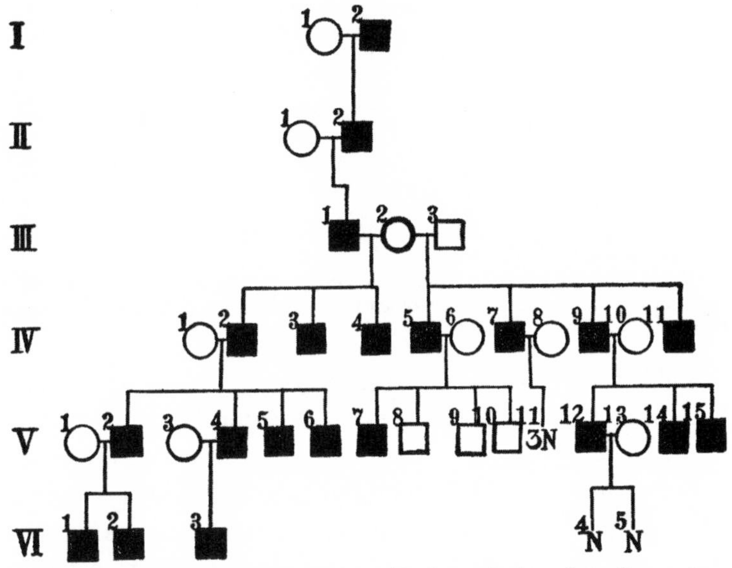

FIG. 151.—Pedigree of hypospadias (black symbols). Inheritance from affected males and unaffected females, III, 2. LINGARD, 1884.

c. **Prolapsus of the Uterus and Sterility.**—Corresponding in a way with incomplete development of the male reproductive organs is the prolapsus of the uterus in the female. This is also definitely inherited but the trait is never transmitted by affected females since they are sterile (Fig. 152).

40. SKELETON AND APPENDAGES

Since the size and form of the bodily frame are greatly influenced by the skeleton the heredity of these features is

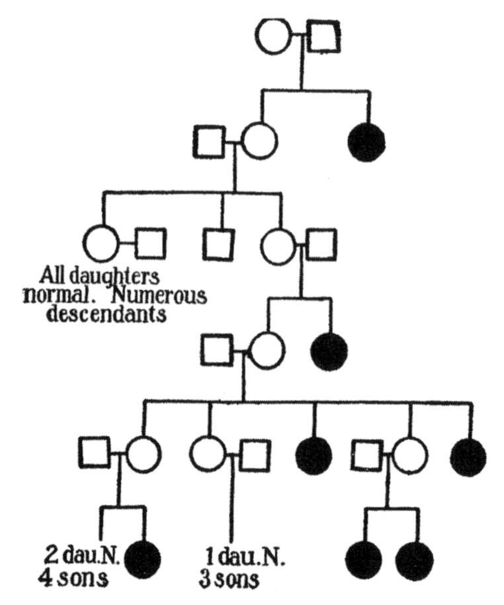

FIG. 152.—Pedigree of a family showing prolapsus of the uterus (females) and sterility. Inherited like the *absence* of a character, with probable consanguinity in marriage. BRONARDEL, 1900.

usually due to an inheritance in the processes that go to determine the form and size of the skeleton.

a. Achondroplasy is characterized by relatively short limbs, a condition in man like that in the Ancon sheep, dachshund and some bull-dogs. The condition is rare and so we have few if any full pedigrees but enough is known to indicate that it is inherited, as in the case cited by Pouchet and Leriche (1903), Fig. 153, and it is probably due to an abnormal positive factor.

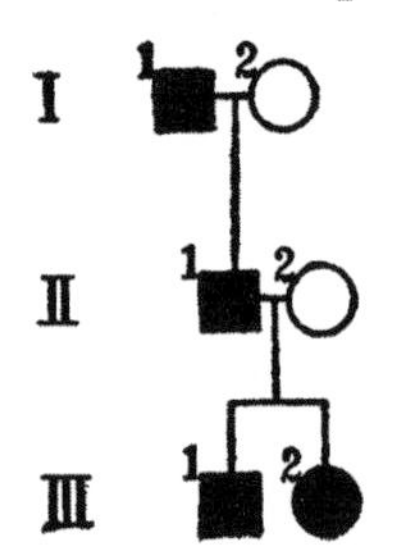

FIG. 153.—Pedigree of achondroplasy (black symbols). POUCHET and LERICHE, 1903.

b. Scoliosis.—The dissymmetry of the trunk accompanied by a curved "spine" is a fairly common condition. That there is an hereditary tendency to it cannot be doubted in view of its frequent occurrence two or more times in one family. Either father or mother of an affected child may be affected; or they may have symmetrical spines themselves but have an affected brother or sister. The offspring are born with an hereditary laxness and weakness of the constituent parts of the spinal column and its ligaments, so that the column easily falls into lateral curves under the influence of secondary causes.

c. Exostoses.—Upon the long bones there occasionally develop osseous outgrowths known as exostoses. The method

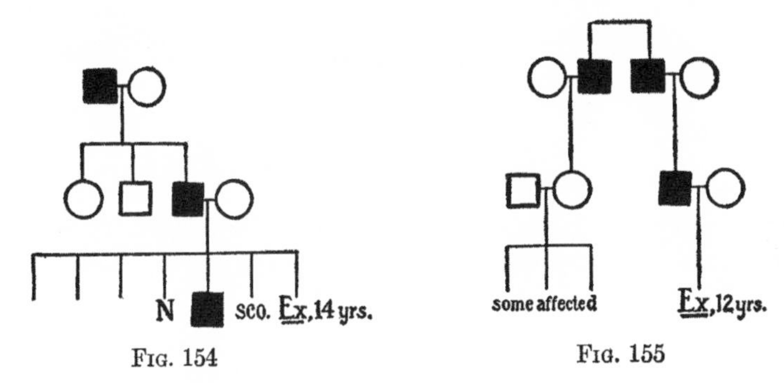

Fig. 154 Fig. 155

Fig. 154.—Pedigrees of exostoses on the long bones. Affected individuals represented by black symbols. *Ex*, exostoses, sex unknown; *sco*, scoliosis or spinal curvature. Teissier and Denecham, 1905.

Fig. 155.—Part of a pedigree of exostoses on the long bones that have been traced through 6 generations. *Ex*, exostoses, sex unknown. Mery and Metayer, 1905.

of inheritance of the tendency to produce such growths is indicated by pedigrees given in Figs. 154, 155.

d. Absence of Clavicles.—The collar bones, or clavicles, are occasionally imperfectly developed and the tendency to this result shows itself in several members of one family. This is well illustrated by a case described by Carpenter (1899) Fig. 156. The high incidence of the abnormal condition in this family suggests that the defect is due to a positive inhibitor.

e. Congenital dislocation at the thigh bone—pelvis joint.—This is a peculiarity that usually runs in families. It is doubtless due to a laxness in the ligaments by which attach-

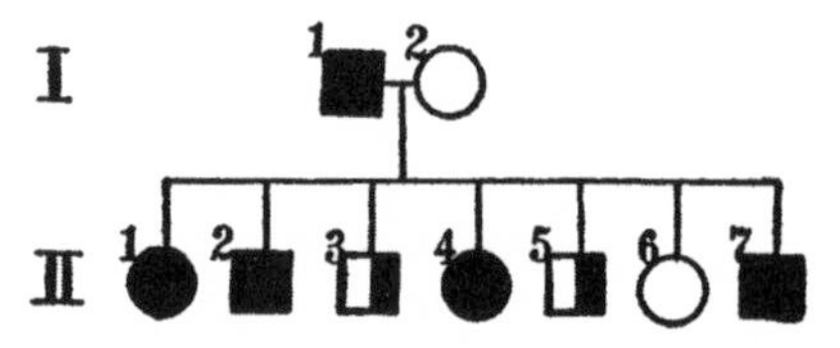

Fig. 156.—Pedigree of absence of clavicles. The father, I, 1, has deformed clavicles. By a normal wife he has 7 children affected as follows: II, 1, has a slightly deformed clavicle; II, 2, has a deformed right clavicle; II, 3, has normal clavicles but a prominent transverse process of the last cervical vertebra; II, 4, has clavicles nearly absent and also the clavicular portion of the great chest muscle; II, 5, has a peculiar kink in the clavicles; II, 6, is normal; II, 7, has a deformed right clavicle. Carpenter, 1899.

ment is made. Several pedigrees have been worked out by Nareth (1903) of which one is reproduced here (Fig. 157).

No evidence appears as to the amount of consanguineous marriage except in one case. The pedigree looks like one

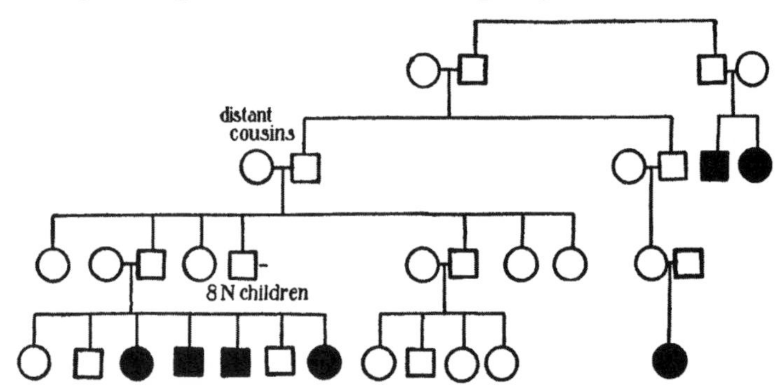

Fig. 157.—Pedigree of a family showing congenital dislocation of the hip. Affected persons (black symbols) descend from unaffected, suggesting that the condition is due to a defect. Senator and Kaminer, 1904.

of albinism and suggests that congenital dislocation is a defect. In that case the marriage of related persons, even though normal, is to be discouraged, but an affected person by marrying into new blood may expect normal offspring.

f. Polydactylism.—The peculiarity of supernumerary fingers and toes is one that is inherited in nearly typical fashion. I have worked extensively on polydactylism in fowls and there can be little doubt that the character behaves in the same way in man. The extra toe is due to an addi-

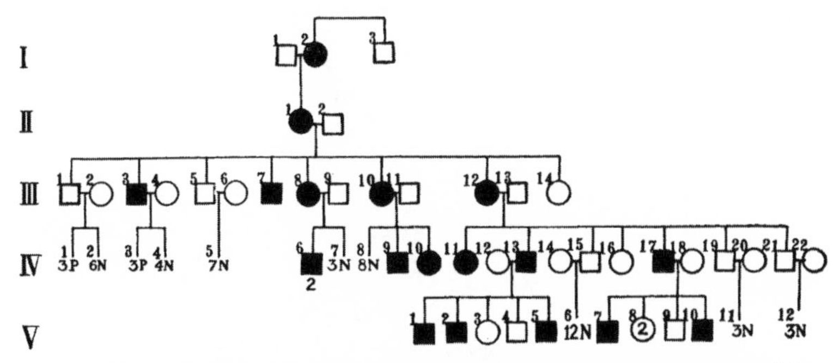

Fig. 158.—Pedigree of polydactylism. Affected persons represented by black symbols. III, 3, has six toes on each foot; III, 8, has six toes on each foot; III, 10, extra fingers on each hand; III, 12, extra fingers on each hand; V, 1, five fingers and thumb on each hand; V, 2, supernumerary digits on both hands and feet; V, 5, extra toes, both feet; V, 7, harelip, cleft palate, web between each big toe; V, 10, 5 fingers and thumb on each hand, 6 toes on each foot, web between all toes. Lucas, 1880.

tional unit so that when one parent has the extra toe the children will also have it. However, it sometimes happens that the offspring fail to produce the extra toe; but such persons, becoming in turn parents, may produce the polydactyl condition again (Fig. 158).

The method of inheritance of polydactylism is well represented by Lucas' case, given in Fig. 158. Here only when one parent was polydactyl were there polydactyl offspring, excepting in the progeny of the oldest son of the third generation. This son is said not to be polydactyl and is recorded as normal. If the record is correct his case is one of failure to dominate of the polydactyl determiner.

The eugenical conclusion is: polydactyl persons will have at least half of their children polydactyl. Those quite free

FIG. 159.—A case of polydactylism. The boy's father has 12 fingers and 12 toes, but the extra fingers are boneless. Besides the boy figured, who is like his father, there is 1 son with extra toes, 1 with extra toes and an extra finger on the left hand only. One sister has extra toes only. The other 5 children were normal in respect to the number of toes and fingers they bear. Through the kindness of Professor C. A. SCOTT.

from the trait, though of the polydactyl strain, will probably have only normal children.

g. Syndactylism.—The union of the bones and tissues of two or more digits into one mass is found in many animals including man. I have studied it in hundreds of fowl. It is inherited there, as no doubt also in man, in such fashion as to permit the conclusion that syndactylism is due to a factor that extends the web *paripassu* with the development of the digits. On this hypothesis the normal hand or foot lacks the factor and two normal persons (even of a syndactylic strain) will not show the abnormality in their offspring. This expectation is indeed realized in most of the pedigrees published; as for instance in that of Parker and Robinson (1887, Clin. Soc. Trans., Vol. XX., p. 181), Fig. 160.

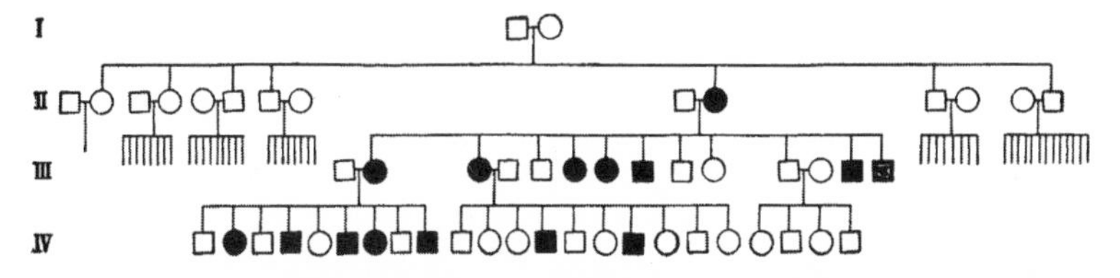

FIG. 160.—A pedigree of syndactylism, or "split foot." All affected persons are from an affected parent; hence the trait is a positive one. Little is known about the condition of the digits in the first generation. PARKER and ROBINSON, 1887.

The general conclusion is that, while a syndactyl individual will transmit his trait, normals from a syndactyl strain have little chance of doing so.

h. Brachydactylism.—This is a condition of shortened digits due to the presence of only two segments to the digit—so that all fingers are like thumbs. The middle phalanx is usually a more or less rudimentary bone attached to the base of the distal phalanx. Inheritance follows the laws of syndactylism. Two normal parents produce only the normal condition; no generation is skipped.

i. Other deformities of the hands.—From time to time other digital peculiarities have been recorded and these are usually strongly inherited. Thus Dobell has described a family in which the hands are double jointed, all joints thick, ring and little finger crooked from the last joint. The peculiarity is distinguishable at birth. The law of inheritance is the same as for syndactylism; viz., normal parents have no offspring with the defect; but one affected parent tends to transmit the defect to half (rarely all) of his offspring (Fig. 161). The tendency of the great toe to grow under the others occurs in at least one family strain (Fig. 162) and is apparently inherited like double jointedness. Another case of family deformity of the digits is given by Carson (Keating's Ency. III, 935). Here there is an absence of the

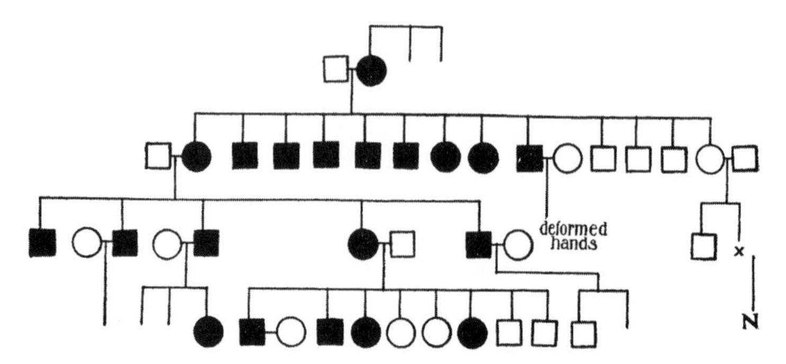

Fig. 161.—Pedigree of family with double jointed hands, all joints thick, ring and little fingers crooked from the distal joint. Affected persons marked by black symbols. Dobell.

distal phalanx and part of the median phalanx from all fingers of both hands, the thumbs being normal. Here again the defect had not skipped a generation, i. e., was not transmitted by normals. It has been known in the family

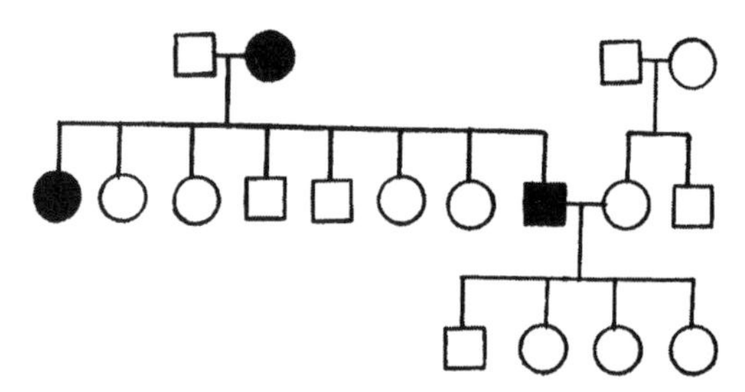

Fig. 162.—Pedigree of tendency of great toe to grow under others (black symbols represent affected persons). F. R.; Ov.

for over a century. Foot (Difformités des Doigts, p. 80) tells of a family in which for three generations the peculiarity has appeared of possessing only the fifth finger. The second and third fingers are represented in these individuals by the metacarpal bone only and the other two fingers are entirely missing. This is, of course, a case of syndactylism, with inheritance of a specific type. In a case cited by Marshall (Trans. Soc. Stud. Disease in Children, III) in which for

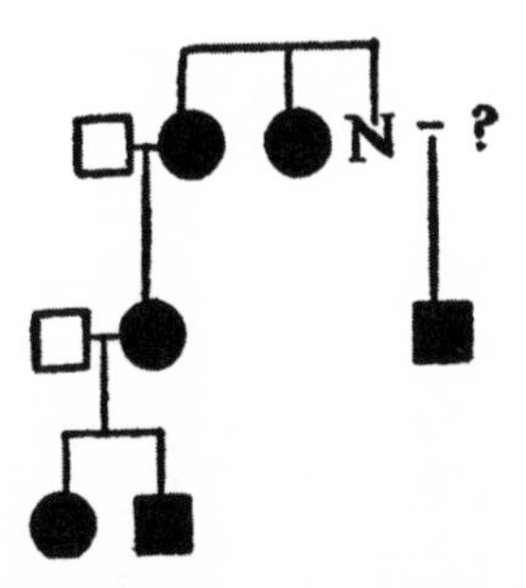

Fig. 163.—Fragment of a pedigree of a family showing hereditary club-foot in 3 generations. So far as it goes this pedigree suggests that the condition is due to a positive character. Drew, 1905.

five generations this peculiarity appeared, each finger stopped short at the proximal phalanx and the thumb was ill developed. Drew has recorded a case of club-foot in three

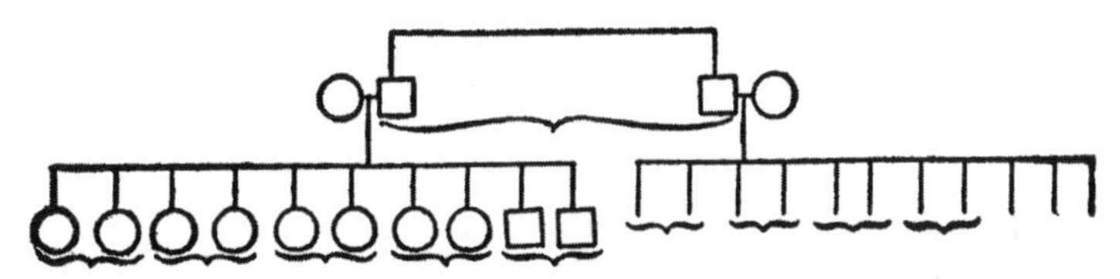

Fig. 164.—Pedigree of a family of twins. Two twin brothers married. The first had 10 children, all born as twins; 4 pair were daughters and 1 pair were sons. Seven of the daughters are married and 4 have produced twins at the first birth, nothing is known of the others. One of the sons is married and has 3 single children. The second brother (first generation) had 8 children born as twins and 3 born singly. Stocks, 1861.

generations (Fig. 163). It is astonishing what a variety of inheritable variations, that are often minute, are shown by the hand and foot. The data are too limited to give assurance as to the law of inheritance in each case.

41. Twins

It is well known that twin production may be an hereditary quality. Thus the Dorset race of sheep is characterized by the tendency to bear twins. In man, too, strains are known where plural births are the rule. Remarkable cases are re-

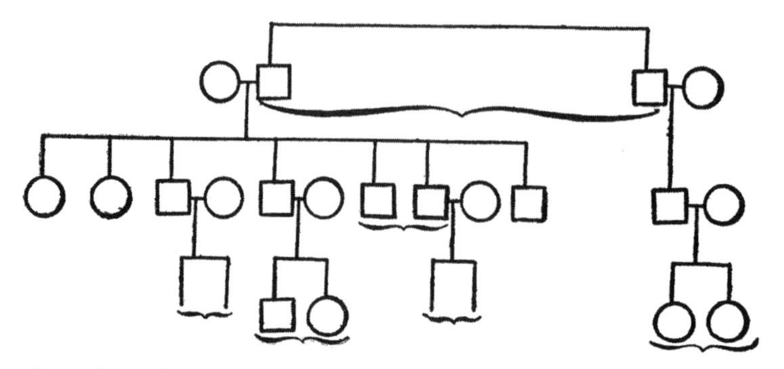

Fig. 165.—Of 2 twin sons one has a pair of twin sons and 5 single born children; the other had 1 son. The former has, through his sons, 3 pair of grandchildren; the latter 1 pair. Wakley, 1895.

corded by Stocks (1861, p. 78), see Fig. 164, and by Wakley (1895, p. 1289). See Fig. 165.

In the foregoing cases inheritance of the twinning capacity is through the males only, and this is true in some strains of sheep. However, other human strains are known with the tendency to twin-production passing along the female line.

CHAPTER IV

THE GEOGRAPHICAL DISTRIBUTION OF INHERITABLE TRAITS

1. The Dispersion of Traits

Traits occur in individuals and the same traits in related individuals. Individuals occupy at any one moment a particular place. Could we take a sort of bird's eye view of the continent and were each individual that bears a given trait conspicuously marked, we should have a perfect picture of the geographic distribution of the trait. Had we such a picture for each day of the hundred thousand odd days since America began to be settled and were they to pass in review as in a cinematograph, then we should see the reproduction and dissemination of the family trait in question. Such a view would show us the traits coming across the ocean from European centres, settling in a place or flitting from point to point, reproducing themselves at a place and continuing to increase there for generations while throwing off individuals to move far athwart the face of the country and to settle down as new proliferating centres. We should see two persons with the same defect coming together as a married couple and proliferating in a few years a number of new individuals with the same negative characters. Or we should see an individual with the defect uniting with a person without it and ending there the trail of the defect. Or, on the other hand, a positive trait, like cataract, hemophilia, or Huntington's chorea, would move about, settle in a spot, multiply itself into many individuals either all of one sex or of both sexes, as the case may be; and these individuals, moving apart, would form new proliferating centres. In the multiplication of negative and positive traits we would see this plain difference—that negative traits multiply most in long established and stable communities where much inbreeding occurs, while positive traits are increased by emigration, as a fire is spread by the wind that scatters firebrands. If, on the other hand, the negative traits be scattered the chance of mating with the same defect is diminished and the trait is not reproduced. Conversely, a country characterized by much inbreeding will have a population that is affected prevailingly by negative traits with a slight tendency for positive traits to increase; while a country that is settled by a restless people will show a small percentage of negative traits and a high percentage of positive ones.

That the picture of the dissemination of traits that I have drawn is not exaggerated but corresponds to the empirical facts is proved by the evidence of many studies. Thus Alexander Graham Bell (1889) finds that not only the deaf mutes of Martha's Vineyard but "groups of deaf mutes who have never been near Martha's Vineyard, trace up to" the blood of James Skiff. A genealogist with unusual intelligence and breadth of interest has traced a "bleeding" tendency from a Hannant who came from Norfolk, England, and whose progeny settled in Sullivan County, Pennsylvania, and created there a colony of bleeders; and by emigration has started new colonies in Minnesota, South Dakota, and California. Students of Huntington's chorea find many of their widely scattered cases tracing back through Delaware County, New York, to the sources of its early population at East Hampton, Long Island, or to that sister settlement of the New Haven Colony, Fairfield County, Connecticut. Even students of crime have traced the disturbing element of a large area to a single focal point; "the Jukes" were traced back to Max living in a lonely mountain valley and the "Ishmaelites" of Indiana were traced back through Kentucky to Virginia and probably to the cutthroats and prostitutes which England spewed out upon, and against the protests of, the Virginia colony in the latter half of the seventeenth century (Butler, 1896). So too a family in New Jersey of over 600 persons, more than three-fourths of them defectives have been derived, by Goddard and his field-workers, from a single pair. These are examples, merely, of a universal fact, that the more strikingly inheritable traits may be followed back generation after generation to a few focal points.

And the focal points of this country have been transported here from abroad. A settlement worker in New York City inquired into the meaning of a particularly unruly and criminalistic section of his territory and found that the offenders came from one village in Calabria—known as the "home of brigands." Of the weary but hopeful thousands of immigrants who weekly (almost daily) enter the port of New York how many are destined to bring in traits for good or evil, that are to proliferate and to affect the future of this country for better or worse! For we must not forget the good. The germ plasm of an Austrian who migrated to the United States three generations ago has produced a race of yacht builders who enable this country to maintain its supremacy in the sport of yachting. From the germ plasm (in part) of an extraordinarily talented but erotic woman who migrated to America in the early part of the seventeenth century have arisen statesmen, college presidents, men of science, great philanthropists from New England to California in extraordinary numbers. From an Irish pair who came to the wilderness of Virginia nearly two centuries ago have descended vice presidents, cabinet officers, admirals, generals, governors, senators and congressmen in great numbers. In these cases the good was not "interred with their bones."

2. Consanguinity in Marriage

The customs of civilized nations oppose certain limits to marriage, almost universally bar the marriage of nearest kin, and have given to the word incest a connotation so loathsome and so emphatic that it is appreciated by practically every normal civilized person. It will be interesting to consider for a moment how wide-spread is this taboo.

First of all it must be said that the union of brother and sister or of parent and child as recognized spouses is not unknown. Various reputable observers report that among the Weddas of Ceylon, probably on account of the sparsity of the population and the isolation of families, the marriage of brother and younger sister is permitted by local custom (Virchow, 1881). In ancient times the marriage of parent and child was not opposed by custom in Persia (Heath, 1887, p. 65) and perhaps in other Eastern countries.

Such customs are to-day, however, highly exceptional and against social ideals. But the line between permissible and non-permissible unions is variously drawn. Thus we are told (Nelson, 1899) that the Eskimos of Behring Strait favor the union of first cousins or even closer relatives on the general ground that in time of stress and hunger the blood tie will be found stronger than the marriage tie to hold the family together. Among other natives of North America a paternal uncle and niece might marry but not a maternal aunt and nephew. However, the North American Indian, on the whole, has strong sentiments against close intermarriage. Also among Africans and the South Sea Islanders cousin marriages are, in general, taboo; and among the Malays "consanguinity, even the remotest, constitutes an important obstacle to marriage." We read of the Islanders making voyages to other islands and carrying off maidens for wives. In India and China marriage of persons within the patronymic is against social ideals.[1] European ideals are largely a legacy of Roman law. Here the purely formal and legal relations constituted as much of an obstacle as blood relationship. A stepchild should not marry his mother nor a father-in-law his daughter-in-law. Only recently has a relic of these legal and non-biological interdictions been removed in England by the repeal of the law prohibiting a man from marrying his deceased wife's sister.

Such wide-spread social barriers to close intermarriage, even among the children of nature—one might almost say *especially* among them—indicates if not an instinctive repugnance to, at least an apprehensiveness toward, such marriages. We have still to inquire if there is any biological basis for such apprehensiveness. The answer to this question has been furnished in many places in the earlier part of the book. Defects in the germ plasm tend to reveal themselves in the offspring of cousin marriages but tend to disappear entirely in the children that are derived from outmatings. On the other hand, undesirable positive traits that are absent from both parents will not reappear in the offspring even though the parents be cousins. One can easily imagine a strain without any important defect, so that a consanguineous marriage would, for generations, be uninjurious to the offspring; but such strains are doubtless rare. We are told that in the family of the Ptolemies and in the royal family of the Incas the marriage of brother and sister repeatedly occurred but, as a friend of mine says, "Where are the Ptolemies and Incas now?" The conclusion seems

[1] The foregoing summary of marriage limitations is based chiefly upon the compiled data of Ploss-Bartels: Das Weib.

FIG. 166.—Rows of maize, each from a single ear of corn. The central row (labeled) is from a 16 row-to-ear race *self-fertilized for five years.* Row to left of center, self-fertilization prevented for six successive years. Row to right, a first cross between long self-fertilized strains.

clear that, while in certain strains consanguineous marriage may not lead to defective offspring, in most families it will, at least after a few generations. This is well illustrated in corn-breeding where self-fertilization leads to rapid loss of productivity and vegetative vigor (Figs. 166, 167).

Let us now consider some of the statistical results gained from a study of consanguineous marriages in a large population. In 1858 Dr. Bemiss reported to the American Medical Association on a collection of 833 consanguineous marriages producing 3,942 children or an average of 4.6 children per marriage. Of these children 28.7 per cent are said to be defective, 3.6 per cent are deaf mutes, 2.1 per cent blind, 7 per cent idiots, 1 per cent insane, 1.5 per cent epileptic, 2.4 per cent deformed, 7.6 per cent "scrofulous" (i. e., probably

FIG. 167.—The piles of ears of corn on the right and left are from seed ears which had been self-fertilized; the pile in the middle from a seed ear in which self-fertilization had been prevented. This figure and the preceding were contributed by Dr. G. H. SHULL.

tubercular) and 22 per cent are said to have "died young." In some data gathered by Dr. Howe (1853) 17 consanguineous marriages produced 50 per cent idiots; in the data of Dr. Mitchell (1866) 7.5 per cent were insane, and 1.4 per cent deaf mutes. Other observers record consanguineous marriages without deaf mutism, others without idiocy, others with less than 1 per cent of insanity. Voisin (1865) tells of the isolated community of Batz where 5 marriages of first cousins and 31 of second cousins has occurred without a case of mental disease, deaf mutism, albinism, retinitis pigmentosa or malformation appearing. These varied results are to be expected. Consanguineous marriage *per se* does not create traits; it permits the defects of the germ plasm, that may not appear in the parents, to reveal themselves in the offspring.

If there is no insanity or albinism in the stock consanguineous marriage will not bring it out; and, strictly, it is not at all consanguinity that brings the trait out but the increasing liability that consanguinity affords to the mating of two similarly defective germ cells.

The variety of the product of consanguineous marriage is well brought out when we compare localities. Thus consanguinity on Martha's Vineyard results in 11 per cent deaf mutes and a number of hermaphrodites; in Point Judith in 13 per cent idiocy and 7 per cent insanity; in an island off the Maine coast the consequence is "intellectual dullness"; in Block Island loss of fecundity; in some of the "Banks" off the coast of North Carolina, suspiciousness, and an inability to pass beyond the third or fourth grade of school; in a peninsula on the east coast of Chesapeake Bay the defect is dwarfness of stature: in George Island and Abaco (Bahama Islands) it is idiocy and blindness (G. A. Penrose, 1905). There is thus no one trait that results from the marriage of kin; the result is determined by the specific defect in the germ plasm of the common ancestor.

The question is often asked, How common are consanguineous marriages? What proportion of marriages are between kin? This question is so ill-defined that a reply is hardly possible. When we recall the enormous number of our ancestors resulting from the fact that the number (theoretically) doubles in each earlier generation, so that there are more than a million in the twentieth ascending generation, and more than a billion in the thirtieth, then we see that *some* degree of consanguinity in the parents is to be expected. There are hardly two persons of European origin who are more distantly related than thirtieth cousin —or who do not have a common ancestor of the time of King William I of England. Indeed, how improbable it is that there are many persons of "pure" European stock whose line of descent has not received contributions from Ethiopia within the last millenium—when we stop to consider the slaves, not only white and yellow but also brown and black, that were brought to Rome, became free there and contributed elements to the population of Italy and to all Europe.

Returning from this digression, we may recognize that, however vague scientifically the term consanguineous may be, popularly, it means related as first or possibly as second cousin. This is, of course, from the standpoint of modern heredity, an absurd limitation of the term since fifth or tenth cousins may carry the same ancestral traits. Our question may then be transformed in this fashion: What proportion of the population marries within the grade of fifth (or tenth) cousin? The answer to this question for the United States as a whole would require a special census, and the proportion, expressed in a single figure would have little significance. Much more important is it to know for each of several small communities the grades of relationship of consorts; and the association of degree of consanguinity with physiographic and other barriers.

3. Barriers to Marriage Selection

Barriers, indeed, to free and wide marriage selection favor consanguineous marriages, and for the same reason they favor the formation of races of men with peculiar traits, even as it has long been recognized that they facilitate the formation of races of plants and animals, by permitting newly-arisen traits to infect, as it were, the entire population and thus to form a new species. The barriers may be classified as physiographic and social.

A. *Physiographic Barriers*

Physiographic barriers are for man, a land animal, stretches of water, such as parts of the ocean, sounds and bays that separate from the mainland, and even broad rivers; also mountain ridges or heights of land. All such barriers restrain exogamy, or marriage outside the family, and favor consanguineous marriage or endogamy.

a. Barrier of Water.—Of oceanic islands the Canaries, Azores, Bermuda, the Bahamas and the Lesser Antilles are examples. In the case of the South Sea Islands the half aquatic nature of the inhabitants has reversed the usual order and made the sea a means of intercommunication. On our own coast we have striking examples of semi-oceanic islands with evidence of consanguineous marriage (Fig. 168).

At Miscou Island on the Northeast coast of New Brunswick there is said to be much intermarriage. The population "is partly English and partly Arcadian French and each race has kept pretty much to itself so they are closely intermarried within the same race."

The islands off the Maine coast show much consanguineous marriage. Thus in Small's (1898) History of Swan's Island it is stated that the amount of intermarriage of persons of the same name in Mount Desert Island, Gott's Island and Swan's and Deer Islands makes genealogy confusing. For example, take the Gott family as shown in Fig. 169; or a family from Swan's Island (Fig. 170). Even more marked examples are furnished by outer Long Island and the islands opposite Jonesport, Maine.

One sees how little opportunity is afforded in such pedigrees for the coming in of new blood. Little wonder that among these descendants of some ancestor who probably carried inferior mentality are some intellectually dull ones.

At western Martha's Vineyard Dr. Alexander Graham Bell (1889, p. 53) has made a careful genealogical study of the inhabitants. "I found," he says "a great deal of intermarrying and a great many consanguineous marriages." Concerning this locality Dr. Withington (1885, p. 26) says:

Fig. 168.—Coast of eastern North America, showing the broken coast line, with islands and peninsulas, each of which is, more or less, a center of consanguineous marriages. Such centers can be picked out by looking at the map.

"The inhabitants are farmers and fishermen of average intelligence and good character, not addicted to drunkenness. A lack of enterprise, associated doubtless with the

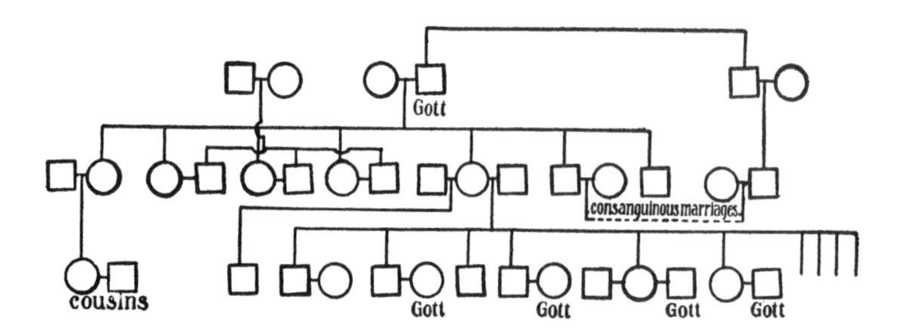

FIG. 169.—Pedigree of a portion of the GOTT family of the Maine Islands, illustrating frequency of cousin marriages in an isolated community.

nature of their occupations, seems to be the cause of their intermarrying." In this locality deaf mutism is the striking trait. In 1880 there was a proportion of 1 to 25 of the whole

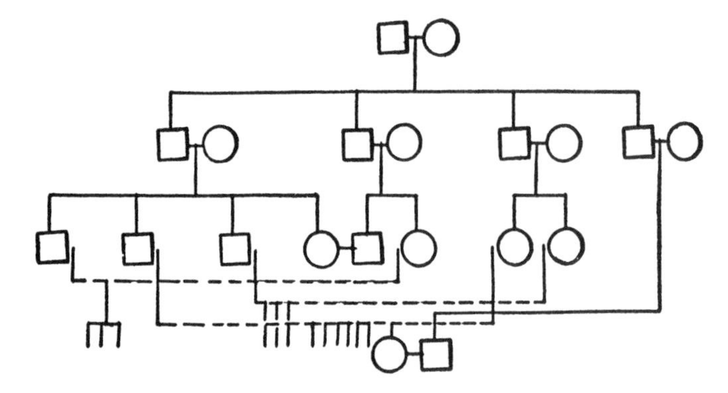

FIG. 170.—Pedigree of a family inhabiting Swan's Island, Maine, illustrating frequency of consanguineous marriage in a restricted and isolated community. The dotted lines connect cousins who have married each other.

population affected (Bell, 1889). Dr. Withington and Dr. Bell report cases of hermaphroditism also from this same locality.

Block Island, comprising about 10 square miles, lies about 40 miles both from Newport, Rhode Island, and from Montauk Point. There are some fine old family names including Ball, Cobb, Dodge, Hall and Littlefield, which constitute a large part of the population of 1,500 souls. The limited area has, however, led those branches of the family who remain on the island to intermarry closely, as

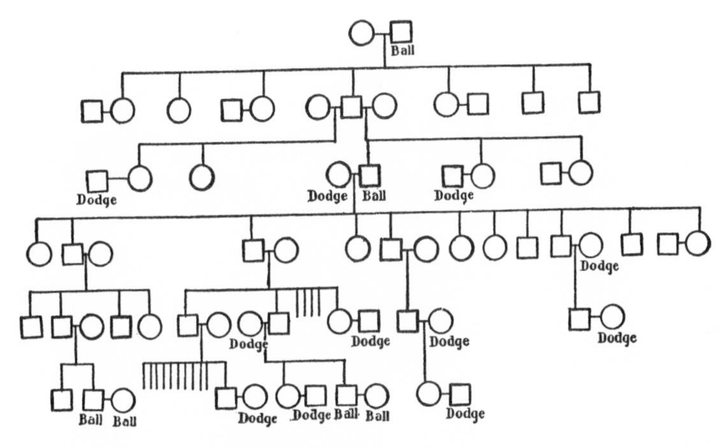

Fig. 171.—Portion of pedigree of the Ball family of Block Island showing frequency of marriage with Dodge and with Ball; a consequence of limited marriage selection in a small island.

illustrated in Fig. 171 based on Ball (1891). The result has not been good. There are families in which all the children are mentally deficient and many marriages that are childless.

As we go south along the Atlantic coast, beaches or "banks" replace offshore islands. When they are so far from the mainland, as at Pamlico Sound, as to make intercommunication difficult, consanguineous marriages occur in extraordinary frequency. A wide-spread trait that may be ascribed to such inbreeding is suspicion and mental dullness; and a relative high frequency of insanity. Even some of the islands of Chesapeake Bay show numerous marriages of kin. Thus Arner (1908, p. 16) states that in Smith's Island, separated from the peninsula of Maryland by twelve miles of water "consanguineous marriages have been very frequent until now nearly all are more or less interrelated. Out of a hundred or more families of which I obtained some record, at least five marriages were between cousins." Over 30 per cent of the inhabitants bear one surname (Evans) and they with Bradshaw, Marsh and Tyler comprise about 59 per cent of the population. The resident physician, here, had noted in 3 years in the community of 700 persons no case of idiocy, insanity, epilepsy or deaf mutism. At the tropics, islands appear again. In some parts of the Bahamas there is a record of consanguineous marriages. C. A. Penrose (1905, pp. 409–414) has described the condition at George Island near Eleuthera Island and at Hopetown, Abaco Island. In George Island close intermarriage occurs, and there is a large proportion of eye diseases, including cataract, and dwarfs with low mental acumen. At Hopetown there are about 1,000 whites. In 1785 a woman, Wyanne Malone, came from Charlestown, South Carolina, with her four children to Hopetown. Three of them married and settled there, a granddaughter marrying a Russell. "From this stock most of the present inhabitants of Hopetown have descended and the names of Malone and Russell are constantly met with throughout the settlement." At Hopetown consanguineous marriage is accompanied by deaf mutism, idiocy, insanity (melancholia) and abnormal appendages.

The island of Bermuda shows the usual consequence of island life. A correspondent writes: "In some of the Parishes (Somerset and Paget chiefly) there has been much intermarriage, not only with cousins but with double first cousins in several cases. Intermarriage has chiefly caused weakness of character leading to drink, not lack of brains or a certain amount of physical strength, but very inert and lazy disposition."

The foregoing studies will suffice to demonstrate, first, the importance of the barrier of water in tending to increase consanguineous marriage and second, the consequences of such consanguineous marriages.

In addition to islands, peninsulas also are more or less isolated and might be expected to yield the same results as islands. There is much evidence that this is so. Cape Cod is a good illustration of a peninsula. Thus Twining (1905, p. 12, note) after giving the pedigree of the descendants of Isabel Twining of Yarmouth who married Francis Baker says, "The frequency of intermarriage between Baker, Chase and Kelly in these records is distinctly observable; it is especially true of the first four generations, confined to the narrow limits of the Cape." Other data proving consanguinity in parentage of Cape families are not difficult to find. Thus Rich (1883, p. 525) tells of William and Mary Dyer, first cousins and Quaker immigrants from England and married. William Dyer (their son?), born 1653, came to Barnstable and married, in 1686, Mary Taylor. Their offspring all married and settled around him and soon became among the most influential people of the town—a position they maintain to this day. "At a recent visit to the Congregational Sunday School, I noticed," says the author, "all officers, many teachers, organist, ex-superintendent, and pastor's wife all Dyers. A lady at Truro united in herself 4 quarters Dyer; father, mother and both grandmothers Dyers." Whether consanguineous marriages at Cape Cod have led to an unusual frequency of any "defects" I cannot say.

Another peninsula of whose marriages there is a record is that of Point Judith. Withington (1885, pp. 14, 15) mentions five marriages of first cousins and two of second cousins. In these marriages insanity (manic-depressive?) and apoplexy were common.

Passing south the peninsulas projecting into Chesapeake Bay often offer extremely isolated situations. A physician of one of the extreme points of Dorchester County, seventeen miles from the railroad, writes me that most of the marriages of that locality—"in fact I may say all, were between relatives and usually of the same name, and with the usual result, dwarfed stature or born crippled, blunted intellect or born idiots." This statement seems to me probably exaggerated—what is meant doubtless is that an exceptional proportion were thus affected.

Finally at Carteret County, North Carolina, we have another example of peninsular conditions which have led to an extreme frequency of consanguineous marriages. Perhaps three-fourths of the inhabitants of the county bear one of four names, and mental deficiency is found in many of the children.

There are other points on our coast which I have not had time to inquire into. It is safe to assume that, in the absence of peculiar, disturbing conditions, all small, inhabited islands off the coast and most of the more isolated peninsulas will show numerous consanguineous marriages and a large proportion of some one of a variety of defects. You can pick out such localities by looking on the map.

b. Barrier of Topography.—A most important barrier is a height of land. How important it is is clear to anyone who has lived in a valley and noted the freedom with which movements of the population take place along the valley as contrasted with movements up the hills to an elevation of even 200 to 500 feet. The valley forms a social center and acquaintances are made and marriages arranged there. Hemmed in by the barriers of the hills and a human inertia that objects to raising the weight of the body, the valley becomes an endogamous center. Such a tendency is much exaggerated in the great valleys of the Appalachian chain. The cradle of the Jukes, however, was in a small valley hemmed in by steep hills only 300 feet high. The valleys of the Taconic Range, of the Catskills, of the Ramapo Mountains of New York are, or have been, regions of much inbreeding and not a little incest, and the product has been much feeble-mindedness, criminality and albinism (Fig. 172). As the mountains rise to the southwestward so do inbreeding, pauperism, and defect, reaching their fullest fruition in the mountain fastnesses of western Virginia and eastern Kentucky and Tennessee. But the story of the effect of this mountain range and its valleys upon consanguineous matings, defect, and crime in America has still to be written.

FIG. 172.—A portion of the U. S. Geological Survey topographic map of the region on the border of the center of the home of the Jukes, showing long, well watered valleys with relatively steep slopes; scale 1: 62,500. Contour interval, 20 feet.

In other countries, longer settled, the influence of mountain barriers is better appreciated. Very famous are the cretins and the imbeciles of the Alps. And from the Chin Hills of Burmah, the Rev. H. East writes about that place as follows (American Naturalist, 1909): "Rau Vau village has been isolated for about seven generations. It contains about sixty houses and possibly two hundred inhabitants. Of these, ten are idiots, many are dwarfs and some hydrocephalic. A number of cases of syndactylism and brachydactylism occur."

B. Social Barriers

The second set of barriers is *social.* These barriers are extremely numerous and complex. There is the barrier of

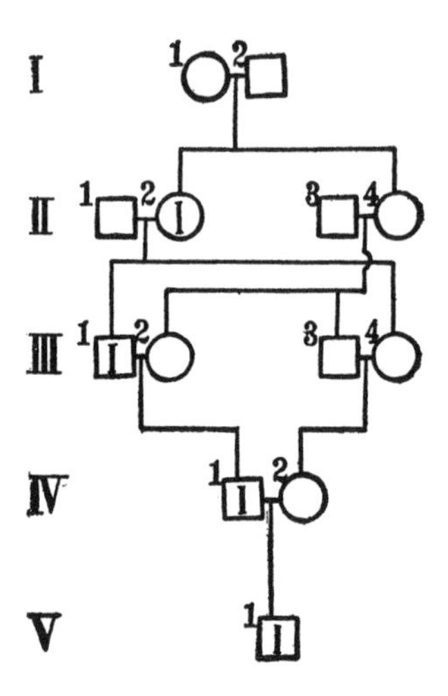

Fig. 173.—Inheritance of a neuropathic taint in a highly inbred family. I, 1, 2, Ferdinand and Isabella of Spain. II, 2, 4, two daughters, Joanna who was insane and Mary; II, 1, 3, their respective consorts, Philip, a weak man and Emanuel also weak; III, 1, is Charles V a great ruler but eccentric, cruel, and subject to melancholia; III, 2, is Isabel; III, 3, is John III of Portugal, a weak man; III, 4, Catherine; IV, 1, is Philip II, morose, sluggish, cruel; IV, 2, is Mary; V, 1, is Don Carlos, "one of the most despicable and unfortunate specimens of humanity in modern history." I (within the symbols) insane. Woods, 1906, pp. 145, 146.

the clan and pride of blood, the barrier of language, the barrier of race, and the barrier of religious sect.

a. The Barrier of the Clan with its pride of blood leads to self-satisfaction and not infrequently to a desire to concentrate wealth and power. This is the barrier that has led the royal families of Europe to inbreed with such disastrous effect, as illustrated by the house of Spain (Woods, 1902, p. 3), Fig. 173. The barrier of the clan is causing the downfall of more than one of America's grand families. The words of Mr. Francis N. Balch are apt here: "I tell you signs are not wanting that if the fine old New England blood despises the ignorant foreigner and stands aloof from him, there will soon be another interesting example of a fine old stock—and our Planters' stock *is* a fine old stock, and a *sturdy* stock,—making a pathetic and unedifying end" (Balch, 1905, p. 22).

b. The Barrier of the Social Status.—This is important where one social class forms a small portion of the community, represented by only a few families. I have in mind a group of persons in a small section of Massachusetts affected by albinism. Probably on this account, together with a mental inferiority, they seem to have been socially ostracized by their neighbors and so were obliged to marry each other. In another instance two families standing above the others in the community in progressiveness and wealth have intermarried extensively; almost exclusively. The effect on consanguineous marriage of an isolated position is well shown by the community of Fort Mardick concerning which a valuable monograph has been written by L. and G. Lancry. They say: "Four families constitute the origin (1670) of the population of Fort Mardick." "This small nucleus was implanted alongside of a population speaking another tongue, having other customs and other occupations than its own, being even more or less hostile to it." To-day, of 300 families 38 bear the name of Everard, of which 9 are Everard-Everard, 36 Hars, 27 Zoonekindt, 24 Benard, and so with the other surnames. To avoid inevitable confusion sobriquets are frequently applied, such as Gros-os, Gros-dos, Bosco, etc. In this community the striking character is sterility. Thus, consanguineous marriages are more than twice as apt to be sterile as non-consanguineous (7.5% : 16%); a single child is 2½ times as common with consanguineous as non-consanguineous marriages and the closer the relationship of the couple the greater the chance of sterile marriage.

In this category may be placed the barrier of life in an institution. A public institution brings together men and women so intimately that marriage frequently occurs after leaving the institution. Thus two persons with the same trait become parents. This is not, strictly, consanguineous marriage but it has much of the essential element of such marriage—viz., the marriage of persons with the same defects. Certainly almshouses in which segregation of the sexes is imperfect yield numerous depauperate and imbecile offspring and there is reason for suspecting that sanatoria and hospitals for the "curable" insane do likewise. That institutions for the deaf mutes lead to intermarriage of persons of this class is notorious. Thus Bell (1884, p. 4) says: "I desire to direct attention to the fact that in this country *deaf mutes marry deaf mutes*. An examination of the records of some of our institutions for the deaf and dumb reveals the fact that such marriages are not the exception but the rule," and later (p. 46) he cites as a cause for this preference "segregation for the purposes of education."

c. **The Barrier of Language** is extremely important in promoting consanguineous marriages or the matings of persons with the same defect. Thus with regard to deaf mutes Bell (1884, p. 44) says: "The practice of the sign language hinders the acquisition of the English language; it makes deaf mutes associate together in adult life, and avoid the society of hearing people; it thus causes the intermarriage of deaf mutes and the propagation of their physical defect." The importance of this barrier is seen among recent immigrants. These tend to herd together largely because of desire to be with people who speak the same language. Thus immigration instead of directly tending to promote matings of dissimilar and unrelated blood, under modern conditions at

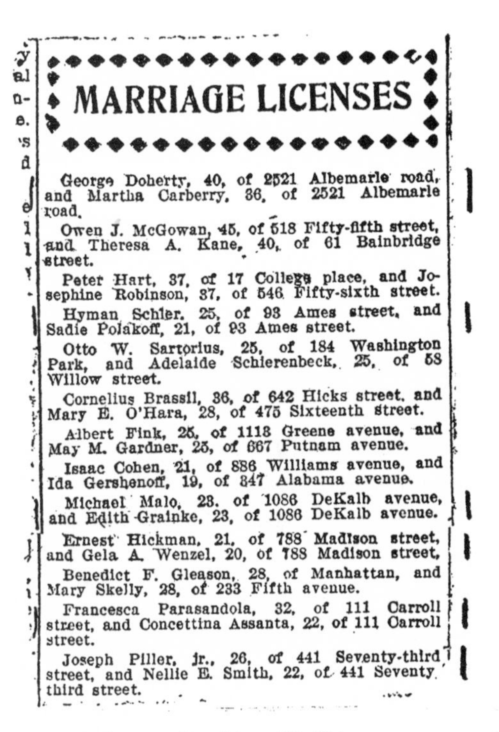

MARRIAGE LICENSES

George Doherty, 40, of 2521 Albemarle road, and Martha Carberry, 36, of 2521 Albemarle road.

Owen J. McGowan, 45, of 518 Fifty-fifth street, and Theresa A. Kane, 40, of 61 Bainbridge street.

Peter Hart, 37, of 17 College place, and Josephine Robinson, 37, of 546 Fifty-sixth street.

Hyman Schler, 25, of 93 Ames street, and Sadie Polakoff, 21, of 93 Ames street.

Otto W. Sartorius, 25, of 184 Washington Park, and Adelaide Schierenbeck, 25, of 58 Willow street.

Cornelius Brassil, 36, of 642 Hicks street, and Mary E. O'Hara, 28, of 475 Sixteenth street.

Albert Fink, 25, of 1118 Greene avenue, and May M. Gardner, 25, of 667 Putnam avenue.

Isaac Cohen, 21, of 886 Williams avenue, and Ida Gershenoff, 19, of 847 Alabama avenue.

Michael Malo, 23, of 1086 DeKalb avenue, and Edith Grainke, 23, of 1086 DeKalb avenue.

Ernest Hickman, 21, of 788 Madison street, and Gela A. Wenzel, 20, of 788 Madison street.

Benedict F. Gleason, 28, of Manhattan, and Mary Skelly, 28, of 233 Fifth avenue.

Francesca Parasandola, 32, of 111 Carroll street, and Concettina Assanta, 22, of 111 Carroll street.

Joseph Piller, jr., 26, of 441 Seventy-third street, and Nellie E. Smith, 22, of 441 Seventy-third street.

FIG. 174.—Clipping from a Brooklyn (N. Y.) newspaper, spring of 1911, showing frequency of marriages between persons from the same address. In the case of recent immigrants this frequently implies that the pair have come from the same home village and are, very likely, somewhat closely related.

first has an exactly opposite effect. The marriage licenses of a large city frequently show bride and groom from the same house—this means frequently, if not usually, that they speak the same dialect, come, very likely, from the same town in the old country, and are probably cousins of some degree (Fig. 174). Even in the well-established populations a barrier of language may cause segregative marriage selection and, if the population is small, lead to consanguinity. Thus at Miscou Island part of the population speaks French and part English and this intensifies the liability to consanguineous marriage.

d. The Barrier of Race is of the very greatest importance in promoting marriage of kin—especially if one race be in a marked minority as the negroes are in New Hampshire and the whites are in the Mississippi River bottom around Vicksburg or in parts of the West Indies. As a striking instance of consanguinity in a colored population in the north may be cited the "Jackson-White" clan of the Ramapo mountain region.

e. Finally, the **barrier of religious sect** has been erected again and again to insure the intermarriage of the faithful only. This is illustrated by the teachings of the Society of Friends and smaller sects such as the Dunkers, Shakers and Amish. Of the Dunkers, Gillen (1906) states: "In their early history marriage out of the church was punishable by expulsion (Chronicon Ephraterise, pp. 96, 346f). It is still frowned upon, but the process of liberalization now in progress has modified the attitude of the Church. In some congregations families intermarry generation after generation. But the degree of kinship is not so close that any evil results appear in the offspring." Nevertheless one sees the danger that any small sect with such tenets runs. A critical study of the Amish of southeastern Pennsylvania with much marriage of kin shows a sufficient frequency of epilepsy and crippled children to serve as a warning that a defect is in the blood of some of the strain that in time will affect the entire sect who remain in that part of the country. It is difficult to see how any religious sect would have a tenet so opposed to the laws of Nature and God as practically to compel consanguineous marriage.

Many other sects are in a worse condition biologically than the Amish. Indeed, the smaller the sect the more apt are its adherents to be thrown closely together and so to become intimately acquainted with one another exclusively; and it is easy to see that in a few generations cousin marriage will be the rule in such sects. From this point of view the Special Report of the Census upon Religious Bodies (1906) becomes of great biological interest. In this report we read of the Duck River Baptists, one-third of whom (2,181) are in the Duck River Association; of the General Six Principle Baptists with 90 per cent of its membership in Rhode Island; of the Amana Society, all (about 1,700) located in Iowa County, Iowa; of the Braederhoef Mennonite Church of Bonhomme County, South Dakota, with 275 members, of the Reformed Presbyterian Church (covenanted) with 17 members, all at North Union, Pa.; and of the 725 Schwenkfelders of Eastern Pennsylvania. In some of these sects it is probable that the tenet of marriage inside the sect does not obtain, but without such a tenet the result tends to follow and we can but regard such small sects as eugenically unfortunate.

CHAPTER V

MIGRATIONS AND THEIR EUGENIC SIGNIFICANCE

1. Primitive Migrations

The human species has come to occupy the entire habitable globe. This fact is mute testimony of man's migratory capacity and tendencies. Just as the Norwegian lemming has been observed, in consequence of several years of favorable conditions for breeding in its mountain home, to spread over the surrounding territory in great bands seeking less crowded breeding-grounds; even as the army worm and the grasshopper swarm from their native territory; so man, also, under the pressure of crowded conditions, poverty and oppression or lured by brighter prospects elsewhere, may move in hordes to other lands that seem to offer better opportunities. Thus Asia seems to have debouched her surplus population upon Europe in the shape of the Huns during the fourth and fifth centuries of our era and the Turks during the fourteenth and fifteenth centuries. So the Anglo-Saxons and the Normans successively swarmed upon England. So, among savages, the Masai of Africa moved upon the neighboring tribes and established themselves over much of southeastern Africa. So in the last three centuries the Americas and Australia have witnessed the greatest migrations that the world has ever seen, hundreds of thousands annually coming from over-crowded Europe and Asia to the "New World."

2. Early Immigration to America

For us in America the phenomena of migration should have a special interest. Excepting for the few scores of thousands of Indians, there was a continent devoid of a population—a clean slate upon which history was to be written and where the effect of "blood" in determining that history might be traced. Fortunately, almost from the beginning, records were made and many have been preserved, despite fire, energetic housecleaners and rats, so that many materials for such a study are still available. It would be a grand contribution to scientific, biological history to show how *traits* of the individual immigrants, no less than conditions, political and other, determined the deeds of communities. For a community is the sum of its constituent individuals, and what each individual does depends on his innate sensitiveness and the vigor and kind of his reactions to the stimuli of conditions. With a given set of conditions the idiosyncrasies of response of the constituent individuals determine the details of history; and these idiosyncrasies depend quite as much on inheritable traits as on training and experience; for just what effect training and experience shall have on the individual depend upon the nature of his protoplasm. Into this grand but unworked historical field we cannot hope to enter here, but a hasty survey of the subject will be attempted.

It would be very difficult now to construct the wave of immigration to the territory of the present United States from 1607 to 1776. The census of 1790 gave a population of nearly 4,000,000; and making every allowance for the high net fecundity of the early immigrants, it is clear that at least a hundred thousand persons must have come in ships from Europe to North America during those 170 years. A concrete idea of the numbers may be gained by the statement (Fiske, 1905, pp. 77, 155, 197) that starting about 1615 Virginia had acquired in 4 years a population of 4,000 souls; between 1630 and 1640, 20,000 persons came to New England [1] but during the following century immigration practically ceased, having been discouraged; and from 1681 to 1684 Pennsylvania gained 8,000 inhabitants. The estimated arrivals from 1776 to 1820 number 250,000 and about 28,000,000 more to 1910.

Since the first few scores of thousands of immigrants had the greatest influence on the ideals of the colonies they established and since their blood has had the longer time to show its effects, and since their traits have had the greatest chance to disseminate widely, they deserve special consideration. The great interest taken in these "forefathers" by their descendants is justified even from the biologic-historic point of view, for their families were large, the pedigrees of their families were often carefully kept and are, for the most part, reliable, and we know much about the characteristics of many of the males who reached maturity. We observe, also, in the colonies the same tendency of persons similar in origin and tastes to segregate that is observed among modern immigrants.

On the James River the first settlers consisted chiefly of "discredited idlers and would-be adventurers," [2] more than half of them "gentlemen" of good family but untrained in labor, trusting for a change of fortune in the new land. Later, men, women and children were *sent* by the London Company to colonize the new land and that company was not particular as to quality. Even felons, murderers and women of the

[1] "It is positively known that early in the spring of 1630, eleven vessels left England for New England with 1700 passengers, arriving at the port of Salem, Mass. in June of that year. Fifty of these families settled in Lynn. In the same year the Massachusetts Bay Co. sent over 16 ships—all arrived safe in New England at the port of Salem." HARRIET R. COOKS, The Driver Family, N. Y. 1889, p. 26.

[2] WILSON, History of the American People, I., p. 45.

streets were at times sent over from London to relieve the city of them; and the governor, who was a pure euthenist, and seemed to think the better environment would cure their evil ways, welcomed all. However, in the middle of the seventeenth century, protests went out from the colony against being made a penal settlement, and in 1670 the House of Burgesses passed an act prohibiting the importation of convicts, but such importations did not wholly cease until declared illegal in Virginia in 1788. Perhaps 20,000 "convicts" altogether, by no means all immoral when judged by our present standards, were imported into the Virginia Colony (Butler, 1896).

But a better blood soon crowded into Virginia to redeem the colony. Upon the execution of Charles I (1649) a host of royalist refugees sought an asylum here, and the immigration of this class continued even after the Restoration. By this means was enriched a germ plasm which easily developed such traits as good manners, high culture, and the ability to lead in all social affairs,—traits combined in remarkable degree in the "first families of Virginia." From this complex and the similar complex of Maryland has come much of the bad blood that found the retreats of the mountain valleys toward Kentucky and Tennessee to its liking, and that spread later into Indiana and Illinois and gave rise, in all probability, to the Ishmaelites, a family of which hundreds have been supported in the almshouses and jails of Indiana. From this complex came also some of America's greatest statesmen and military leaders; the Randolphs, the Marshalls, the Madisons, the Curtises, the Lees, the Fitzhughs, the Washingtons and many others born with the instinct to command. Such are the descendants of the high-spirited cavaliers. It might have been predicted that the future state would be the Mother of Presidents and that in a civil war the hardest fought battles should be fought on her soil.

Further north, at Manhattan Island, a settlement was being made by another sort of people; a band of Dutch traders. The fur trade with the Indians waxed profitable. They maintained friendly relations with the Indians, as the main source of their wealth, and under their protection established trading posts up the North River even as far as the present site of Albany and along the valley of the Mohawk; while others went east as far as the Connecticut River. Little wonder that such blood, under the favorable environment of an admirable location, has created the commercial center of the western world.

On the bleak coasts of New England were being founded settlements of idealists, men who were willing to undergo exile for conscience' sake. They included many scholars like the pastor Robinson, Brewster who, while self-exiled at Leyden, instructed students at the University, John Winthrop "of gentle breeding and education," John Davenport whom the Indians named "So-big-study-man." [1] Little wonder that the germ plasm of these colonies of men of deep convictions and scholarship should show its traits in the great network of its descendants and establish New England's reputation for conscientiousness and love of learning and culture. As it was almost the first business of the founders of the colonies of Massachusetts Bay and New Haven to found a college, so their descendants—the families of Edwards, Whitney, Dwight, Eliot, Lowell, Woolsey and the rest have not only led in literature, philosophy and science but have carried the lamps of learning across the continent, lighting educational beacons from Boston to San Francisco. Nor is it an accident that on the soil tilled by these dissenters from the Established Church of England should be spilled the first blood of the American Revolution.

Later, to the shores of the Delaware, Penn led his band of

[1] Cotton Mather, Magnolia III, 56.

followers, consisting of men and women whose natures were attracted to his principles of thrift, absence of show, and non-resistance. The germ plasm of his followers soon peopled Penn's woods and it is not due solely to chance that Pennsylvania has the largest number of homes owned and free from debt of any state and that the "powers that prey" prowl here so unmolested.

Thus the characteristics of each commonwealth were early determined by the traits of the persons who were attracted toward it. These traits still persist in their dwindling descendants who strive to secure the preservation in the state of the ideals inculcated by their forefathers.

One common characteristic these early immigrants had, which led them to leave family and friends, to undergo the trials of the long sea voyage in small ships and to settle in a rigorous climate among unreliable savages, and that was a willingness to break with tradition, to exchange the old for the new and better. This trait, that amounts in extreme cases to a "Wanderlust," is illustrated by the history of many a pioneer. For example, Simon Hoyt landed in Salem, Mass., in 1628, went in the first company of settlers to Charleston (1629); went to Dorchester (1630) with the first company of settlers there; joined the church at Scituate (1635) and built a house there; then, probably in the spring of 1636, migrated to Windsor, Connecticut colony, which he helped found. In 1649 he was granted land at Fairfield and in 1657 he died at Stamford. Thus in the space of thirty years Simon Hoyt lived in seven villages in America and was a founder of at least three of them—a truly restless spirit like many another settler, and the parent of a restless progeny.

Still another example is that of Hans Jorst Heydt of Strasburg. He fled to Holland when his native town was seized by Louis XIV, married there Anna Maria DuBois, a French Huguenot refugee from Wicres; came with her to

America and settled at New Paltz on the Hudson about 1710. Schismatic dissensions having broken out in the new colony, Heydt, with others, left and settled about 1717 in Philadelphia County not far from Germantown where he acquired several hundred acres of land, established a colony, built mills and entered upon various commercial enterprises. In 1731, having acquired a grant of 40,000 acres of land in the Shenandoah Valley, he migrated thither, became known as Baron Hite, and died there in 1760. One of his friends, Van Metre, who originally settled at New Paltz, had moved first to Somerset Co., New Jersey, then to Salem County in the same colony, later to Prince George's County, Maryland, and, finally, to Orange County, Virginia (Smyth, 1909). These are examples, merely, of the restlessness,—of the enterprising restlessness—of the early settlers.

This trait of restlessness and ambitious search for better conditions shows itself in the frequent migrations of the descendants of the early settlers. The abandoned farms of New England point to the trait in our blood that entices us to move on to reap a possible advantage elsewhere. "I don't know a farmer in Illinois," said a friend that has traveled over the state extensively, "who wouldn't sell his farm tomorrow and go to a distant state if he could be sure of bettering himself financially by doing so." This restlessness affects whole states. Thus from 1900 to 1910 the population of Iowa decreased because so many thousands of her people moved to the newly opened lands of Canada, Washington and Oklahoma. There was an ambitious tendency in the germ plasm out of which the forefathers developed that lured them from Europe and it is in the same germ plasm yet and shows itself in these later generations.

A shorter but not less pregnant migration is that to the metropolis from the surrounding rural districts. One after another, as they grow up, many or most of the young men and many of the young women also leave the farm for the office, shop and factory.

Now all of these migrations have a profound eugenic significance. The most active, ambitious and courageous blood migrates. It migrated to America and has made her what she has become; in America another selection took place in the western migrations and what this best blood—this crême de la crême—did in the west all the world knows. Great cities like Chicago, with its motto "I will," arose in a generation or two to the front rank of world metropolises, and New England, the early home of the sewing machine and the cotton gin, has yielded the palm to the central west, the home of the harvesting machine and the aëroplane.

And when the best and strongest migrated, the weaker minds were left behind to breed in the old homestead. A recent British Committee on Physical Deterioration[1] contains the testimony of Dr. C. R. Browne about conditions in the west of Ireland. He says: "The sound and the healthy—the young men and young women—from the rural districts emigrate to America in tremendous numbers, and it is only the more enterprising and the more active that go, as a rule." And Dr. Kelly, the Roman Catholic Bishop of Ross testified: "For a considerable number of years it has been only the strong and vigorous that go—the old people and the weaklings remain behind in Ireland." And even in New England we see signs of decadence of the old stock and men speak of racial deterioration. But the race as a whole has not deteriorated but only the New England representatives—the "left-behinds" of the grand old families, whose stronger members went west. Likewise in the rural and semi-rural population within a hundred miles of our great cities we find a disproportion of the indolent, the alcoholic, the feeble-

[1] Inter-departmental Committee on Physical Deterioration, Vol. I, p. 37, 1904.

minded, the ne'er-do-weel. I know intimately several such localities and have seen in one family after another, how the ambitious youth leave the parental roof-tree to try their fortunes in the city while the weakest young men stay behind, supported by their parents, or earning only enough to buy the liquor their defective natures crave, and are finally often forced to marry a weak girl and father her imbecile offspring. Such villages, depleted of the best, tend to become cradles of degeneracy and crime. Thus our great cities lure to themselves the best of the rural protoplasm, surround it with conditions that discourage reproduction, either by creating a disinclination to marriage or making it inconvenient and expensive to have children. So our great cities act anti-eugenically, sterilizing the best and leaving the worst to reproduce their like.

3. Recent Immigration to America

We have seen that the early immigrants to America were men of courage, independence, and love of liberty; and many of them were scholars or social leaders. Are these the characteristics of the immigrants at this later day? Let us examine the matter of immigration to America during the past hundred years. We shall find great differences from the immigration of the 17th and 18th centuries. Thus where the annual immigration was formerly a few thousand it is now hundreds of thousands. The wave of immigration is shown in Plate II. From 1820 to 1824, inclusive, the annual immigration was less than 10,000 but it has never fallen below that limit since. From 1825 to 1844 (with one exception) it has remained below 100,000, but in 1845 it passed that number and (excepting for 1862, in the depth of our Civil War) it has not since fallen below that limit. In 1905 it passed the 1,000,000 mark. The general population meanwhile rose from over 9,000,000 to 90,000,000, or only one-tenth as fast.

The wave of immigration shows great fluctuations in height. Referring to this the Commissioner General of Immigration (Keefe, 1910, p. 10) says: "This periodical rise and fall well represents the relative prosperity of the country, while the gradual increase from decade to decade may be taken as a fairly accurate index of the country's development and growth and its capacity to employ larger numbers of alien laborers."

It may be added that, on account of the departure of aliens, the net increase is less than the totals shown on the chart. Thus there were over 200,000 emigrants in the year ending June 30, 1910, leaving a net increase of something over 800,000. Even that is enormous, and no patriotic American can contemplate this vast annual addition to our kinds of germ plasm without inquiring as to the sort of potential traits they carry and the probable eugenic effect on our nation of this constant influx of new blood.

a. The Irish.—The consequences of the immigration of the earlier half of the period of 91 years are already seen. In 1846 there was a severe famine in Ireland and during the next five years over a million souls, or one-eighth of her population, emigrated thence to the United States, and Ireland has remained one of the most persistent sources of our foreign population. The traits that the great immigration from the south of Ireland brought were, on the one hand, alcoholism, considerable mental defectiveness and a tendency to tuberculosis; on the other, sympathy, chastity and leadership of men. The Irish tend to aggregate in cities and soon control their governments, frequently exercising favoritism and often graft. The young women were formerly much employed as household servants, but more recently have become shop girls and factory hands. Many of the Irish, most strikingly those of the northern part of that island, were among the nation's most intrepid frontiersmen and

PLATE II

WAVE OF IMMIGRATION into the United States

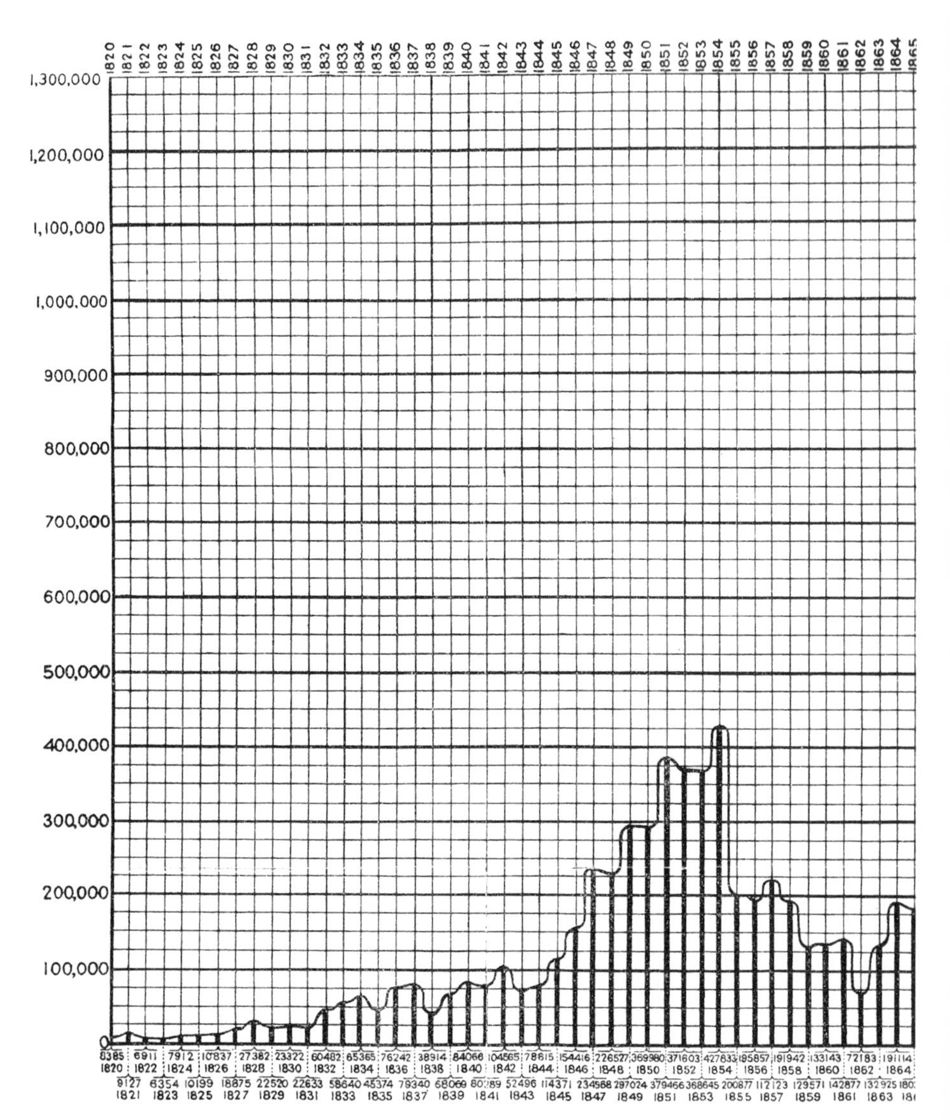

ARRIVALS 1820 to 1910 27,894,293

Figures denoting immigration for the years 1832, 1843, 1850, 1857, represent respectively 15 month, 9 month, 13 month, and 6 month periods, while 12 month periods for those years have been approximated in the graphic representation.

FROM ALL COUNTRIES, during the past 91 YEARS

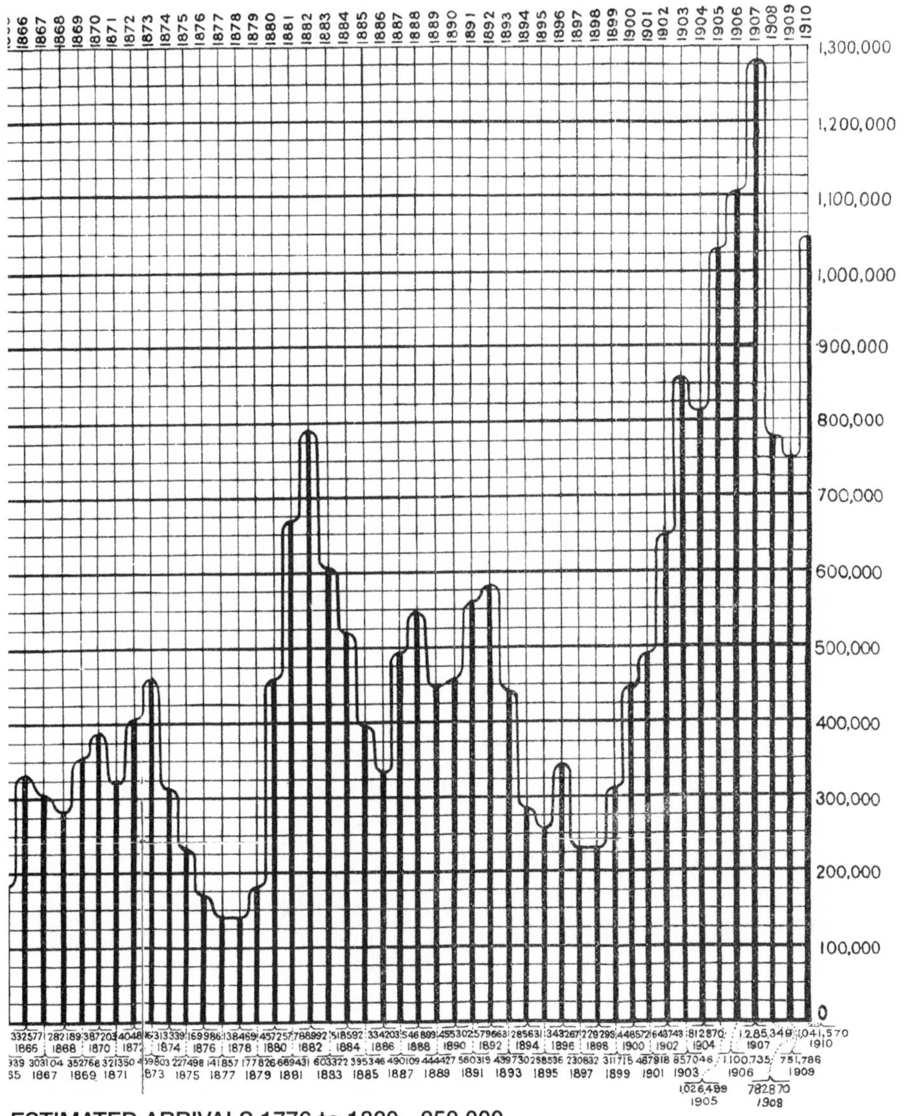

ESTIMATED ARRIVALS 1776 to 1820 250,000

their descendants have served the nation in many important positions.

b. The Germans.—The year 1845 marked the rapid rise of the liberal spirit in Germany and a revolt against the attempt of the ruling class to weaken representative government. Then followed a great increase in immigration to America, advancing to over 140,000 a year for the three years 1852–54. The German immigrants of this period were lovers of freedom, full of courage and daring, and furnished the Union Army during the Civil War with many of its best officers. More recently the Protestant Germans have come to us as unskilled laborers and, after working for a time as farm hands, save enough to buy a place of their own. Great numbers, however, settle in the cities, make useful clerks and often rise to positions of trust. Germans are, as a rule, thrifty, intelligent and honest. They have a love of art and music, including that of song birds, and they have formed one of the most desirable classes of our immigrants.

c. The Scandinavian immigration first assumed considerable proportions in 1866 at the close of our Civil War, reached a maximum (105,000) in the prosperous year 1881, and has since declined somewhat, being now about 50,000 a year. Our Scandinavian population is found chiefly in the central west and northwest, above all in Minnesota, Wisconsin and Iowa. It tends to group itself into colonies; for example, 32 per cent of the entire population of Chisago Co., Minnesota, consisted, in 1900, of immigrants from Sweden; similarly, 26.5 per cent of the population of Traill Co. consists of persons who sailed to this country from Norway. In this tendency to form colonies the Scandinavian immigration of a decade ago shows much resemblance to that of the early English of the 17th century. Such colonization is bound to stamp the impress of the "national traits" upon the community. These national traits include a love of independence in thought and action, chastity, self-control of other sorts, and a love of agricultural pursuits. The latter is less marked in the Swedes than the Norwegians, for of the former only one-third, while of the later more than half, are engaged in farming.

d. **Austro-Hungary.**—The immigration from Austro-Hungary was the next to assume large proportions. It first became considerable with 17,000 in 1880; rose to 77,000 in 1892, and to 338,000 in 1907. It now consists of diverse races; Germans, Slavonians, Croatians and Dalmatians, Bohemians, Magyars, Slovaks, Ruthenians, Roumanians. The latter races are brunet in skin, hair and eye color and of average to short stature. The Bohemians that have migrated to the United States are engaged prevailingly in agriculture. Colonies are found in the prairie states of the upper Mississippi Valley, and in Nebraska and Texas. The Report of the Commissioner-General of Immigration gives Illinois as the intended home of 26 per cent of the immigrant Bohemians and Moravians, New York of 19 per cent, Ohio of 9 per cent and Texas and Pennsylvania each of 7 per cent. In both rural and urban conditions they show prevailing traits of self-respect and pertinacity. The Slovaks in America (to whom nearly 8,000 were added in 1910) are agricultural laborers, not farm owners, but they have founded a few colonies, like that at Slovaktown, near Stuttgard, Ark. Most of those in the East become miners, especially of bituminous coal, and have settled largely in Pennsylvania.

e. **Hebrews** have formed a marked proportion of the population of North America from an early period; even in prerevolutionary times they penetrated to the frontier as peddlers. But the great immigration began with that from Germany and has continued from that country, from Austro-Hungary and Russia in ever increasing numbers.

For the most part they have settled in our large cities, and their frequency is roughly proportional to the size of the city, yet with a preponderance in the East. Though it is superficial to attempt to name the traits of even so relatively homogeneous a company as the Hebrews, yet a sort of average or prevailing condition may be recognized. As the Abstract of the Report of the Immigration Commission on Recent Immigration in Agriculture says, p. 41, "The Hebrew on the land is peaceable and law abiding, but he does not tamely submit to what he believes to be oppression and he has a highly developed sense of personal rights, civil and economic." Probably with few changes this statement would stand for the Hebrews of the cities where the mass of recent Hebrew immigrants occupy a position intermediate between the slovenly Servians and Greeks and the tidy Swedes, Germans and Bohemians. In earning capacity both male and female Hebrew immigrants rank high and the literacy is above the mean of all immigrants. Statistics indicate that the crimes of Hebrews are chiefly "gainful offenses," especially thieving and receiving stolen goods, while they rarely commit offenses of personal violence. On the other hand, they show the greatest proportion of offenses against chastity and in connection with prostitution, the lowest of crimes. There is no question that, taken as a whole, the hordes of Jews that are now coming to us from Russia and the extreme southeast of Europe, with their intense individualism and ideals of gain at the cost of any interest, represent the opposite extreme from the early English and the more recent Scandinavian immigration with their ideals of community life in the open country, advancement by the sweat of the brow, and the uprearing of families in the fear of God and the love of country.

f. The Italian immigration first passed the 10,000 mark

in 1881. That from Southern Italy has always been five or six times as great as from Northern Italy. Immigrants from the former country are darker and doubtless have derived part of their blood from Greece and Northern Africa. It is these South Italians that we generally have in mind when we speak of Italians. Eighty per cent of those who come are males and a quarter of them return each year to their homes. In America they become, prevailingly, general laborers, relatively few specifically farm laborers; yet they are going into agriculture to a considerable extent and buying land as they save the money. Of the agricultural Italians many are truck farmers near large cities, and a few isolated settlements have been made like that at Hammonton or at Vineland, New Jersey. Others are found in central New York State, and a few colonies have been established in the South where they compete with negro labor. Apparently North Italians are to a certain extent influenced in locating in this country by topography like that of their homes. "While sentiment often has much to do with the choice of a location," says Cance (1911, p. 23) "it can not be said that the success of the settlement at Genoa, Wis., is due to the Alpine aspect of the topography rather than to the excellence of the soil and the favorable markets; nor that the fine North Italian settlers of Valdese, N. C., would not have made more progress every way had they settled nearer markets and on level land where there was more fertility and less Swiss scenery." The traits of the Southern Italians are thus expressed: "The Italian has not the self-reliance, initiative resourcefulness nor self-sufficing individualism that necessarily marks the pioneer farmer." "On the whole the Italian farmer compares well with other foreign farmers in his neighborhood in industry, thrift, careful attention to details, crop yields and surplus returns from his farm. His strength lies in his patience, unflagging industry and capacity for hard, monotonous labor." Aside from his tendency to crimes of personal violence the average Italian has many excellent characteristics, not one of the least of which is his interest in his work, even as a day laborer. He assimilates fairly rapidly, especially in rural districts; not a few Irish girls marry Italian husbands when both are Catholics; and this assimilation will add many desirable elements to the American complex.

g. The Poles are distributed under their political affiliations as German, Austrian, Russian and so on. The race constitutes one of the largest contributors to the American population. The cause of this emigration of a large proportion of the European Poles is doubtless the political disabilities under which they have labored. Poles first began to form colonies in the United States in 1885 (in Texas), from 1895 they came in numbers to Wisconsin and Michigan, and later to Indiana and Illinois. More than any other recent immigrants, except the Italians, they become general laborers, largely in rural districts, and as they save money they buy farms. The Poles are independent and self-reliant though clannish. They love the land and work hard to gain a piece of it. They are able to make pay the farms of New England which the sons of the early settlers have abandoned. We may welcome this freedom-loving people whose blood is bound largely to replace that of the old New England stock.

h. The Portuguese are among our more recent immigrants, since their numbers did not exceed 2,000 per year until 1889 and first reached 5,000 in 1902. They are classified either as white (largely from the Azores) or dark, from the Cape Verde Islands. The former become farm laborers, general laborers, mill hands, and farmers, and are steady, reliable, and efficient. In Rhode Island they form a notable colony of potato planters; in Massachusetts their headquarters are at New Bedford and from this city they have spread through the "Old Colony" region and into Cape Cod. The Black Portuguese are the principal cranberry pickers employed on the Massachusetts bogs. "They are largely recruited from the ranks of dock laborers near New Bedford and neighboring cities. Five-sixths of them are men or boys, many of them single or without families in the United States." The cranberry pickers of Massachusetts are illiterate and neither resourceful nor intelligent; but this has the less eugenic significance since few settle permanently in this country.

Summarizing this review of recent conditions of immigration it appears certain that, unless conditions change of themselves or are radically changed, the population of the United States will, on account of the great influx of blood from South-eastern Europe, rapidly become darker in pigmentation, smaller in stature, more mercurial, more attached to music and art, more given to crimes of larceny, kidnapping, assault, murder, rape and sex-immorality and less given to burglary, drunkenness and vagrancy than were the original English settlers. Since of the insane in hospitals there are relatively more foreign-born than native it seems probable that, under present conditions, the ratio of insanity in the population will rapidly increase.

As to the question of increasing dependence and credulity among recent immigrants it appears that "the immigrant to the United States in a large measure assists as well as advises his friends in the Old World to emigrate." Next to this "the propaganda conducted by steamship agents is undoubtedly the most important immediate cause of emigration from Europe to the United States," especially in Austria, Hungary, Greece and Russia. While America will be slow to relinquish her position as the home of the oppressed of all nations, she may well oppose any practice that tends to lure persons here by raising false hopes of an easy acquisition of riches.

4. Control of Immigration

It has long been recognized in this country that it is a national duty to regulate immigration. Our present immigration laws recognize this right and duty. Section 2 of the Immigration Act has the following eugenic provisions:

> "That the following classes of aliens shall be excluded from admission into the United States: All idiots, imbeciles, feeble-minded persons, epileptics, insane persons, and persons who have been insane within five years previous; persons who have had two or more attacks of insanity at any time previously; paupers; persons likely to become a public charge; professional beggars; persons afflicted with tuberculosis or with a loathsome or dangerous contagious disease; persons not comprehended within any of the foregoing excluded classes who are found to be and are certified by the examining surgeon as being mentally or physically defective, such mental or physical defect being of a nature which may affect the ability of such alien to earn a living; persons who have been convicted of or admit having committed a felony or other crime or misdemeanor involving moral turpitude; polygamists, or persons who admit their belief in the practice of polygamy, anarchists, or persons who believe in or advocate the overthrow by force or violence of the Government of the United States, or of all government, or of all forms of law, or the assassination of public officials; prostitutes, or women or girls coming into the United States for the purpose of prostitution or for any other immoral purpose; persons who procure or attempt to bring in prostitutes or women or girls for the purpose of prostitution or for any other immoral purpose."

Now while few dispute the right and the duty of this country to control immigration there is a difference of opinion as to the degree and nature of that control. There are those who think that the present restrictions are sufficient and beyond them immigration should be encouraged; there are others who believe that immigration should be much further restricted by requiring educational, property and other qualifications. This difference of opinion is based partly on differences of needs and ideals. Those who would keep the door open are largely employees of labor who need most of it to "develop" or exploit the resources of the country. Those who wish to restrict belong partly to the class of laborers and low-grade artisans who desire to keep wages high and partly to the old families who fear the consequences of this copious infusion of South-eastern European blood. This difference of opinion must, as is generally the case, be ascribed to ignorance. If we knew the probable consequences upon our national life we would probably be agreed what to do.

To a biologist it seems that the economic aspects of the immigration problem will take care of themselves, just because immigration is, from this side, self-regulatory. When wages fall immigration diminishes to a third or a quarter of the volume that it has in times of prosperity and high wages. Moreover, it is (isn't it?) a rather selfish policy to keep out those who are qualified to become good citizens that we may fatten the faster on their destitution. But on its biologic side the problem is real and urgent. How can we keep out defective germ plasm while we admit that which is strong? The attempt to do this by examination of the immigrant is as unscientific as it is inadequate. A person who by all physical and mental examinations is normal may lack in half of his germ cells the determiner for complete mental development. In some respects such a person is more undesirable in the community than the idiot (who will probably not reproduce) or the low-grade imbecile, who will be recognized as such and be selected against in marriage, or be sent by his neighbors to an institution where he may be kept from reproducing. Nor can the immigration problem be solved by excluding on the ground of race or native country. No one has suggested excluding the natives of Switzerland, yet a normal woman from the neighborhood of

Tenna, Canton Graubunden, may become a focus of hemophilia in this country. On the other hand, the exclusion of one Hungarian family of my acquaintance would have deprived American Universities of three of their best scientific professors. The fact is that no race *per se*, whether Slovak, Ruthenian, Turk or Chinese, is dangerous and none undesirable; but only those individuals whose somatic traits or germinal determiners are, from the standpoint of our social life, bad. While all somatically defective may well be excluded at once, it is, within limits, hazardous to admit any person permanently to this country because he has no undesirable somatic trait—for no one transmits to his progeny his somatic traits but rather the determiners in his germ plasm. The proper way to classify immigrants for admission or rejection is on the basis of the probable performance of their germ plasm. In other words, immigrants are desirable who are of "good blood"; undesirable who are of "bad blood."

Since "blood" cannot be judged by inspection of the individual what practicable method remains for separating the sheep from the goats? Experience indicates the one best way. Before any one person is admitted to citizenship let something be learned concerning his family history and his personal history on the other side of the ocean. How can this be done? By means of field workers performing a service similar to that which they are doing in this country, visiting the relatives of the person in question and learning his personal and family history. Is this feasible? Governments might interpose an objection, but it seems probable that the matter could be put before them so that they would not. Experience indicates that few families approached in the proper spirit would decline to give information. It is then only a matter of money to pay for the required studies. How much money? It appears that about 200,000 declarations of intention to become naturalized are filed annually in the United States. It seems probable that field workers by properly sorting their families geographically could each report on the average on ten persons a week or, say, 500 a year. This average is the more reasonable since brothers sometimes make declaration simultaneously so that the history of two persons can be got in one visit. At this rate 400 field workers would be required. At the low price of living abroad the cost of each field worker's salary and traveling expenses would not exceed $1,200, or $480,000 for all. With 10 district inspectors at $2,000, including traveling expenses, and a central office at $10,000, the total cost would be $510,000 a year, and this amount should furnish our government with a report on practically every applicant for naturalization, which would serve as a proper basis for judging of his desirability. Compared with the annual expenditure of over $100,000,000 in this country to take care of our defectives this amount seems small and would be well invested, for, within a decade, the annual saving to our institutions would pay for the work. Moreover, an increase of 50 cents in the head-tax of immigrants would supply funds enough for the entire undertaking.

With a control such as is outlined above we may, it seems to me, face the addition annually of 200,000 Europeans to our citizenship with equanimity. Despite the tendency of encouraged immigration to bring in a less independent and self-reliant class, a significant selection is still exercised. This is clearly expressed in the Report on Emigration Conditions in Europe, published by the Immigration Commission, p. 11.

> The present-day emigration from Europe to the United States is for the most part drawn from country districts and smaller cities or villages and is composed largely of the peasantry and unskilled laboring classes. This is particularly true of the races or peoples from countries furnishing the newer immigration, with the conspicuous exception of Russian Hebrews, who are city dwellers by compulsion. Emigration being mainly a result of economic conditions, it is natural that the emigrating spirit should be

strongest among those most seriously affected, but notwithstanding this the present movement is not recruited in the main from the lowest economic and social strata of the population. In European countries, as in the United States, the poorest and least desirable element in the population, from an economic as well as a social standpoint, is found in the larger cities, and as a rule such cities furnish comparatively few emigrants. Neither do the average or typical emigrants of to-day represent the lowest in the economic and social scale even among the classes from which they come, a circumstance attributable to both natural and artificial causes. In the first place, emigrating to a strange and distant country, although less of an undertaking than formerly, is still a serious and relatively difficult matter, requiring a degree of courage and resourcefulness not possessed by weaklings of any class. This natural law in the main regulated the earlier European emigration to the United States, and under its influence the present emigration represents the stronger and better element of the particular class from which it is drawn.

A most potent adjunct to the natural law of selection, however, is the United States immigration act, the effect of which in preventing the emigration, or even attempted emigration, of at least physical and mental defectives is probably not generally realized. The provisions of the United States immigration law are well known among the emigrating classes of Europe, and the large number rejected at European ports, or refused admission after reaching the United States, has a decided influence in retarding emigration, and naturally that influence is most potent among those who doubt their ability to meet the law's requirements.

If increasing attention is paid to the selective elimination at our ports of entry of the actually undesirable (those with a germ plasm that has imbecile, epileptic, insane, criminalistic, alcoholic, and sexually immoral tendencies); if agents in Europe learn the family history of all applicants for naturalization; if the luring of the credulous and suggestible by steamship agents abroad and especially in the south-east of Europe be reduced to its lowest limits, then we may expect to see our population not harmed but improved by this mixture with a more mercurial people.

CHAPTER VI

THE INFLUENCE OF THE INDIVIDUAL ON THE RACE

As one stands at Ellis Island and sees pass the stream of persons, sometimes 5,000 in a day, who go through that portal to enter the United States and, for the most part, to become incorporated into it, one is apt to lose sight of the potential importance to this nation of the individual, or, more strictly, the germ plasm that he or she carries. Yet the study of extensive pedigrees warns us of the fact. Every one of those peasants, each item of that "riff-raff" of Europe, as it is sometimes carelessly called, will, if fecund, play a rôle for better or worse in the future history of this nation. Formerly, when we believed that factors blend, a characteristic in the germ plasm of a single individual among thousands seemed not worth considering: it would soon be lost in the melting pot. But now we know that unit characters do not blend; that after a score of generations the given characteristic may still appear unaffected by the repeated unions with foreign germ plasm. So the individual, as the bearer of a potentially immortal germ plasm with innumerable traits becomes of the greatest interest. A few examples will illustrate this law and its practical importance.

1. Elizabeth Tuttle

From two English parents, sire at least remotely descended from royalty, was born in Massachusetts Elizabeth Tuttle. She developed into a woman of great beauty, of tall and commanding appearance, striking carriage, "of strong will, extreme intellectual vigor, of mental grasp akin to rapacity, attracting not by a few magnetic traits but repelling" when she evinced an extraordinary deficiency of moral sense.

"On November 19, 1667, she married Richard Edwards of Hartford, Connecticut, a lawyer of high repute and great erudition. Like his wife he was very tall and as they both walked the Hartford streets their appearance invited the eyes and the admiration of all." In 1691, Mr. Edwards was divorced from his wife on the ground of her adultery and other immoralities. The evil trait was in the blood, for one of her sisters murdered her own son and a brother murdered his own sister. After his divorce Mr. Edwards remarried and had five sons and a daughter by Mary Talcott, a mediocre woman, average in talent and character and ordinary in appearance. "None of Mary Talcott's progeny rose above mediocrity and their descendants gained no abiding reputation."

Of Elizabeth Tuttle and Richard Edwards the only son was Timothy Edwards, who graduated from Harvard College in 1691, gaining simultaneously the two degrees of bachelor of arts and master of arts—a very exceptional feat. He was pastor of the church in East Windsor, Connecticut, for fifty-nine years. Of eleven children the only son was Jonathan Edwards, one of the world's great intellects, preeminent as a divine and theologian, president of Princeton College. Of the descendants of Jonathan Edwards much has been written; a brief catalogue must suffice: Jonathan Edwards, Jr., president of Union College; Timothy Dwight, president of Yale; Sereno Edwards Dwight, president of Hamilton College; Theodore Dwight Woolsey, for twenty-five years president of Yale College; Sarah, wife of Tapping Reeve, founder of Litchfield Law School, herself no mean lawyer; Daniel Tyler, a general of the Civil War and founder of the iron industries of north Alabama; Timothy Dwight, the second, president of Yale University from 1886 to 1898; Theodore William Dwight, founder and for thirty-three years warden of Columbia Law School; "Henrietta Frances, wife of Eli Whitney, inventor of the cotton gin, who, burning the midnight oil by the side of her ingenious husband, helped him to his enduring fame; Merrill Edwards Gates, president of Amherst College; Catherine Maria Sedgwick of graceful pen; Charles Sedgwick Minot, authority on biology and embryology in the Harvard Medical School, and Winston Churchill, the author of *Coniston.*"[1] These constitute a glorious galaxy of America's great educators, students and moral leaders of the Republic.

Two other of the descendants of Elizabeth Tuttle through her son Timothy, have been purposely omitted from the foregoing catalogue since they belong in a class by themselves, because they inherited also the defects of Elizabeth's character. These two were Pierrepont Edwards, who is said to have been a tall, brilliant, acute jurist, eccentric and licentious; and Aaron Burr, Vice-President of the United States, in whom flowered the good and the evil of Elizabeth Tuttle's blood. Here the lack of control of the sex-impulse in the germ plasm of this wonderful woman has reappeared with imagination and other talents in certain of her descendants.

The remarkable qualities of Elizabeth Tuttle were in the germ plasm of her four daughters also: Abigail Stoughton, Elizabeth Deming, Ann Richardson and Mabel Bigelow. All of these have had distinguished descendants of whom only a few can be mentioned here. Robert Treat Paine, signer of the Declaration of Independence, descended from Abigail, the Fairbanks Brothers, manufacturers of scales and hardware at St. Johnsbury, Vt., and the Marchioness of

[1] From a manuscript furnished by a reliable genealogist. The statements have not all been checked.

Donegal were descended from Elizabeth Deming; from Mabel Bigelow came Morrison R. Waite, Chief Justice of the United States, and the law author, Melville M. Bigelow; from Ann Richardson proceeded Marvin Richardson Vincent, professor of Sacred Literature at Columbia University, the Marchioness of Apesteguia of Cuba, and Ulysses S. Grant and Grover Cleveland, presidents of the United States.[1] Thus two presidents, the wife of a third and a vice-president trace back their origin to the germ plasm from which (in part) Elizabeth Tuttle was also derived, but of which, it must never be forgotten, she was not the author. Nevertheless, had Elizabeth Tuttle not been this nation would not occupy the position in culture and learning that it now does.

2. The First Families of Virginia

This remarkable galaxy arose by the intermarriage of representatives of various English aristocratic families. The story of these early matings is briefly as follows: Richard Lee, of a Shropshire family that held much land and many of whose members had been knighted, went, during the reign of Charles I, to the Colony of Virginia as Secretary and one of the King's Privy Council. "He was a man of good stature, comely visage, enterprising genius, sound head, vigorous spirit and generous nature." He gained large grants of land in Virginia. His son Richard married, in 1674, Laetitia, daughter of Henry Corbin and Alice Eltonhead. The Corbins were wealthy and extensive landowners in England for 14 generations, and the Eltonheads were also an aristocratic family and extensive landowners of Virginia, holding high offices in the colony. Richard and Laetitia had six sons and one daughter (Fig. 175). Their daughter Ann married Colonel William Fitzhugh, a descendant of the English barons of that name who took prominent parts in

[1] From the genealogist's manuscript, deposited at the Eugenics Record Office.

political and military movements of the day and occupied seats in parliament generation after generation. Their eldest son, Henry Fitzhugh, married Lucy Carter. One of their granddaughters married a Randolph; one of their sons, William Fitzhugh, a near neighbor and trusted friend of Washington, married Anne Randolph. *Their* daughter Anne married Judge William Craik; their daughter Mary married George Washington Parke Custis and became the mother of Mary Anne Randolph Custis and the grandmother of Robert E. Lee; and their son William Henry Fitzhugh married Anna Goldsborough.

Richard Lee, son of Richard and Laetitia (Corbin) Lee, married an English heiress, Martha Silk, and had several children of whom one married a Fairfax, another a Colonel Corbin and a third Major George Tuberville of an ancient English family, himself Justice, Sheriff and Clerk.

Philip Lee, another son of Richard, married a daughter of Hon. Thomas Brooke and Barbara Addison and their children married well. Thomas, brother of Philip, was a member of the House of Burgesses, member, and later president of the Council and later Acting Governor of the Colony. He married Hannah, daughter of Colonel Philip Ludwell, a descendant of a brother of Lord Cattington, a prominent statesman and diplomat of the reign of Charles II. One of the sons of Thomas and Hannah was Richard Henry Lee, a representative to the Continental Congress, who prepared the resolutions for independence; and another son was Francis Lightfoot Lee, a member of Congress; still another, Thomas, was a judge of the General Court.

Finally there was Henry Lee, son of Richard and Laetitia, who lived quietly at the ancestral Lee Hall. He married Mary, daughter of Colonel Richard Bland, descendant of Sir Thomas Bland, of ancient and honorable family, created baronet by Charles I. Mary Bland's grandfather, Theod-

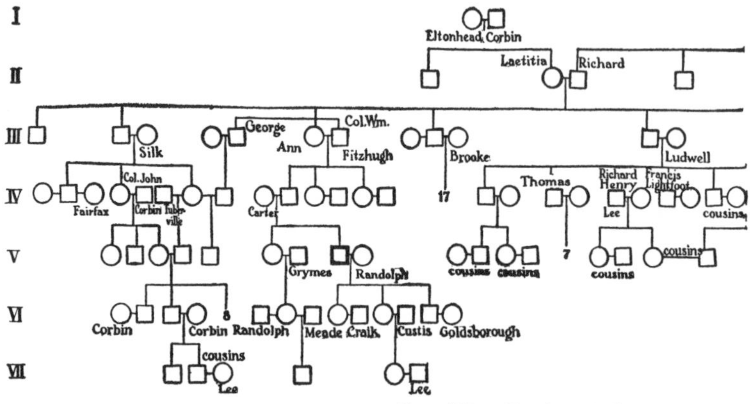

FIG. 175.—Portion of the LEE family

rick Bland, was speaker of the House of Burgesses, a member of the Council, inferior to none in his time. Of the three sons of Henry Lee and Mary Bland, John was a clerk of courts and a member of the House of Burgesses; Richard, was in the house of Burgesses and the House of Delegates; Henry, in the House of Burgesses, Conventions, and the State Senate. Such is a sample, merely, of the intermarriages of the first families of Virginia and their product—statesmen and military men, the necessary consequence of the determiners in their germ plasm.

3. THE KENTUCKY ARISTOCRACY

Nearly two centuries ago John Preston of Londonderry, Irish born though English bred, married the Irish girl Elizabeth Patton, of Donegal, and to the wilderness of Virginia took his wife and built their home, Spring Hill. "Of this union there were five children, Letitia, who married Colonel Robert Breckinridge; Margaret, who married the Rev. John Brown; William, whose wife was Susannah Smith; Anne, who married Colonel John Smith; and Mary, who married Benjamin Howard." From them have come the most conspicuous of those who bear the name of Preston, Brown,

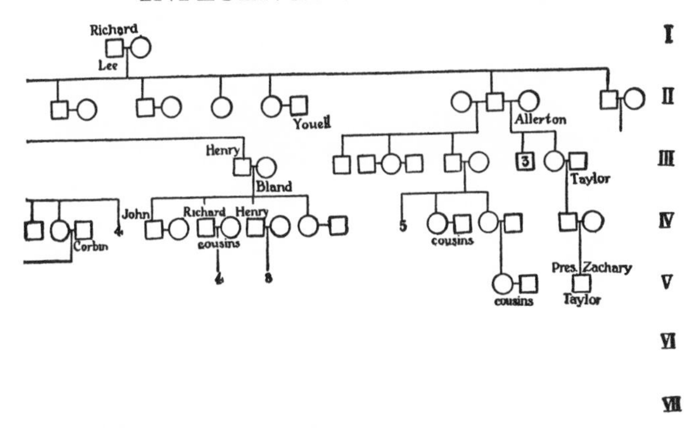

of Virginia, showing intermarriages.

Smith, Carrington, Venable, Payne, Wickcliffe, Wooley, Breckinridge, Benton, Porter and many other names written high in history.

"They were generally persons of great talent and thoroughly educated; of large brain and magnificent physique. The men were brave and gallant, the women accomplished and fascinating and incomparably beautiful. There was no aristocracy in America that did not eagerly open its veins for the infusion of this Irish blood; and the families of Washington and Randolph and Patrick Henry and Henry Clay and the Hamptons, Wickliffes, Marshalls, Peytons, Cabells, Crittendens, and Ingersolls felt proud of their alliances with this noble Irish family.

"They were governors and senators and members of Congress, and presidents of colleges and eminent divines, and brave generals from Virginia, Kentucky, Louisiana, Missouri, California, Ohio, New York, Indiana, and South Carolina. There were four governors of old Virginia. They were members of the cabinets of Jefferson and Taylor and Buchanan and Lincoln. They had major-generals and brigadier-generals by the dozen; members of the Senate and House of Representatives by the score; and gallant officers in the army and navy by the hundred. They furnished three of the recent Democratic candidates for Vice-president of the United States. They furnished the Union Army General B. Gratz Brown, General Francis P. Blair, General Andrew J. Alexander, General Edwin C. Carrington, General Thomas C. Crittenden, Colonel Peter A. Porter, Colonel John M. Brown, and other gallant officers. To the southern army they gave Major-General John C. Breckinridge, Major-General William Preston, General Randall Lee Gibson, General John B. Floyd, General John B. Grayson, Colonel Robert J. Breckinridge, Colonel W. C. P. Breckinridge, Colonel William Watts, Colonel Cary Breckinridge, Colonel William Preston Johnson, aide to Jefferson Davis, with other colonels, majors, chaplains, surgeons, fifty of them at least the bravest of the brave, sixteen of them dying on the field of battle, and all of them, and more than I can enumerate, children of this one Irish emigrant from the county of Derry, whose relatives are still prominent in that part of Ireland, one of whom was recently mayor of Belfast."

Overlooking the pardonable rhetoric and family pride in the last sentence, that neglects the hundreds of other ancestors of these famous men, the quotation has a scientific value in comparison with the product of Elizabeth Tuttle. The New England family glows with scholars and inventors, the Virginia and Kentucky families with statesmen and military men. The result is not due to the differences in the characteristics of Elizabeth Tuttle and Richard Edwards, Richard and Laetitia Lee, John and Elizabeth Preston, respectively, but to the different traits of the New England settlers as a whole and Virginia cavalier-colonists as a body. The initial person becomes a great progenitor largely because of some fortunate circumstance of personal gift or excellent reputation that enables his offspring to marry into the "best blood."

4. The "Jukes"

On the other hand, we have the striking cases of families of defectives and criminals that can be traced back to a single ancestor. The case of the "Jukes" is well known. We are first introduced to a man known in literature as Max, living as a backwoodsman in New York State and a descendant of the early Dutch settlers; a good-natured, lazy sot, without doubt of defective mentality. He has two sons who marry two of six sisters whose ancestry is uncertain but of such a nature as to lead to the suspicion that they are not full sisters. One of these sisters is known as "Ada Juke," also as "Margaret, the mother of criminals." She was indolent and a harlot before marriage. Besides an illegitimate son she had four legitimate children. The first, a son, was indolent, licentious and syphilitic; he married a cousin and had eight children all syphilitic from birth. Of the 7 daughters 5 were harlots and of the others one was an idiot and one of good reputation. Their descendants show a preponderance of harlotry in the females and much consanguineous marriage. The second son was a farm laborer, was industrious and saved enough to buy 14 acres of land. He married a cousin and the product was 3 stillborn children, a harlot, an insane daughter who committed suicide, an industrious son, who, however, was licentious, and a pauper son. The first daughter of "Ada" was an indolent harlot who later married a lazy mulatto and produced 9 children, harlots and paupers, who produced in turn a licentious progeny.

Ada had an illegitimate son who was an industrious and honest laborer and married a cousin. Two of the three sons were licentious and criminalistic in tendency and the third, while capable, drank and received out-door relief. All of the three daughters were harlots or prostitutes and two married criminals. The third generation shows the eruption of criminality. Excepting the children of the third son, none of whom were criminalistic, we find among the males 12 criminals, 1 licentious, 5 paupers, 1 alcoholic and 1 unknown; none were normal citizens. Among the females 3 were harlots, 1 pauper, 1 a vagrant and 2 unknown; none were known to be reputable. Thus it appears that criminality lies in the illegitimate line from Ada and not at all in the legitimate—doubtless because of a difference in germ plasm of the fathers.

The progeny of the harlot Bell Juke is a dreary monotony of harlotry and licentiousness to the fifth generation. Two in the fourth generation there are and two in the fifth against whom there is nothing and their progeny mostly moved to another neighborhood and are lost sight of. Very likely they have married into stronger strains and are founders of reputable families.

The progeny of Effie Juke and the son of Max (a thief) show to the fifth generation a different aspect. Some larceny and assault there is and not a little sexual immorality, but pauperism is the prevailing trait.

Thus, in the same environment, the descendants of the illegitimate son of Ada are prevailingly *criminal;* the progeny of Bell are *sexually immoral;* and the offspring of Effie are *paupers.* The difference in the germ plasm determines the difference in the prevailing trait. But however varied the forms of non-social behavior of the progeny of the mother of the Juke girls the result was calculated to cost the State of New York over a million and a quarter of dollars in 75 years—up to 1877, and their protoplasm has been multiplied and dispersed during the subsequent 34 years and is still marching on.

5. The Ishmaelites

Another example of a great family tracing back to a single man may be taken from "the Tribe of Ishmael" of Central

Indiana, as worked out under the direction of the Rev. Oscar C. McCulloch of the Charity Organization Society, Indianapolis. The progenitor of this tribe, Ben Ishmael, was in Kentucky as far back as 1790, having come from Maryland through Kentucky. One of his sons, John, married a half-breed woman and came into Marion County, Indiana, about 1840. His three sons who figure in this history married three sisters from a pauper family named Smith. They had altogether 14 children that survived, 60 grandchildren and 30 great-grandchildren living in 1888. "Since 1840 this family has had a pauper record. They have been in the almshouse, the House of Refuge, the Woman's Reformatory, the penitentiaries and have received continuous aid from the townships. They are intermarried with the other members of this group,—and with over two hundred other families. In this family history are murderers, a large number of illegitimacies and of prostitutes. They are generally diseased. The children die young. They live by petty stealing, begging and ash-gathering. In summer they "Gipsy" or travel in wagons, east or west. We hear of them in Illinois about Decatur and in Ohio about Columbus. In the fall they return. They have been known to live in hollow trees on the river bottoms or in empty houses. Strangely enough, they are not intemperate to excess."

"A second typical case is that of the Owens family, also from Kentucky. There were originally four children, of whom two have been traced, William and Brook. William had three children, who raised pauper families. One son of the third generation died in the penitentiary; his two sons in the fourth generation have been in the penitentiary; a daughter in the fourth generation was a prostitute with two illegitimate children. Another son in the third generation had a penitentiary record and died of delirium tremens." An illegitimate half-breed Canadian woman enters this family. There have been several murders and a continuous pauper and criminal record. There is much prostitution, but little intemperance.

"Brook had a son John, who was a Presbyterian minister. He raised a family of 14 illegitimate children. Ten of these came to Indiana, and their pauper record begins about 1850. Of the ten, three raised illegitimate children in the fifth generation."

The families with which the Ishmaelites intermarried (30 in number) came mostly from Kentucky, Tennessee, and North Carolina. "Of the first generation—of 62 individuals—we know certainly of only three. In the second generation we have the history of 94. In the third generation, we have the history of 283. In the fourth generation (1840–1860) we have the history of 644. In the fifth generation (1860–1880) we have the history of 57. Here is a total of 1,750 individuals. Before the fourth generation (from 1840–1860), we have but scant records. Our more complete data begin with the fourth generation, and the following are valuable. We know of 121 prostitutes. The criminal record is very large,—petty thieving, larcenies, chiefly. There has been a number of murders. The first murder committed in the city was in this family. A long and celebrated murder case known as the 'Clem' murder, costing the State immense sums of money, is located here, nearly every crime of any note belongs here." What a vivid picture has McCulloch drawn of the influence on a community of its "bad blood," forming an intergenerating, self-perpetuating, anti-social class—anti-social because possessed of such traits as feeble-mindedness, wandering mania, eroticism, and "moral imbecility." How slow the community is to protect itself by adopting some method of preventing their reproduction!

6. The Banker Family

The examples given above are extreme, to be sure; they were selected just because they are extreme. But it is just as true that every family whose early ancestors showed some striking trait reveals that trait now and again in the offspring. One can find evidence of this in almost any intelligently compiled genealogical history. Take, for example, the Banker family. There were two Dutchmen who were early settlers in New York State: Gerrit, who settled about 1654 in Albany, and Laurens, who settled some years later in Tarrytown. They were, apparently, not related and their descendants have not intermarried. The two lines present some striking contrasts.

"Gerrit appears to have been well educated for that time and was a very successful merchant and Indian trader, accumulating a considerable property. His descendants were largely merchants, although many become farmers." In general they maintained a high degree of culture and social rank. Several of them attained to positions of prominence in the affairs of the Colony before and during the Revolution. For example, the first Treasurer of the State and the first Speaker of the Assembly were both from this family, while several held commissions in the Revolutionary Army. Since that period they have been less prominent in public affairs, although maintaining a position of high social standing and respectability."

Laurens, on the other hand, had no education, could not write his name, at least when a young man, and was a laborer and farmer. His descendants "may be said in some ways to have started at the bottom. The family prior to the Revolution was obscure, its members were chiefly laborers, farmers, and artisans with only limited opportunities for education and acquiring but little of this world's goods. In the Revolution they actually furnished more soldiers than the Gerrit Banker family, but none of them held rank above a corporal. They were, in fact, as often described in legal documents, yeomen, and yeomen under a semi-feudal system. With the organization of the new nation a larger opportunity opened. To-day many of this family have reached places of high social standing while a few have been brought into a considerable degree of public prominence." [1] In this instructive example we see the persistence of an initial difference with a final tendency to approach a common level. Because in the absence of caste, and the desire to marry as well as possible, new and strong characters are introduced into the germ plasm.

[1] Compare Banker, 1909.

CHAPTER VII

THE STUDY OF AMERICAN FAMILIES

Nowhere else is a genealogical interest keener than in America. The possibility of tracing one's pedigree back to the first ancestor of the name in the country has inspired thousands of genealogical researches, and the demand for assistance in working out pedigrees has created the professional genealogist. Still the amateur's work, like most labors of love, is usually to be preferred because of the personal element involved.

1. The Study of Genealogy

The study of genealogy, under the stimulus of our modern insight into heredity, is destined to become the most important handmaid of eugenics. The conscientious and scientific genealogist records a brief biography of each person of the pedigree and such a biography should be an analysis of the person's traits; an inventory of his physical and mental characteristics; his special tastes and gifts as shown by his occupation and especially his avocations. It would be well, so far as possible, to go further than that, if not for publication at least for record.[1] It will be desirable to get a statement of physical weaknesses, diseases to which there was liability and causes of death. There are none of these classes of data that are not included in some genealogies; it

[1] The Eugenics Record Office has an isolated fire proof vault at Cold Spring Harbor, N. Y., in which it will receive and keep safe and confidential any records that genealogists will deposit there. All genealogical data is indexed on cards so as to be made accessible to properly qualified persons who wish to use it for justifiable purposes.

would be well if all were included in all genealogies. Another desideratum is abundant photographs of the persons whose biographies are given; especially, strictly full-face and profile, to facilitate comparisons; and two or three photographs at successive ages would be still better than one.

Attention should be paid to the form of the pedigree. The commonest form is that which begins with the first known male ancestor bearing the surname. His children are given, but in the later generations only the offspring of males are named. Few genealogies attempt either to trace the lines going through females or to give the ancestry of the consorts. A second form of pedigree begins with the author or some other one person and gives an account of all of his direct ancestors in ever expanding number toward the earlier generations. This method is scarcely more valuable than the other from a scientific point of view, based as it is upon the exploded idea that inheritance is from parents, grandparents, etc.

The ideal genealogy, it seems to me, starts with a (preferably large) fraternity. It describes fully each member of it. It then describes each member of the fraternity to which the father belongs and gives some account of their consorts (if married) and their children. It does the same for the maternal fraternity. Next, it considers the fraternity to which the father's father belongs, considers their consorts, their children and their grandchildren and it does the same for the fraternities to which the father's mother belongs. If possible, earlier generations are to be similarly treated. It were more significant thus to study in detail the behavior of all the available product of the germ plasms involved in the makeup of the first fraternity than to weld a chain or two of links through six or seven generations. A genealogy constructed on such a plan would give a clear picture of heredity, would be useful for the prediction of the characteristics of the generations yet unborn, and would, indeed, aid in bringing about better matings. It is to be hoped that the time will come when each person will regard it as a patriotic duty to coöperate in the compilation of such genealogical records even to the statement of facts which are, according to the (often false) conventions of the day, not considered "creditable."

2. Family Traits

The results of such genealogical studies will be striking. Each "family" will be seen to be stamped with a peculiar set of traits depending upon the nature of its germ plasm. One family will be characterized by political activity, another by scholarship, another by financial success, another by professional success, another by insanity in some members with or without brilliancy in others, another by imbecility and epilepsy, another by larceny and sexual immorality, another by suicide, another by mechanical ability, or vocal talent, or ability in literary expression. In some families the members are prevailingly slender, in others stout; in some tall, others short; some blue-eyed, others dark-eyed; some with flaxen hair, others with black hair; some have diseases of the ear, others of the eye, or throat or circulation. In some nearly all die of consumption; in others there is no weakness of the mucous membranes but a tendency to apoplexy; others die prevailingly of Bright's disease or valvular disease of the heart, or of pneumonia. In some families nearly all die at over 80, in others all die under 40 years of age. Stammering, hirsuteness, extra dentition, aquiline nose, lobeless-ears, crooked digits, extra digits, short digits, broad thumbs, ridged nails,—there is hardly an organ or the smallest part of an organ that has not its peculiar condition that stamps a family.

Said a lady to me, "I was traveling in Egypt and met a

man who was introduced to me as Mr. Osborn. I said to him 'My mother was an Osborn. I wonder if we are related.' He replied, 'Let me see if you have the Osborn thumb,'" and she was able to show the family trade-mark. How often a peculiar laugh, a trick of speech or gesture will serve to identify the family of a stranger. Once in a city where my family was well known but where I was a stranger I needed to get a check cashed and went to an office where my father and brother had done business. On explaining my need to the head of the firm he supplied it without hesitation, saying: "Though I have never seen you before I would know anywhere that you were a Davenport." So wonderfully are details of facial muscles, form of skull bones and nose cartilage stamped in the family blood. Such features as these deserve full treatment in the philosophical family history.

Many works on genealogy, as I have said, give a little account of family traits. A few of those have been excerpted from the published works and are reproduced here chiefly to illustrate the specificity of human families. Of course, except where there is much consanguineous marriage, not all traits will appear in all or even most individuals of the family, and new traits are being introduced by marriage. But certain characteristics because of their special nature or the frequency with which they occur in certain branches of the family will come to be known as "family traits."

Allerton (Allerton, 1888). The great majority of the family to-day, as always, are farmers; have never showed a tendency to city life. Next to farming, machinist is the most favored occupation. Mostly large framed, few blondes, slender and lithe in youth; fleshy in old age. A quick-tempered race; decided, uncommunicative, reserved.

Balch (MSS.). "Balch spelling" said to be a recognizable trait.

Bascom (Harris, 1870). Stout, compact form, head well set back upon the shoulders, dark skin, dark gray eye, massive head and round, high, full forehead.

Banning (Banning, 1908). Determination and will-power almost to point of stubbornness; faithful to friends and families, fairness to enemies; clannishness, ability for hard, reliable work, firmness of mouth.

Breed (Breed, 1892). As a rule, positive, determined, industrious and persevering in business and careful of their income.

Brinckerhoff (Brinckerhoff, 1887). Blue eyes, Roman features, magnetic and generous; ofttimes impulsive, sometimes absolutely wrong in actions and convictions but true and steadfast in the wrong. Usually can whistle a tune or sing a song without any apparent effort.

Buck (Buck, 1893). Quickness and activity in movement; fast walkers. One could seize with his right hand the toe of his left boot and whilst so holding it and standing erect jump with his right foot backwards and forwards over his left leg. Fluency in conversation and aptness for acquiring languages.

Cole (Cole, 1887). Asa Cole was a man of immense physical strength and endurance; he suffered a paralytic stroke. His son, John Cole, was a man of fine physique, and died from a stroke of apoplexy; a second cousin, Salmon Cole, was almost a giant in strength.

Colegrove (Colegrove, 1894). Strong individuality of character, often called peculiar or secretive, very self-reliant.

Doolittle (Doolittle, 1901). Large, robust physique, florid complexion, high spirit, jovial disposition.

Dwight (Dwight, 1874). Moderate sized families; longevity not high, commonly well-to-do and inclined to liberal culture; much military talent.

Humphreys (Humphreys, 1883). Self-reliance, readiness

of acquisition; professional men, few tradesmen and mechanics; artistic temperament, good talkers and eloquent speakers; benignity and quietness.

Johnsons of Harpswell, Maine (Sinnett, 1907). Hospitality, story-telling.

Kimball (Morrison, 1897). Powerful memory; few politicians.

Lemen (Lemen, 1898). Strongly accentuated mental and moral traits; a "family habit" of slight despondency; some gift for poetry.

Lindsay (Lindsay, 1889). Cheerfulness, hospitality.

Mell (Mell, 1897). Social, genial, fun-loving temperaments.

Mickley (Mickley, 1893). No lawyers, but other professions; nearly all in comfortable circumstances.

Neighbor or Nachbar (Neighbor, 1906). Not restive; neighborly, temperate.

Reed of Massachusetts (Reed, 1861). Few die of pulmonary complaints. Generally live to old age, 85 or 90 or even 100 years being nothing unusual. Capable of great endurance. Taller than average. One custom has prevailed among them to some extent; that of marrying relatives. "Consequences have been injurious; many of the offspring of such marriages dying in infancy, early youth or middle age, few living to advanced years, to say nothing of cases where effect has been still more melancholy."

Riggs (Wallace, 1901). A large proportion are governed by strong religious convictions and are active in religious thought and work. Many daughters of the family have married Presbyterian ministers and in due time became mothers of Presbyterian ministers themselves.

Root (Root, 1870). Eight sons of Samuel were tall (with two exceptions), quick, subject to frequent attacks of headache; general family trait a prominent (frequently aquiline) nose, light complexion, blue eyes, somewhat commanding presence and vivacity of manner.

Sinclair (Morrison, 1896). Fond of athletic sports and feats of strength and skill, much mechanical knowledge, practical, loving activities and experiences of frontiersman better than books or studies of scholars and of professional life. Love of military life.

Slayton (Slayton, 1898). Musical, especially vocally. Large families, twenty pairs of twins and one set of triplets recorded.

Tapley (Tapley, 1900). Quick and nervous movements, fondness for music, short stature, genial disposition. Men of affairs rather than of professions.

Tiffany (Tiffany, 1903). Complexion dark, eye bright with expression changing rapidly with mood indicating health, sympathy, grief, determination or anger with quickness and unerring certainty; "a Tiffany mark."

Twining (Twining, 1905). Broad-shouldered, dark hair, prominent nose, nervous temperament, temper usually quick, not revengeful. Heavy eyebrows, humorous vein and sense of ludicrous; lovers of music and horses.

Varick (Wheeler, 1906). A colored family, very light in complexion, some members pass for white.

Zahniser (Zahniser, 1906). Tall, many 6 feet or over, heavy black hair, rarely falling out, face broad, cheek-bones prominent, eyebrows protruding. Type becoming rarer in recent generations.

The traits named in the foregoing list have a very dissimilar value and significance as inheritable characters. But some, at least, have the same value as the famous "Hapsburg lip." Were our population so closely inbred as European royalty it would show hundreds of characteristics with the same family value. But our families are constantly outmarrying and a definite trait becomes disseminated into scores of family names so that its family signification becomes lost.

The facts that we have been considering above lead to a conclusion quite in line with modern experimental work in heredity and with the interpretation of varieties. The white race as seen in America to-day is made up of thousands, yes, hundreds of thousands of kinds of protoplasm which differ by the possession of at least one determiner for a peculiar, differentiating trait. The potential strains that are constituted by these different kinds are not, however, real strains because they are constantly crossed into other strains. Only when there is a high degree of consanguineous marriage, as in small islands, or mountain valleys, is this potentiality realized. Otherwise the traits soon become dissociated from the family names of those who brought them to this country and they become disseminated into many related families. But the potentiality for the production of a strain or race remains.

Now the fact of the existence of such strains in this country has an important bearing upon studies made on man. For example, our text-books on anatomy give an account of structure that is based on the finding of numerous autopsies. The original author of such a work records for each organ and part the condition in which he has found it in the material that he has dissected. If he goes into enough detail he has to state in connection with each description that it does not hold universally but that, on the contrary, in one cadaver or another this and that modification has been found. The name of the family to which the cadaver belongs, its ancestral history, is usually not given (and indeed it frequently cannot be obtained), but it is important that it should be ascertained, if possible, for the same reason that it is important to know if the cadaver were of a Caucasian or a Chinaman. Indeed, as a textbook of Human Anatomy must be rewritten for the Chinese, for the Ethiopians, and for the Eskimos, so must it be rewritten for the Rumanian, for the North Italian, for the Norwegian and for the Spaniard. Nor will the same description of structure of the human body serve, in all details, for the Lees of Virginia, the Ishmaelites of Indiana and the Edwards family of New England. Similarly the text-books of pathology are not universally applicable. There are hundreds of diseases listed that you and I could no more have than we could have extra fingers or a retina without pigment. Even the symptoms of a disease will differ in different strains; for the symptoms of a disease like typhoid fever are not due only to the typhoid germ but to the reaction of the particular living body to those germs. In not a few cases the prognosis, or prospect of the course of the disease, should read: The prognosis can be got by asking the head of the family "What is the usual course of the disease in this family?" Indeed, the classification and diagnosis of a disease is often got better by a comparison of the brother and sister of the patient than by reference to a book of symptoms. "I knew a family of four sisters," said Dr. E. E. Southard to me, "three of whom had manic-depressive insanity; the fourth had a mental disorder that had been classified quite otherwise by another physician. But a comparison of the sisters showed that the mental disorder was of the same type in all." Bleeders in different families differ in the ease with which hemorrhage is induced and the difficulty in stopping it; and in the Sullivan County bleeders the disorder runs a peculiar course so that they are called "nine-day bleeders." Of imbecility there are, as we have seen, all grades and all usually incurable; but the great "moron" or simpleton family of New Jersey is peculiar in that mental development is not permanently arrested but only much retarded.

So albinism varies much in degree and certain families are recognized as containing partial albinos; others, nearly complete albinos; still others, complete albinos.

Pathologies describe some diseases as common, others as rare; yet, within limits, this must depend on the geographical location of the author. At the east end of Long Island Huntington's chorea is not a rare disease as it seems to be in Eastern Massachusetts. Deaf mutism was found in 4 per cent of the population of Chilmark, in 1880, and the practitioner of that place would gain an impression of its frequency which would differ from that of a hospital surgeon in New York City. Hospital surgeons in great cities believe they get a better *average* view because they get random samples out of a great mixture; but in just so far they lose sight of the essential feature of the specificity of the different strains of human germ plasm and too often gain the impression that the sporadic examples of a disease that come to their hands prove the purely accidental nature of its incidence. The metropolitan hospital with its random sampling is the last place to get a proper idea of the relation of disease to germ plasm. It is the venerable country doctor in a long settled and stable community who can tell tales of hereditary tendencies.

It was stated above that coöperation in putting on record one's family history should be regarded as a patriotic duty. I might go further and say that, just as the traits of criminals and defectives go on public or semi-public records, with even more reason a record should be kept of our best families and of their traits. Enlightened communities preserve records of births, marriages and deaths and of various business transactions, especially in land. It is not less important to keep a record of innate capacities and valuable traits. For it is not too much to say that the future of our nation depends on the perpetuation by reproduction of our best protoplasm in proper matings and we cannot have proper matings unless our best protoplasm is located and known. The day may come when in intelligent circles a woman will accept a man without knowing his biologico-genealogical history with as much hesitation as a stock-breeder will accept as a sire for his colts or calves an animal without a pedigree. Since restriction of the number of children seems, for better or worse, to be the fashion with our older families, let every effort be put forth to secure that each child shall be of the best quality in respect to inborn capacities.[1]

3. The Integrity of Family Traits

We often hear persons who are impressed by the multiplicity of one's ancestors make light of family pride in some preëminent forbear. They ask of what significance can such an ancestor be whose blood is diluted to one part in a thousand? This way of looking at heredity is a relic of a former view that a trait when mated to its absence produced a half trait in the progeny as skin color was considered to do, and which gave rise to the conception of quadroons, octaroons, etc., with successive lightening of the skin to ¼, ⅛ and so on. Now that we know that even skin color may segregate out in the ancestral full grades we are ready to accept as practically universal the rule that unit characters do not blend; that apparent blends in a trait are a consequence of its composition out of many units. Since this is so, a unit character (especially a negative character) which a remote ancestor possessed may reappear, after many generations have passed, in its pristine purity. A germ plasm that produced a mathematical genius only

[1] The need for a full Family Record is, we may hope, about to be filled by Dr. J. Madison Taylor of Philadelphia. Meanwhile those who wish a copy of the Family Records of the Eugenics Record Office may obtain it on application.

once, a century ago, may produce another not less noteworthy again.

A feature of positive unit characters, which from their very nature tend to reappear in each generation is that of *anticipation.* This means that the trait appears at an earlier age in each generation. Nettleship (1910, pp. 23–25) has referred to some striking cases of this. Thus he gives three pedigrees of hereditary glaucoma and diabetes illustrating this law. In one case the average known age in successive generations for the incidence of glaucoma is 66 and 48 years; in another family 71, 45, and 23 years; in still another, 47 and 20. In the case of diabetes deaths occurred, on the average, at 69, 35 and 26 years. Nettleship explains this result "by assuming that certain defects, taints or vices of the system, say of the blood, are not only hereditary in the true or germinal sense, but able to produce toxic agents in the embryo which have an evil influence upon all its cells, and thus so lower their power of resistance that the innate hereditary factor has freer play and is likely to manifest itself earlier."

The law of segregation of traits, the disproof of the blending hypothesis, is of the utmost importance since it shows how a strain may get completely rid of an undesirable trait. If the undesirable character is a positive one, like polydactylism, it will disappear if the normal children alone have offspring. If it is a negative character its complete and certain elimination is not so easy to be assured of, but offspring without the undesirable trait are easily secured if marriage be always with germ plasm that is without the defect. Thus a simpleton married into a mentally strong strain will probably have mentally well endowed offspring. Here is where the beneficence of heredity clearly appears.

But do traits never arise *de novo* is often asked. If you deny it, how do you account for the presence of great men from obscure origin? For example, Mohammed, Napoleon, Lincoln. First of all, in seeking for an explanation of the origin of such "sports" of which history is full, we must inquire if the putative paternity is the real one. Not infrequently a weak woman has had illegitimate children by the wayward scion of a great family. The oft repeated story that Abraham Lincoln was descended on his mother's side from Chief Justice John Marshall of Virginia, whether it has any basis or not, illustrates the possibility of the origin of great traits through two obscure parents. In the second place we have seen that many elements of genius are negative characters and, as such, they may be transmitted without influencing the soma of the transmitter.

Thus two parents without mathematical genius might bring together germ cells whose union would favor a mathematical prodigy; and the same is true of many other traits. Indeed, as many of our pedigrees show, genius frequently, if not usually, appears in families with mental defects, insanity, or at least neurotic tendencies. It is just these sturdy, stolid communities of which not a few are found in Eastern Pennsylvania that, I am informed, produce few insane persons as well as few geniuses. The connection between genius and mental defect or aberration has been often referred to, especially by Lombroso and his followers, and as often scoffed at. But, apart from the significant association of the two conditions in pedigrees, there is no *a priori* objection to the view that the flights of the imagination, one of the most constant features of genius, should be associated with that flightiness that is a symptom of insanity, or that the absence of complete mental development should be associated with the absence of one or more of these inhibitors that marks the man or woman of great talent.

CHAPTER VIII

EUGENICS AND EUTHENICS

1. Heredity and Environment

Admitting, as we must, the importance of hereditary tendencies in determining man's physical traits, his behavior and his diseases, we cannot overlook the question that must occur to all—What relation have the facts of heredity to those of environmental influence, to the known facts of infection and bad conditions of life? Indeed, were we to accept the teachings of some, environment alone is important, good training, exercise, food, and sunlight can put anybody in a "normal" condition.

So long as we regard heredity and environment as opposed so long will we experience endless contradictions in interpreting any trait, behavior or disease. The truth seems to be that for human phenomena there is not only the external or environmental cause but also an internal or personal cause. The result is, in most cases, the reaction of a specific sort of protoplasm to a specific stimulus. For example, the controversy as to the inheritableness versus the communicableness of "the itch" receives a simple solution if we recognize that there is an external agent, probably a parasite, that can, however, develop only in persons who are non-immune. Since such persons are rather uncommon and the absence of immunity is inheritable, the disease tends to run in families and can rarely be caught even through inoculation, by persons outside such families. Even in cases where the hereditary factor is universally admitted as in manic-depressive insanity, the onset of the symptoms may be delayed by very favorable conditions of life. But though such symptoms may be diminished and the patient be discharged from the hospital as "cured," yet the weakness in his germ plasm is not removed and it will, unless he be fitly mated, show itself in his children when they, in turn, experience an unusual stress. Even the fugue tendency of the child of three years (page 89) might not have expressed itself so acutely had he lived in the country with freedom to wander widely at will instead of being restrained within the confines of city houses and narrow streets, In extreme cases, however, of which complete albinism is an example, the trait seems to be due to the entire absence in both of the united germ cells of any determiner for the character. Under these circumstances not even the best of environmental conditions can bring about pigmentation. Albinism is a protoplasmic "accident" as independent of environment as drowning by the overturning of an ocean steamship is independent of heredity.

With few exceptions, the principle that the biological and pathological history of a child is determined both by the nature of the environment and the nature of the protoplasm may be applied generally. It is an incomplete statement that the tubercle bacillus is the cause of tuberculosis or alcohol the cause of delirium tremens or syphilis the cause of paresis. Experience proves it, for not all that harbor the tubercle bacillus show the dread symptoms of tuberculosis (else there were little hope of escape for any of us); nor do all drunkards have delirium tremens, nor are all who are infected by syphilis paretic, else our hospitals for the insane would be fuller than they are. Rather, each of these diseases is the specific reaction of the organism to the specific poison. In general, the causes of disease as given in the pathologies are not the real causes. They are due to inciting conditions acting on a susceptible protoplasm. The real cause of death of any person is his inability to cope with the disease germ or other untoward conditions.

How prone we are to neglect the personal side of the result! We explain that Mr. A. has gone insane from business losses or overwork. Yet hundreds suffer great losses and work hard and show no signs of nervous breakdown. It would be more accurate to say A. went insane because his nervous mechanism was not strong enough to stand the stresses to which it was put. As a matter of fact insanity rarely occurs except where the protoplasm is defective. Also epilepsy, which is so often ascribed to external conditions, is, like imbecility, determined chiefly by the conditions of the germ plasm; and the trivial circumstance that first reveals the defect is as little the true cause as the touching the electric button that opens an exposition is the motive power of its vast engines. "Father," says the young hopeful, "may I go skating?" "So far as I am concerned; but you had better ask your mother," replies the father. "No, indeed," puts in the mother, "for I read in the paper the other day of a boy who fell on the ice and had an epileptic fit." Thus does the untrained mind confuse contributing and essential causes.

2. Eugenics and Uplift

The relation of eugenics to the vast efforts put forth to ameliorate the condition of our people, especially in crowded cities, should not be forgotten.

Education is a fine thing and the hundreds of millions annually spent upon it in our country are an excellent investment. But every teacher knows that the part he plays in education is after all a small one. In the same class will be two boys who have had the same school training. One catches ideas almost before they are expressed, makes knowledge his own as soon as it is acquired, and passes with swiftness and thoroughness to the limit of the teacher's capacity to impart. Another comprehends slowly, advances only by constant drill and hammering, and seems as little plastic as a piece of wood. Another may be slow in most work but rapid in mathematics, and still another may be first in English composition and incapable of acquiring algebra. The expert teacher can do much with good material; but his work is closely limited by the protoplasmic makeup—the inherent traits—of his pupils.

Religious teachers do a grand work and the value to the state of properly developed and controlled emotions is incalculable. Yet how dependent, after all, are religious or moral teachings upon the nature of those who receive them. I have heard ministers express regret that they preached only to those who least needed their ministrations, but they forgot that to others their ministrations would be of little avail. Religion would be a more effective thing if everybody had a healthy emotional nature: and it can do nothing at all with natures that have not the elements of love, loyalty and devotion.

Of the importance of fresh air, good food, and rest in curing tuberculosis I have no doubt, yet how often have I seen persons brought up in the best of hygienic conditions, with every need supplied, forced to live in a camp in the Adirondacks or in Southern Arizona and, despite the best of trained nursing, gradually fade away. That cleaner milk, more air and sunlight will still further reduce the death rate of infants in New York city cannot be denied; yet there are infants who do not succumb to infantile diarrhea even in the slums. The personal side must not be overlooked in properly estimating the value of prophylaxis.

3. The Elimination of Undesirable Traits

The practical question in eugenics is this: What can be done to reduce the frequency of the undesirable mental and bodily traits which are so large a burden to our population? This question has often been asked. It has been answered in diverse ways, and, indeed, there are several methods of stopping the reproduction of undesirable traits.

There is, first, the method of surgical operation. This prevents reproduction by either destroying or locking up germ cells. There are two principal methods of surgical interference. One is castration, which removes the reproductive gland and destroys sexual desire. The other is vasectomy which prevents the escape of the germ cells to the exterior but does not lessen desire. Neither of these operations is necessarily painful or liable to cause death or much inconvenience to the males. Corresponding operations can be performed on the female but they are more serious in this sex since they involve opening the abdominal cavity.

Concerning the power of the state to operate on selected persons there can be little doubt, not only since the right to the greater deprivation—that of life—includes the right to the lesser deprivation—that of reproduction—but also since these operations are actually made to-day and that of sterilization is legalized, under certain precautions, in six states of the union. There is no question that if every feeble-minded, epileptic, insane, or criminalistic person now in the United States were operated on this year there would be an enormous reduction of the population of our institutions 25 or 30 years hence; but is it certain that such asexualization or sterilization is, on the whole, the best treatment? Is there any other method which will interfere less with natural conditions and bring about the same or perhaps better results? One is struck by the contrast between the haste shown in legislating on so serious a matter compared with the hesitation in appropriating even a small sum of money to study the subject.

First, it may be pointed out that such legislation as is enacted does not square with what we know about heredity. It is based on the old notions that parents transmit their traits to their children. Now we know that traits are transmitted by means of the germ cells and by them alone, and the resemblance of children to parents is due to the fact that both arise from the same material—the father is half-brother to his child. While a feeble-minded person lacks, *ipso facto,* the determiner for normal development in his germ cells, still we do not know that his children will be defective. Such evidence as we have goes rather to show that if, for example, a man whose germ cells have the determiner for normal mentality marry a feeble-minded woman all of the children will be mentally normal or practically so. I can well imagine the marrying of a well-to-do, mentally strong man and a high-grade feeble-minded woman with beauty and social graces which should not only be productive of perfect domestic happiness but also of a large family of normal happy children. Half of the germ cells of such children would, indeed, be defective, but as long as the children married into normal strains the offspring, through an indefinite number of generations, would continue to be normal. Yet in many states of the Union such a marriage cannot be legalized; and, in others, the potential mother might be sterilized.

Secondly, the laws against the marriage of the feeble-minded are unscientific because they attempt no definition of the class. If feeble-mindedness were always as clearly distinct from normality as polydactylism then there would be no objection to the law on this score. But this is by no means the case. If we measure the mentality of 10,000 individuals by a quantitative test, such as that of Binet and Simon, then we shall find that the retardation in mental development for 1 year, 2 years, 3 years, etc., shows nowhere a sharp change indicating where the normal ceases and the abnormal begins. Shall we sterilize or forbid marriage to all children whose mental development is retarded as much as one year? That would include 38 per cent of all children, and one of yours, O legislator! Shall the limit be two years of retardation? That would include 18 per cent of the children. Shall the limit be three years? That will still be over 8 per cent—full one-twelfth of the population to be sterile. Is it not reckless to pass such serious legislation in such loose terms?

Third, have we good ground for denying marriage, generally and under all circumstances, to persons who as school children were even four years behind their fellows? Is it certain that the progeny of such a person will be four years older than their classmates at school, or three years, or two years or even one year? Is it desirable to encourage non-legal and irregular unions to sustain a law passed without inquiry and based on no certain knowledge? Oh, fie, on legislators who spend thousands of dollars on drastic action and refuse a dollar for an inquiry as to the desirability of such action!

Fourth, even if it were desirable to prevent procreation of feeble-minded males of a certain grade, is it certain that vasectomy is to be preferred to castration? It is urged as one of the advantages of vasectomy that it does not interfere with desire nor its gratification but only with paternity. But is it a good thing to relieve the sexual act of that responsibility that it ought to carry and of which it has hitherto not been entirely free? Is not many a man restrained from licentiousness by recognizing the responsibility of possible parentage? Is not the shame of illicit parentage the fortress of female chastity? Is there any danger that the persons operated upon shall become a peculiar menace to the community through unrestrained dissemination of venereal disease? Will the frequency of the crime of rape be diminished by vasectomy? To many it would seem that to secure to a rapist his eroticism and uninhibited lust while he is released from any responsibility for offspring is not the way to safeguard female honor. Castration for rapists would seem preferable to vasectomy. Perhaps Indiana's experiment will give an answer to these questions.

Fifth. Is there any alternative besides sterilization or asexualization? There doubtless is, though it may at first be more expensive. This method is the segregation throughout the reproductive period of the feeble-minded below a certain grade. If, under the good environment of institutional life, they show that their retarded development is a result merely of bad conditions they may be released and permitted to marry. But such as show a protoplasmic defect should be kept in the institution, the sexes separated, until the reproductive period is passed. If this segregation were carried out thoroughly there is reason to anticipate such a reduction in defectiveness in 15 or 20 years as to relieve the state of the burden of further increasing its institutions, and in 30 years most of its properties, especially acquired to accommodate all the seriously defective, could be sold. We have the testimony of Dr. D. S. Jordan (1910) that the cretins who formerly abounded at Aosta in Northern Italy were segregated in 1890 and by 1910 only a single cretin of 60 years and 3 demi-cretins remained in the community. "Soeur Lucie, at the head of the work of the Little Sisters of the Poor, summed up the position in these words 'Il n'y en a plus'"—there are no more. Such then, would seem to be the proper program for the elimination of the unfit—segregation of the feeble-minded, epileptic, insane, hereditary criminals and prostitutes throughout the reproductive period and the education of the more normal people as to fit and unfit matings.

4. The Salvation of the Race through Heredity

Heredity is often regarded as a terrible fact; that we suffer limitations because of the composition of our germ plasm is a blow to pride and ambition. But, on the other hand, with limitation in capacity goes limitation in responsibility. Those who held the hazy doctrine of freedom of the will must have postulated uniformity of capacity for discriminating between right and wrong and uniformity in responsiveness to similar stimuli. Of course such an assumption is false. How we respond to any stimulus depends on the nature of our protoplasm. The nature of the response may be modified by training, by the formation of habits; but the result of training is, within limits, determined by the impressibility of the protoplasm. So I do not condemn my neighbor however regrettable or dangerous he may be.

And while heredity limits capacity in one point it extends it in others. If I have mental limitations, I have also gifts of natural health, of physical vigor, of persistence, and so on. Thus, as there is hardly a strain of human germ plasm that is without some defect or limitation so there is hardly a strain without the determiner of some admirable characteristic. While education and moral and religious instruction may do much to develop one's native traits, heredity can introduce the desirable determiner that will make such training more useful or less necessary. Indeed, while by good conditions we help the individual to make the most of himself, by good breeding we establish a permanent strain that is strong in its very constitution. The experience of animal and plant breeders who have been able by appropriate crosses to increase the vigor and productivity of their stock and crops should lead us to see that proper matings are the greatest means of permanently improving the human race—of saving it from imbecility, poverty, disease and immorality.

5. The Sociological Aspect of Eugenics

Human society, as its exists in these United States in this twentieth century, is complex. How complex it is, is indicated in some degree by the vast number of laws that have been passed and represent the rules of that society. These rules apply generally to all people alike. They tacitly assume that all people are alike; while admitting that there are some who are different and who constitute special classes that must be specially provided for. These special classes are of eugenic interest. Although well defined at one extreme, at the other they merge with the great mass of the population. The individuals composing these special classes are not in all respects distinct, but rather they are more or less peculiar in one or more respects. In fact the special classes which are the concern of the boards and associations of charities and correction consist of individuals with one or more traits that are more or less disturbing to the social organization. These individuals, or rather their traits—cause a disturbance and an expense of time and money quite out of proportion to their numbers in the community—they seem to be the main hindrance to our social progress. Moreover, their numbers seem to be increasing, hence it is a pressing need of the day to find out what is the cause and cure of defectiveness and delinquency.

The diversity of answers to such inquiry shows the depth of our helplessness. Mental defectiveness is ascribed to malnutrition of the fetus, to asphyxiation of the child during the labor of birth, to adenoids, to infection with venereal disease—despite the fact that (excepting mongolism) it usually occurs only in families with the defect on both sides of the house. Likewise criminality is ascribed to poverty, to bad example, to bad or inadequate education, despite the fact of incorrigibility. Even when there is some relation between the alleged cause and the result one feels that all these explanations are based on the logical error: *post hoc ergo propter hoc:* and that the cart is often put before the mule. The very multiplicity of explanations shows their inadequacy. There is a more fundamental explanation for these non-social traits than any of those that are usually ascribed.

First of all we can see clearly that the traits that cause so much trouble are "unfortunate" or "bad" only in relation to our society, i. e., relatively, not absolutely. Lack of speech, inability to care for the person or to respond in the conventional fashion to the calls of nature, failure to learn the art of dressing and undressing, inability to count, entire lack of ambition beyond getting a meal, abject slothfulness, love of sitting by the hour picking at a piece of cloth—these are unfortunate traits for a twentieth-century citizen but they constitute a first-rate mental equipment for our remote ape-like ancestors, nor do we pity infants, who invariably have them. So likewise with crimes:—the acts of taking and keeping loose articles, of tearing away obstructions to get at something desired, of picking valuables out of holes and pockets, of assaulting a neighbor who has something desirable or who has caused pain or who is in the way, of deserting family and other relatives, of promiscuous sexual relations—these are crimes for a twentieth-century citizen but they are the normal acts of our remote, ape-like ancestors and (excepting the last) they are so common with infants that we laugh when they do such things. In a word the traits of the feeble-minded and the criminalistic are normal traits for infants and for an earlier stage in man's evolution. There is an aphorism that biologists use which is apt here—ontogeny recapitulates phylogeny. This means that the individual (ontos) in its development passes through stages like those the race (phylum) has traversed in its evolution. The infant represents the ape-like stage.

Just as certain adult persons show ancestral organs that most of us have lost—such as a heavy coat of hair, an elongated coccyx (tail), an unusually large appendix, a third set of teeth,—so some adult persons retain certain ancestral mental traits that the rest of us have got rid of. And just as the heavy coat of body hair can be traced back generation after generation until we cannot avoid the conclusion that these hairy people represent a human strain that has never gained the naked skin of most people, so imbecility and "criminalistic" tendency can be traced back to the darkness of remote generations in a way that forces us to conclude that these traits have come to us directly from our animal ancestry and have never been got rid of.

The question how these traits ever came to be so rare in mankind is one with the question of human evolution and on this subject there is no historical evidence. It is clear, however, that after the new traits became established and constituted the basis for the new society, those persons who had the old traits stood a good chance of being killed off and many a defective line was ended by their death. We are horrified by the 223 capital offenses in England less than a century ago, but though capital punishment is a crude method of grappling with the difficulty it is infinitely superior to that of training the feeble-minded and criminalistic and then letting them loose upon society and permitting them to perpetuate in their offspring these animal traits. Our present practices are said to be dictated by emotion untempered by reason; if this is so, then emotion untempered by reason is social suicide. If we are to build up in America a society worthy of the species *man* then we must take such steps as will prevent the increase or even the perpetuation of animalistic strains.

6. Freedom of the Will and Responsibility

The consideration of the facts of heredity inevitably raises the ancient question of the freedom of the will, and throws a new light upon it. What is this free will? As I sit here in my study I will that to-morrow I shoot my dog. But when, to-morrow, I approach the dog to carry out my resolution his signs of fondness for me, the *abandon* with which he throws himself in the most helpless position at my feet, make the act impossible for me. I go to a neighbor and say, "My dog is decrepit and enjoys life no longer. I cannot kill him, will you do me the favor of shooting him?" He says, "I will" and does. We both had the will, why the difference in execution? Was he more resolute, more indomitable than I? It does not follow; simply his reaction to the sight of the dog did not overcome his resolution; mine did. There are various ways in which I might bring myself to do such an act. I might shut out the stimulus of the sight of the dog by covering him, or I might train myself to view him with indifference by associating him with some wrong, or I might picture more vividly my duty so that it would be a stronger motive than my affection or sympathy. By these means I might strengthen my "will." But except in some such indirect way my conduct is unmodifiable. Given such and such conditions I am bound to react in such and such ways.

A man of indomitable will is one who pictures so vividly the work he plans to do that other, minor, stimuli are relatively ineffective in opposition to the major stimulus. The man of weak will has usually a less vivid and powerful imagination and hence his actions are more determined by numerous incidental stimuli. "Free will" is predicated in matters of small consequence or concern to the person so that his action is determined by habit or slight stimuli whose source is unperceived. Though a man pride himself on the freedom of his will his every action is determined by his protoplasmic makeup, plus the modification it has received through experience, plus the relative vigor and quality of the stimulus he receives.

Is a man on this view less of a responsible agent? It depends on what is meant by responsible. I am responsible in the sense of answerable to society if I kill a man. If I kill him without intention or knowledge—if, for instance, my foot sets a stone rolling that starts an avalanche—then society decides that there is no evidence that my freedom imperils it and nothing is done. If I kill in self-defense society decides that my reaction is, on the whole, not prejudicial or disadvantageous to it and I am set free. If I kill on sudden anger society decides, whether rightly or wrongly, that my action does not prove that I may not, by training, gain inhibitions such that I shall thereafter react more slowly, giving time for other stimuli to play their part. But if I kill after prolonged premeditation, so that there is no question of merely temporary absence of inhibitions or of chance for numerous other stimuli to act, then society decides that my makeup is fundamentally bad and that the acquisition of a new method of reacting is not to be expected and so, properly enough, cuts me off. My name may indeed become a by-word, since society, rather unreasonably, takes that method of designating the combinations of characteristics that are antisocial. But I am not responsible in the sense of "deserving" pain because of the inadequacy of the determiners in my protoplasm. I am what the determiners in my two fused germplasms have developed into under the culture which they have experienced during their development. I am not responsible for my early culture nor for the reactions determined by it; but that culture is partly determined by my makeup, as when I find pleasure in the society of bad companions, and partly is imposed by the formal "good influences" that society has organized. Now, what I do depends on what I am, on the one hand, and the nature of the stimuli I receive, on the other, and neither what I am nor the nature of the stimuli I receive can be an excuse for adding more than is necessary to society's welfare to the sum of the world's pain. But organized society, on the contrary, has a responsibility towards its members in the sense of a duty to perform under penalty of dire consequences that will follow automatically. That responsibility involves, first, preventing the mating that brings together the antisocial traits of the criminal; second, after this damage is done, in securing the highest development of the good traits and the inhibition of the bad, surrounding the weak protoplasm with the best stimuli and protecting it from harmful stimuli. Here is where society must act to cut off the evil suggestions of immoral theaters, yellow journals and other bad literature. These stimulate those who react violently to this kind of suggestion. "The prisoner was a paranoiac and had a delusion of persecution; but had the play at the theater not been what it was he would not have murdered *that* night."

CHAPTER IX

THE ORGANIZATION OF APPLIED EUGENICS

1. State Eugenic Surveys

The commonwealth is greater than any individual in it. Hence the rights of society over the life, the reproduction, the behavior and the traits of the individuals that compose it are, in all matters that concern the life and proper progress of society, limitless, and society may take life, may sterilize, may segregate so as to prevent marriage, may restrict liberty in a hundred ways.

Society has not only the right, but upon it devolves the profound duty, to know the nature of the germ plasm upon which, in last analysis, the life and progress of the state depend. It has not only the right, but the duty, to make a thorough study of all of the families in the state and to know their good and bad traits. It may and should locate traits of especial value such as clear-headedness, grasp of details, insight into intricate matters, organizing ability, manual dexterity, inventiveness, mechanical ability and artistic ability. It may and should locate antisocial traits such as feeble-mindedness, epilepsy, delusions, melancholia, mental deterioration, craving for narcotics, lack of moral sense and self-control, tendency to wander, to steal, to assault and to commit wanton cruelties upon children and animals. It may and should locate strains with an inherent tendency to certain diseases such as tuberculosis, rickets, cancer, chronic rheumatism, gout, diabetes insipidus, goitre, leuchemia, chlorosis, hemophilia, eye and ear defects and the scores of other diseases that have an hereditary factor. It should know where the traits are, how they are being reproduced, and how to eliminate them. It should locate in each country the centers of feeble-mindedness and crime and know what each hovel is bringing forth. In fact it should let the bright light of knowledge into all matters of the reproduction of human traits, as the most dangerous of its enemies or the most valuable of its natural resources.

We take our census decennially or at more frequent intervals. We learn how many persons there are of military age, their race, birthplace and occupation, and we learn how many are blind and deaf, and it is well. But by a very little additional labor we could gain many not less significant facts, such as how each of our blind and deaf and feeble-minded came to be, so that the laws of their origin can be studied and the defective germ plasm located. It would seem worth while to use the census as a means of securing data on human blood lines and tracing the descent of defects.

A state eugenic survey should be taken in at least the older states. The organization of the survey could be relatively simple; the 630,000 teachers of state and city schools might be used to secure the census of the 24,000,000 children of "school age" and their parents. Through a series of visits on Saturday afternoons or during vacations the parents could be interested to furnish the desired data. The teachers could be instructed how to fill out the schedules by superintendents or at teachers' institutes. They should, of course, receive special compensation, but it would be difficult to think of any other method of making a census so cheaply and effectively; the more so since the teacher through her pupil has ready access to most homes. The schedules of questions should be prepared so as to avoid giving any offense, to secure the required data as to physical and mental family traits, and to get such names and places of birth and residence as would serve to tie families together. After study the data might be used to give particular families advice as to how their children should marry to avoid the recurrence of undesirable traits in the children's children.

Objection will probably be offered to any such survey on the ground that inheritable traits are private and personal matters; but this is surely a narrow and false view. The collective traits of any person constitute a mosaic whose elements have been derived from thousands of germ plasms and parts of which may be passed on to thousands of the persons who will constitute the social fabric of a few generations hereafter. What justification have I, whose elements are derived from the society of the past and will pass into the society of the future, to maintain that the society of to-day has no right to question me—who am merely a sample of this universal germ plasm. No one who looks broadly at the relation his family bears to the commonwealth will hesitate to put on record an account of his family traits.

The objection that such a survey is impracticable can be met by the assertion that in the State of New Jersey such a survey is already well advanced, largely through private initiative. The work has been done by means of field workers attached to various institutions for defectives. Massachusetts, also, has made a good beginning in this direction. The suggestion as to a state survey is merely an extension of such work as is being carried on in a more limited fashion to-day.

2. A Clearing House for Heredity Data

While states should undertake eugenic surveys, it is clear that, in a country like ours where extensive intermigration takes place between States, "blood lines" are not limited by state boundaries. There is need, consequently, of a central clearing house for data concerning family traits in America. This will serve not only as a headquarters for investigation but also for education.

It will be interesting to trace the history of institutions of this sort in America. One was planned in 1881 or 1882 by Mr. Loring Moody of Boston. In his booklet entitled "Heredity: its relations to human development. Correspondence between Elizabeth Thompson and Loring Moody," he tells how he had hoped for aid from a philanthropist. He adds "in the earnest hope and expectation that such persons will soon appear ready for their work, as a colaborer therein and as preliminary steps toward the formation of an

INSTITUTE OF HEREDITY

which shall found a library, establish lectureships with schools of instruction and take in hand the diffusion of knowledge on the subject of improving our race by the laws of physiology, I propose, with the aid of such as may volunteer their patronage and support, to open a school and lecture room in Boston with the nucleus of a library for such conversations, consultations and illustrated lectures as may awaken interest and lead toward a realization of these great and beneficent ends." This plan failed because of the early death of its projector.

About 1887 or 1888 Dr. Alexander Graham Bell founded at Washington, D. C., the Volta Fund which has grown to over $100,000. Out of this was established the Volta Bureau, which collects all valuable information that can be obtained with reference not only to deaf mutes as a class but to deaf mutes individually. In this bureau can be found the names of over twenty thousand deaf and the particulars respecting their history. They are so systematically arranged that without a moment's delay the facts with reference to any of them can be turned to. These valuable manuscripts and indices are placed in a perfectly fire-proof section of the building of the Bureau. The library is rich in New England town histories and genealogies, in addition to works on the deaf.

About 1905 the late Sir Francis Galton contributed to the support of a Eugenics Laboratory at University College, London, under the direction of Professor Karl Pearson, and at his death in 1911 Galton made it his residuary legatee. This laboratory is publishing an important "Treasury of Human Inheritance."

In October, 1910, The Eugenics Record Office was started at Cold Spring Harbor, Long Island, N. Y., in connection with the Eugenics Section of the American Breeders' Association in a tract of 80 acres, with a good house to which has been added a fire-proof vault for the preservation of records. Mr. H. H. Laughlin is its superintendent. At this place the collecting and cataloguing of records goes on apace. It is hoped to establish here a very completely indexed collection of published genealogical and town histories for the United States as well as the manuscript reports of the field investigators. The main work of the office is investigation into the laws of inheritance of traits in human beings and their application to eugenics. Two series of publications are contemplated, an octavo series of Bulletins and a quarto series of Memoirs. Several numbers of the Bulletin are issued or in press. The Eugenics Record Office wishes to coöperate with Institutions and State Boards of Control in organizing the study of defectives and criminalistic strains in each State. It will offer suggestions as to the organization of local societies devoted to the study of Eugenics. It proffers its services free of charge to persons seeking advice as to the consequences of proposed marriage matings. In a word it is devoted to the advancement of the science and practice of Eugenics.

BIBLIOGRAPHY

The following method of citation is adopted. 1. Name of author, in capital letters. 2. Date of publication, used, with the author's name, for reference (in the body of the work) to the publication. 3. Title of the publication. 4. If published in a periodical, name of periodical, in italics, followed by volume number and page. If published as a separate book, the place of publication is given, and sometimes the name of the publisher. *p.* stands for page; *pl* for plate; *v* for volume.

ALLERTON, WALTER S., 1888. A History of the Allerton Family in the United States, 1585 to 1885, and a genealogy of the descendants of Isaac Allerton. N. Y., 166 pp.

ANDERSON, T. MCCALL, 1863. Hereditary Deaf-mutism. *Med. Times and Gazette*, London, II, 247.

APERT, E., 1907. Traité des maladies familiales et des maladies congénitales. Paris, Libraire J. B. Baillière et fils.

ARNER, G. B. L., 1909. Consanguineous Marriage in the American Population. *Studies in Hist., Economics and Public Law.* Columbia Univ., XXXI, No. 3.

ATKINSON, J. E., 1875. Observations upon Two Cases of Fibroma Molluscum. *New York Medical Journal*, XXII, 601–610.

BABINGTON, B. G., 1865. Hereditary Epistaxis. *Lancet*, London, Sept., 1865, II, 362–363.

BAER, TH., 1907. Zuer Kasuistik der Hypotrichosis Congenita Familiaris. *Arch. f. Dermatologie und Syphilis*, LXXXIV, 1 Th., pp. 15–18.

BALCH, W. L. (Secretary), [1905]. First Reunion and organization of the Balch Family Association by the descendants of John Balch one of the "Old Planters" of Naumkeag, now Salem, Beverly and North Beverly, Massachusetts, 52 pp.

BALL, NICHOLAS, 1891. Edward Ball and some of his Descendants. Newport, R. I., *Mercury Print*, pp. 1–15.

BANKER, H. J., 1909. A partial history and genealogical record of the Bancker or Banker families of America and in particular the descendants of Laurens Mattipe Bancker. Rutland, Vt., The Tuttle Co., 458 pp.

BANNING, PIERSON W., 1908. The First Banning Genealogy. Chicago.

BARR, MARTIN W., 1897. Some Studies in Heredity. *Jour. Nerv. and Mental Diseases*, N. Y., XXIV, 155–162.

—, 1904. Mental Defectives: their History, Treatment and Training. Phila., P. Blakiston's Son, 368 pp.

BATESON, W., 1906. Address on Mendelian Heredity and its Application to Man. *Brain*, V. 29, p. 157.

—, 1906. Progress of Genetics since the Rediscovery of Mendel's Papers. *Progr. Rei Bot.*, I, p. 368.

—, 1908. Methods and scope of genetics. Cambridge, Eng., Univ. Press.

—, 1909. Mendel's Principles of Heredity. Cambridge, Univ. Press.

BELL, ALEXANDER GRAHAM, 1884. Memoir upon the Formation of a Deaf Variety of the Human Race. *Mem. of National Acad. of Sciences*, 86 pp.

—, 1889. Royal Commission on the Blind, the Deaf and Dumb, etc.: Minutes of Evidence taken by the Royal Commission. London, Eyre & Spottiswoode.

—, 1906. The Blind and the Deaf, 1900. Special Report to Bureau of the Census. Washington, Gov't Printing Office, ix+264 pp.

BEMISS, S. M., 1858. Report on Influence of Marriages of Consanguinity upon Offspring. *Trans. Am. Med. Ass'n*, Phila., XI, 321–425.

BENTLEY, MADISON, 1909. Mental Inheritance. *Pop. Sci. Mo.*, v. 75, p. 458.

BERNHARDT, M., 1885. Beitrag zur Pathologie der sogenannten "Thomsen'schen Krankheit." *Centralb. f. Nervenh.*, Leipzig, VIII, 122–126.

BERZE, J., 1910. Die hereditären Beziehungen der Dementia Praecox. Beitrag zur Hereditätslehre. Leipzig a. Main.

BONAJUTI, F., 1890. Contributo allo studio della epidermolysis bullosa hereditaria di Köbner. *Il Morgagni*, Milano, I, 770–780.

BOND, C. J., 1905. The Correlation of Sex and Disease. *British Med. Jour.*, Lond., II, Oct., 1094–1095.

BORDLEY, J., Jr., 1908. A Family of Hemeralopes. *Johns Hopkins Hosp. Bull.*, XIX, 278–280, 1 pl.

BOVAIRD, D., 1900. Primary Splenomegaly; Endothelial Hyperplasia of the Spleen; 2 cases in children, autopsy and morphological examination in one. *Am. Jour. Med. Science*, Phila., CXX 377–402.

BRAMWELL, BYROM, 1876. Progressive Pernicious Anemia. *Rep. Proc. Northumb. & Durham M. Soc.*, Newcastle-upon-Tyne, 1876–7, pp. 151–167.

—, 1901. Wednesday Cliniques. Case VIII. Case of Hereditary Ichthyosis of the Palms and Soles. *Clinical Studies*, Edinburgh, I, 1903, pp. 77–80.

—, 1903. Wednesday Cliniques. Case IV. Hereditary Optic Atrophy. *Clinical Studies*, Edinburgh, II, 1904, pp. 44–55.

—, 1906. Wednesday Cliniques. Case XLV. Haemophilia. *Clinical Studies*, Edinburgh, V, 1907, 368–370.

—, 1907. Wednesday Cliniques. Hereditary Webbing of Second and Third Toes of the Left Foot. *Clinical Studies*, Edinburgh, v, 1907, pp. 373.

—, 1907. Wednesday Cliniques. Case XXXVIII. Diabetes Mellitus: strong hereditary history; differential diagnosis of glycosuria and diabetes mellitus. *Clinical Studies*, Edinburgh, VI, 1908, pp. 263–266.

Breed, J. Howard, 1892. A Record of the Descendants of Allen Breed. Phila., Hathaway & Bros.

Brill, N. E., 1901. Primary splenomegaly. *Am. Jour. Med. Sci.*, Phila. and N. Y., April, 1901, CXXI, 377–392.

Brinckerhoff, R., 1887. The Family of Joris Dircksen Brinckerhoff. New York, pp. 1–188.

Broca, Paul, 1866–9. Traité de Tumeurs, Vols. I and II. Paris, P. Asselin.

Bronardel, P., 1900. Le Mariage, Nullité, Divorce, Grossesse, Accouchement. Paris, Libraire J. B. Baillière et fils, pp. 1–452.

Buck, Wm. J., 1893. Account of the Bucks Family of Bucks Co., Pa. Philadelphia, pp. 1–142.

Bulloch, W. and P. Fildes, 1911. Haemophilia. *Treasury of Human Inheritance*, Parts V and VI. London.

Bureau of the Census (Department of Commerce and Labor), Special Reports, 1910. Religious Bodies: 1906, Part I, 576 pp. Summary and General Tables, Part II, 670 pp. Separate Denominations: History, Description, and Statistics. Washington, Gov't Printing Office.

Burger, Eugen, 1900. Ueber Haemophilie mit Geschichte einer Bluterfamilie. Inaugural Diss., Freiburg, pp. 1–30.

Buschan, G., 1894. Die Basedow'sche Krankheit. Eine Monographie. Leipzig u. Wien.

Butler, James D., 1896. British Convicts Shipped to American Colonies. *Am. Hist. Review*, II, 1, Oct., pp. 12–33.

Cance, Alex. E., 1911. Abstract of the Report on Recent Immigrants in Agriculture. Reports of the Immigration Commission. Washington, Gov't Printing Office, 75 pp.

Cannon, G. and A. J. Rosanoff, 1911. Preliminary Report of a Study of Heredity in Insanity in the Light of the Mendelian Laws. *Bull. Eugenics Record Office*, No. 3, 11 pp.

CARPENTER, G., 1899. A case of absence of the clavicles, with an account of various deformities of the clavicles in 5 other members of the same family. *Lancet*, London, Jan., 1899, I, 13–17.

CARSON, W., 1890. Congenital Abnormalities of the Extremities. In Neilson, H. R.: Keating's *Encyclopedia of the Diseases of Children*, III, p. 935. Philadelphia, 1890.

CASTLE, W. E., 1903. Heredity of Sex. *Bull. Mus. Comp. Zool. Harvard*, v. 40, No. 4.

—, 1903. Laws of Heredity of Galton and Mendel and Some Laws Governing Race-improvement by Selection. *Proc. Amer. Acad. Arts and Sci.*, v. 39, p. 223.

—, 1903. Mendel's Law of Heredity. *Science*, N. S., 24, p. 396.

— and others, 1906. Effects of Inbreeding, Cross-breeding and Selection upon the Fertility and Variability of Drosophila. *Amer. Acad. Arts and Sci. Proceed.*, v. 41, No. 33.

— and FORBES, ALEXANDER, 1906. Heredity of Hair-length in Guinea-pigs and its Bearing on the Theory of Pure Gametes. Wash., Carnegie Inst., Wash., Pub. No. 49.

—, 1906. Origin of a Polydactylous Race of Guinea-pigs. Wash., Carnegie Inst. Wash., Pub. No. 49.

—, 1909. Studies of Inheritance in Rabbits. Wash., Carnegie Inst. Wash., Pub. No. 114.

—, 1911. Heredity, N. Y.

CHEADLE, W. B., 1900, Occasional Lectures on the Practice of Medicine, London, 324 pp.

CHURCH, SIR WILLIAM S., and others, 1909. Influence of Heredity on Disease, with Special Reference to Tuberculosis, Cancer and Diseases of the Nervous System: a discussion by Sir W. S. Church, Sir W. R. Gowers and others. London: Longmans, 1909.

CLARKE, Ernest, 1903. Hereditary Nystagmus. *Ophthalmoscope*, London, I, 86–87.

CLUBLE, W. H., 1872. Hereditariness of Stone. *Lancet*, London, Feb. 1872, 204.

COLE, FRANK T., 1887. Early Genealogies of the Cole Families in America. Columbus, Ohio, Hann & Adair, pp. 1–308.

COLEGROVE, WILLIAM, 1894. History and Genealogy of the Colegrove Family in America with Biographical Sketches, Portraits, etc. Chicago, Ill., pp. 1–792.

COOKE, HARRIET R., 1889. The Driver Family. A Genealogical Memoir of the Descendants of Robert and Phebe Driver, of Lynn, Mass. New York. John Wilson & Son, pp. 1–531.

COUCH, J. KYNASTON, 1895. A Family History of Hernia. *Lancet*, London, October 1895, II, pp. 1043.

CUNIER, 1838. Annales Soc. Méd. de Gand.

CUTLER, C. W., 1895. Ueber angeborene Nachtblindheit und Pigment-Degeneration. *Arch. f. Augenheilk*, XXX, p. 92.

DARWIN, C., 1894. The Variation of Animals and Plants under Domestication, 2d Ed. N. Y., D. Appleton.

DAVENPORT, C. B., 1906. Inheritance in Poultry. Carnegie Inst. Wash., Pub., No. 52.

—, 1908. Degeneration, Albinism and Inbreeding. *Sci.*, N. S. 28, p. 454.

—, 1908. Heredity of Some Human Physical Characteristics. *Proc. Soc. Exper. Biol. and Med.*, V, pp. 101–2.

—, 1909. Influence of Heredity on Human Society. *Annals Amer. Acad. Polit. and Soc. Sci.*, v. 34, p. 16 (Race improvement in the United States).

—, 1909. Heredity in Man. Mar. 6. *Harvey Lectures, 1908–09*, pp. 280–90.

—, 1910. The Imperfection of Dominance and Some of its Consequences. *Amer. Nat.*, v. 44, Mar.

—, G. C. and C. B., 1907. Heredity of Eye Color in Man. *Science*, N. S., pp. 589–592, Nov.

—, 1908. Heredity of Hair Form in Man. *Amer. Nat.*, v. 42, p. 341.

—, 1909. Heredity of Hair Color in Man. *Amer. Nat.*, v. 43, No. 508, Apr.

—, 1910. Heredity of Skin-Pigment in Man. *American Naturalist*, XLIV, Nov. and Dec., pp. 642–672, 705–731.

DAVIS, C. H. S., 1870. History of Wallingford, Conn., Meriden, 956 pp.

DE BECK, D., 1886. A Rare Family History of Congenital Coloboma of the Iris. *Arch. of Ophthal.*, XV, p. 8, and ibid., 1894, XXIII, p. 264.

DEBORE, M. and RENAULT, JULES, 1891. Du tremblement hereditaire. *Bull. et Mém. Soc. Méd. des Hôpitaux de Paris*, Paris, VIII, 3d series, July, 1891, pp. 355–361.

DENIKER, J., 1906. The Races of Man. London and N. Y., pp. xxiii + 611.

DEPARTMENT OF COMMERCE AND LABOR, 1911. Immigration, Laws and Regulations of July 1, 1907. Washington, Gov't Printing Office, 97 pp.

DERCUM, F. X., 1897. Three Cases of the Family Type of Cerebral Diplegia. *Jour. Nerv. and Ment. Dis.*, New York, 24, 396–399.

DEVRIES, HUGO, 1906. Species and Varieties, their Origin by Mutation, ed. by D. T. MacDougal. Chicago, Open Court Pub. Co.

DOBELL, HORACE, 1863. A Contribution to the Natural History of Hereditary Transmission. *Med.-Chir. Trans.*, London, XLVI, pp. 25–28.

DOOLITTLE, WM. F., 1901. The Doolittle Family in America. Parts I–VII. Cleveland, Acme Printing Co., pp. 1–730, 1901–8.

DREW, DOUGLAS, 1905. Acquired Club Foot with Marked Hereditary History. *Reports of the Soc. for the Study of Dis. in Children,* London, V, 1904–05, 172–3.

DRINKWATER, H., 1908. An account of a Brachydactylous Family, *Proc. Roy. Soc.*, Edinburgh, 28, p. 35.

DUGDALE, R. L., 1902. The Jukes; a study in crime, pauperism, disease and heredity. 7th edition. N. Y., G. P. Putnam's, viii + 120 pp.

DWIGHT, BENJ. W., 1874. History and Descendants of John Dwight of Dedham, Mass. Vol. I and II. New York, John F. Trow & Son.

ELLIS, H., 1904. A Study of British Genius. London, Hurst, 300 pp.

EUGENICS REVIEW. Vol.—date. Apr. 1909—date.

FARRABEE, W. C., 1905. Inheritance of Digital Malformations in Man. *Papers of Peabody Mus. of Am. Arch. and Ethn.*, Harvard Univ., III, 3, p. 69.

FAY, EDWARD ALLEN, 1898. Marriages of the Deaf in America. Wash. D. C., vii + 527 pp., Volta Bureau.

FEER, E., 1907. Der Einfluss der Blutesverwandschaft der Eltern auf die Kinder. *Jahrb. f. Kinderh.* Berlin, LXVI, 188–219.

FERNALD, WALTER E., 1909. The Imbecile with Criminal Instincts. *Am. Jour. of Insanity*, LXV, pp. 731–749, April.

FISKE, JOHN, 1905. The Discovery and Colonization of North America. Boston, xiv + 224 pp.

FOOT, A. J. A., 1869. Des difformités congénitale et acquise des doigts. Paris.

FREUD, SIGM., 1893. Ueber familiare Formen von cerebralen Diplegien. *Neurol. Centralblatt.*, Leipzig, XII, 512–515; 542.

GALTON, FRANCIS, 1869. Hereditary Genius: an Inquiry into its Laws and Consequences. London. Macmillan.

—, 1889. Natural Inheritance. N. Y., Macmillan, ix + 259 pp.

—, 1892. Finger Prints. London. Macmillan.

—, 1895. English Men of Science; their Nature and Nurture. N. Y., Appleton & Co.

—, and SCHUSTER, EDGAR, 1906. Noteworthy Families (Modern Science); an index to kinships in near degrees between persons whose achievements are honorable, and have been publicly recorded. London, J. Murray.

GARROD, ARCHIBALD E., 1902. The Incidence of Alkaptonuria; a Study in Chemical Individuality. *Lancet*, London, Dec. 13, 1902, 1616–1620.

—, 1908. Inborn Errors of Metabolism (Croonian lectures). *Lancet*, 1908, II, pp. 1, 73, 142, 214.

Gillin, J. L., 1906. The Dunkers, N. Y. (pp. 221, 222).

Goddard, H. H., 1911. Heredity of Feeblemindedness. *Bull. No. 1*, Eugenics Record Office. Cold Spring Harbor, N. Y., pp. 1–14.

—, and Helen F. Hill. 1911. Feeblemindedness and Criminality. *The Training School*, VIII, pp. 3–6, March.

Gossage, A. M., 1907. The Inheritance of Certain Human Abnormalities. *Quarterly Jour. Med.*, Oxford, I, 331–347.

von Graefe, 1869. Beiträge zur Pathologie und Therapie des Glaucoms. *Arch. f. Ophth.*, Bd. XV, p. 228.

Gray, 1907. Memoir on the Pigmentation Survey of Scotland. *Jour. Roy. Anthropological Inst.*, XXXVII, pp. 375, 401, pl. XXVII–XLVII.

Great Britain, 1904. Inter-departmental committee on physical deterioration. Report on physical deterioration. v. 1, Report and appendix.

Guilford, S. H., 1883. A dental anomaly (a man 48 years of age, edentulous from birth, totally lacking the sense of smell and almost devoid of the sense of taste, surface of body destitute of fine hairs, and he has never perspired). *Dental Cosmos*, Phila., 1883, XXV, 113–118.

Gunzberg, F. 1889. Kasuistik der Angeborenen Muskelanomalien. *Klin. Monatsbl. f. Augenheilk*, S. 263.

Gutbier, 1834. Inaug. Diss., Wurzburg. Quoted by Loeb.

Hammerschlag, Victor, 1905. Zur Frage der Vererbbarkeit der Otosklerose. *Wien. Klin. Rundschau*, Wien., XIX, 5–7.

—, 1906. Beitrag zur Frage der Vererbbarkeit der "Otosklerose." *Monats. f. Ohrenh.*, Berlin, XL, 443–464.

—, 1908. Zur Atiologie der Otosklerose. *Wien. Med. Wochensch.*, Wien., LVIII, 566–567.

—, 1909. Zur Kenntnis der hereditär-degenerativen Taubstummheit. *Zeits. f. Ohrenh.*, B. 59, 315–329.

Harrington, Harriet L., 1885. A family record showing the heredity of disease. *Physician and Surg.*, Ann Arbor, Mich., VII, 49–51.

Harris, Edward Doubleday, 1870. A Genealogical Record of Thomas Bascom and his Descendants. Boston, Wm. P. Lunt, pp. 1–79.

Hartmann, Arthur, 1881. Deaf-mutism, and the Education of Deaf-mutes by Lip-reading and Articulation. London, Bailliere, Tindall & Cox, XIV, 1–224.

HERRINGHAM, 1889. Muscular Atrophy of the Peroneal Type Affecting Many Members of a Family. *Brain*, Vol. XI, 230.

HERTEL, E., 1903. Uber Myopie, klinische-statistische Mitteilungen. V. Graefe's, *Archiv. f. Opthal.*, Leipzig, LVI, 326–386.

VON HIPPEL, E., 1909. Die Missbildungen des Auges; in E. Schwalbe. Missbildungen des Menschen. III, Theil., 1 Lief., 2 Abth.

HOLMES, S. J., and H. M. LOOMIS, 1909. The Heredity of Eye Color and Hair Color in Man. *Biol. Bull.*, XVIII, 50–65, Dec.

HOWE, L., 1887. A Family History of Blindness from Glaucoma. *Arch. of Ophthalmology*, N. Y., XVI, 72–76.

—, S. G., 1858. On the Causes of Idiocy; being the supplement to a report by Dr. S. G. Howe . . . . appointed by the governor of Massachusetts to inquire into the condition of the idiots of the commonwealth. Feb. 1848, Edinburg, 1858.

HUMPHREY, FREDERICK, 1883. The Humphreys Family in America. New York, Humphreys Print, pp. 1–1115.

HUNTINGTON, GEORGE, 1872. On chorea. *Med. and Surg. Reporter*, Phila., 1872, XXVI, 317.

HURST, C. C., 1906. Mendelian Characters in Plants and Animals *Report Conf. on Genetics, R. Hortic. Soc.* London, p. 114.

—, 1908. Mendel's Law of Heredity and its Application to Man. *Leicester Lit. Phil. Soc. Trans.*, 12, p. 35.

—, 1908. On the Inheritance of Eye Color in Man. *Proc. Royal Soc.*, B. vol. 80, pp. 85–96.

HUTCHINSON, JONATHAN, 1886. Congenital Absence of Hair and Mammary Glands. *Med.-Chir. Trans.*, London, LXIX, May, 473–477.

HUTCHINSON, JOSHUA, 1876. Brief Narrative of the Hutchinson Family. Boston, Lee & Shepard, 73 pp.

HUTH, A. H., 1887. The Marriage of Near Kin, 2d edition. London and N. Y., x+475 pp.

IMMIGRATION COMMISSION, 1911. Abstract of the Report on Emigration Conditions in Europe. Washington, Gov't Printing Office.

IWAI, T., 1904. La Polymastie au Japon. *Arch. de Medecine Experimentale*, XVI, 489–518.

JACKSON, V. H., 1904. Orthodontia. Phila., 517 pp.

JACOBS, P. J., 1911. A Tuberculosis Directory, containing a list of institutions, associations and other agencies dealing with tuberculosis in the United States and Canada. New York; Nat. Ass'n for Study and Prevention of Tuberculosis, 331 pp.

JENNINGS, H. S., 1909. Heredity and Variation in the Simplest Organisms. *Amer. Nat.*, v. 43, No. 510, June.

JOHANNSEN, W., 1903. Uber Erblichkeit in Populationen und in reinen Linien. Jena, Fischer.

—, 1909. Elemente der Exakten Erblichkeitslehre. Jena, G. Fischer.

JOHNSTON, A., 1887. Connecticut, A Study of a Commonwealth-Democracy. Boston.

JOLLY, F., 1891. Uber Chorea Hereditaria. *Neurol. Centrabl.*, Leipzig, 1891, X, 321.

JORDAN, D. S., 1907. The Human Harvest. Boston, Amer. Unitarian Assoc.

—, 1910. Cretinism in Aosta. *The Eugenics Review*, II, 3, Nov. 247–248.

JÖRGER, J., 1905. Die Familie Zero. *Archiv. für Rassen und Gesellschafts Biologie*, Berlin, 1905, II, 494–559.

KEEFE, D. J., 1910. Annual Report of the Commissioner-General of Immigration to the Secretary of Commerce and Labor, for the fiscal year ended June 30, 1910. Washington, Gov't Printing Office, 248 pp., 2 charts.

KELLICOTT, WILLIAM E., 1911. The Social Direction of Human Evolution. N. Y., 249 pp.

KELLOGG, VERNON L., 1907. Darwinism To-day. N. Y., Henry Holt.

KELLY, A. B., 1906. Multiple Telangiectases of the Skin. *Glasgow Med. Jour.*, Glasgow, LXV, June, 411–422.

LANCRY, LOUIS and GUSTAV, 1890. La Commune de Fort-Mardick près Dunkerque (étude historique, démographique et médicale). Paris, 72 pp.

LAUNOIS and APERT, 1905. Achondroplasie hereditaire. *Soc. méd. des hôpitaux*, 30 juin, 1905.

LEBER, TH., 1871. Ueber anomale Formen der Retinitis pigmentosa, *Arch. f. Ophth.*, XVII, 1, S. 314.

LEE, EDMUND J., 1895. Lee of Virginia, 1642–1892, Biographical and Genealogical Sketches of the Descendants of Colonel Richard. Philadelphia, Franklin Printing Co., pp. 1–586.

LEICHTENSTERN, 1878. Ueber das Vorkommen und die Bedeutung Supernumerärer (accessorischer) Brüste und Brustwarzen. Virchow's Archiv., Berlin, LXXIII, 222–256.

LEMEN, FRANK B., 1898. History of the Lemen Family of Illinois, Virginia and elsewhere. Two Parts. Collinsville, Ill., pp. 1–643.

LEWIS, G. G., 1904. Hereditary Ectopia Lentis with Report of Cases, *Archives of Ophth.*, XXXIII, No. 3, p. 275.

LEWIS, T. and EMBLETON, D., 1908. Split-hand and Split-foot Deformities, their Types, Origin, and Transmission. *Biometrika*, 6, p. 26.

LINDSAY, MARGARET T., 1889. The Lindsays of America: A Genealogical Narrative and Family Record. Albany, Joel Munsell's Sons, pp. 1–275.

LINGARD, A., 1884. The Hereditary Transmission of Hypospadias and its Transmission by Indirect Atavism. *Lancet*, London, 1884, i, 703.

LOCY, WILLIAM A., 1908. Biology and its Makers. N. Y., Henry Holt.

LOEB, CLARENCE, 1909. Hereditary Blindness and its Prevention. St. Louis, 1909. *Annals of Ophthalmology*, Jan.-Oct.

LONDON COUNTY COUNCIL, 1909. Report of the Medical Officer (Education) for the 12 Months ended December 31, 1909, pp. 96.

LORENZ, OTTOKAR, 1898. Lehrbuch der gesammten wissenschaftlichen Genealogie. Berlin, W. Hertz.

LOSSEN, 1905. Die Bluterfamilie Mampel in Kirchheim bei Heidelberg. Deutsche Zeitschr. f. Cherurgie, LXXVI, 1.

LUCÆ, A., 1907. Die chronische progressive Schwerhorigkeit, ihre Erkenntnis und Behandlung. Berlin, 403 pp., 2 pl.

LUCAS, R. CLEMENT, 1880. On a Remarkable Instance of Hereditary Tendency to the Production of Supernumerary Digits. *Guy Hosp. Repts.*, London, XXV, 417–419.

LYDSTON, G. FRANK, 1904. The Diseases of Society. Philadelphia, 1904, pp. 1–626. J. B. Lippincott Co.

MCCULLOCH, REV. OSCAR C., 1888. The Tribe of Ishmael: a study in social degradation. *Proc. of 15th National Conf. Char. and Correction*, Buffalo, July.

MACDONALD, A., 1908. Juvenile Crime and Reformation. Washington, Gov't Printing Office, 339 pp.

MCQUILLEN, J. H., 1870. Hereditary Transmission of Dental Irregularities. *Dental Cosmos*, Phila., XII, Feb., pp. 73–75.

MARSHALL, L. 1903. Deformity of the Hands and Feet Transmitted through Five Generations. *Reports Soc. Stud. Disease in Children*, III, 222--225.

MARTIN, F. 1888. Ueber Microphthalmus. Inaug. Diss. Erlangen.

MASON, L. D., 1910. The Etiology of Alcoholic Inebriety, with special reference to its true status and treatment from a medical point of view. *Monthly Cyclopedia and Med. Bull.*, Phila., III, Sept., 521–532.

MELL, DR. and MRS. P. H., 1897. Genealogy of the Mell Family in the Southern States. Auburn, Ala., 61 pp.

MENDEL, GREGOR. Versuche über Pflanzen-Hybriden. Brünn, G. Gastl, 1866.

MERZBACHER, L., 1909. Gesetzmässigkeiten in der Vererbung und Verbreitung Verschiedener Hereditär-familiärer Erkrankungen. *Archiv. f. Rassen u Ges. Biologie*, VI, 172–198, May.

MICKLEY, MINNIE F., 1893. Genealogy of the Mickley Family of America. Newark, N. J., Advertiser Printing House, pp. 1–182.

MITCHELL, A., 1866. Blood-Relationship in Marriage, Considered in its

Influence on the Offspring. *Anthropolog. Soc. of London*, II, pp. 402–456.

MOLENES, PAUL, 1890. Sur un cas d'alopecie congenitale. *Ann. de Derm. et Syph.*, Paris, 3d series, I, 548–557.

MOORE, ANNE, 1911. The Feeble-Minded in New York. A report prepared for the Public Education Association of New York. Published by the State Charities Aid Association. Special Committee on provision for the feeble-minded. N. Y. United Charities Bldg., 111 pp., June.

MORGAN, T. H., 1910. Chromosomes and Heredity. *Amer. Nat.*, v. 44. p. 449, Aug.

MORRISON, LEONARD A., 1896. History of the Sinclair Family in Europe and America. Boston, Mass., Damrell & Upham, pp. 7–453.

—, 1897. History of the Kimball Family in America from 1634 to 1897 and of its Ancestors. The Kemballs or Kemboldes of England. Boston, Damrell & Upham, Vol. I and II.

MOTT, F. W., 1905. A Discussion on the Relationship of Heredity to Disease. *Brit. Med. Jour.*, London, Oct., 1086–1091.

NARETH, 1903. Beiträge zur Luxatio coxæ Congenitalis. Wien u. Leipzig.

NEIGHBOR, LAMBERT B., 1906. Descendants of Leonard Neighbor, Immigrant to America, 1738. Dixon, Ill., 48 pp.

NELSON, E. W., 1899. The Eskimo about Bering Strait. *18th Ann. Rept. Bureau of American Ethnology*, pp. 289–291.

NETTLESHIP, E., 1905. On Heredity in the Various Forms of Cataract. Additional cases of hereditary cataract. *Report Roy. Lond. Ophth. Hosp.*, v. 16, p. 1.

—, 1905. Cases of Color-Blindness in Women. *Ophth. Soc. Trans.*, v. 26.

—, 1906. On Retinitis pigmentosa and Allied Diseases. *Report Roy. Lond. Ophth. Hosp.*, v. 17, pts. I, II, and III.

—, 1907. History of Congenital Stationary Night-blindness in Nine Consecutive Generations. *Ophth. Soc. Trans.*, v. 27, 269–293.

—, 1908. Three New Pedigrees of Eye Disease. *Ophth. Soc. Trans.*, v. 28, p. 220.

—, 1910. Some Points in Relation to the Heredity of Disease. *St. Thomas's Hospital Gazette*, March, 1910, pp. 37–65.

—, and OGILVIE, F. M., 1906. Peculiar Form of Hereditary Congenital Cataract. *Ophth. Soc. Trans.*, v. 26.

NEW YORK INSTITUTION FOR THE INSTRUCTION OF THE DEAF AND DUMB, 1853. *35th Annual Report and Documents.* Albany, 1854, pp. 95–120.

NICOLLÉ, C., and HALIPRÉ, A., 1895. Maladie Familiale Caracterisée par des Alterations des Cheveux et des Ongles. *Ann. de Derm. et Syph.*, Paris, 3d series, VI, 1895, 804–811.

OSLER, WM., 1901. On a Family Form of Recurring Epistaxis Associated with Multiple Telangiectases of the Skin and Mucous Membranes. *Johns Hopk. Hosp. Bull.*, XII, 333.

OSWALD, AGNES B., 1911. Hereditary Tendency to Defective Sight in Males only of a Family. *Brit. Med. Jour.*, London, Jan., 1911, p. 18.

PARDOE, GEORGE MOLYNEUX, 1894. Genealogy of Wm. Molyneux and Descendants. Sioux City, Iowa, 1894, 1–24.

PARKER, R. W., and ROBINSON, H. B., 1887. A Case of Inherited Congenital Malformation of the Hands and Feet: . . . . with a Family Tree. *Clin. Soc. Trans.*, Vol. XX, London, April, pp. 181–189.

PEARSON, KARL, 1909. Note on the Skin Color of the Crosses between Negro and White. *Biometrika*, V. 6, pt. 4, p. 348.

PELIZAEUS, FR., 1885. Ueber eine Eigenthümliche Forms pastischer Lähmung mit Cerebralerscheinungen auf hereditärer Grundlage (Multiple Sklerose). *Archiv. f. Psych.*, XVI, 698–710.

PENROSE, C. A., 1905. Sanitary Conditions on the Bahama Islands, pp. 387–416; in G. B. Shattuck: The Bahama Islands. N. Y.

PLOSS, H., and M. BARTELS, 1905. Das Weib in der Natur und Volkerkunde. 8 Aufl. Leipzig, Th. Grieben's Verlag, Bd. I u. II.

POLITZER, ADAM, 1907. Geschichte der Ohrenheilkunde. Stuttgart, F. Enke, Bd. I and II.

POTAIN, C., 1870. Anemie. *Dict. Encycl. des Sci. Méd.*, Paris, IV, 1st series, 327–406.

POUCET, M. A., and LERICHE, M. R., 1903. Nains d'aujourd'hui et nains d'autrefois. Nanisme ancestral. Achondroplasie ethnique. *Bull. de l'Acad. de Med.*, Paris, L, 3d series, Oct., 174–188.

POULTON, E. B., and others, 1909. Fifty Years of Darwinism: Modern Aspects of Evolution. N. Y., Henry Holt.

PUNNETT, R. C., 1905. Mendelism. Cambridge, Macmillan.

—, 1908. Mendelism in Relation to Disease. *Proc. Roy. Soc. Med.*

RADCLIFFE-CROCKER, H., 1903. Diseases of the Skin, their Description, Pathology, Diagnosis and Treatment. Philadelphia, Blakiston's Son & Co., pp. 1–1439.

REBER, W., 1895. Six Instances of Color Blind Women Occurring in Two Generations of One Family. *Medical News*, Phila., 1895, LXVI, 95–97.

REED, JACOB W., 1861. History of Reed Family in Europe and America. Boston, John Wilson & Son, pp. 1–588.

REZEK, 1877. Hereditäre Herzfehler. Allg. Wien-med. Zeitschr. Wien, XXXVII, Sept., 338–339.

Rich, Shebnah, 1883. Truro, Cape Cod or Land Marks and Sea Marks. Boston, D. Lothrop & Co., pp. 1–580.

Ripley, Wm. Z., 1899. The Races of Europe, a sociological study. New York, D. Appleton & Co., XXXII, 624.

—, 1908. The European population of the United States. *Jour. Royal Anthrop. Inst. of Great Britain and Ireland*, XXXVIII, 221–240, also *Ann. Rept. Board of Regents of The Smithsonian Inst.*, for 1909, pp. 585–606.

Rischbeth, H., 1909. Hare Lip and Cleft Palate. *Treasury of Human Inheritance*, Part IV. *Eugenics Laboratory Memoirs*, XI. London, pp. i–vii, 79–126, 8 plates.

Roberts, Charles, 1878. A Manual of Anthropology or a Guide to the Physical Examination and Measurement of the Human Body. London, J. & A. Churchill, 115 pp.

Root, James P., 1870. Root Genealogical Records, 1600–1870, comprising the General History of the Root and Roots Families in America. N. Y., R. C. Root, Anthony & Co., pp. 1–533.

Rose, Felix, 1907. Obesité familiale. *L'encephale*, II, 299–303.

St. Hilaire, E., 1900. La surdi-mutité. Paris, 1–300. Maloine, Editeur.

Saleeby, Caleb Williams, 1909. Parenthood and Race Culture; an Outline of Eugenics. London, Cassell & Co.

Schamberg, Jay Frank, 1908. Diseases of the Skin and Eruptive Fevers. Phila. and London, W. B. Saunders Co., 1–534.

Senator, H. and S. Kaminer, 1904. Health and Disease in Relation to Marriage and the Married State. N. Y., 1257 pp.

Silcox, A. G., 1892. Hereditary Sarcoma of eyeball in 3 generations. *Brit. Med. Jour.*, I, p. 1079.

Simon, Chas. E., 1903. Heredity in Ménière's Disease. *Johns Hopkins Hosp. Bull.*, Baltimore, IV, pp. 72–84.

Sinnett, Rev. Chas. N., 1907. Jacob Johnson of Harpswell, Maine, and his Descendants, East and West. Concord, N. H., pp. 1–132.

Slayton, Asa W., 1898. History of the Slayton Family, Biographical and Genealogical. Grand Rapids, Mich., pp. 1–330.

Small, H. W., 1898. History of Swan's Island, Me. Ellsworth, Me., pp. 1–244.

Smyth, S. Gordon. 1909, Hans Joest Heydt, The Story of a Perkiomen Pioneer. *The Pennsylvanian-German*, July, pp. 11.

Spokes, Sidney, 1890. Report at Monthly Meeting of Odontological Society. *Trans. Odont. Soc.*, London, XXII, 229–232.

Stelwagon, Henry W., 1907. Treatise on Diseases of the Skin. Philadelphia, W. B. Saunders Co., 1150 pp.

Stocks, A. W., 1861. Sterility in Twin Sisters. *Lancet*, Lond., July, 1861, II, 78.

STREATFIELD, J. F., 1858. Coloboma Iridis. Heredity and rare cases. *Roy. London Ophthal. Hospital Reports*, I, p. 153.

SWIFT, C. F., 1888. Genealogical Notes of Barnstable Families, being a reprint of the Amos Otis Papers originally published in the Barnstable Patriot, Vols. I and II. Barnstable, Mass.

TAPLEY, HARRIET S., 1900. Genealogy of the Tapley Family. Danvers, Mass., pp. xix, 1–256.

THOMSEN, A. 1885. Zur Thomsen'schen Krankheit. *Centralbl. f. Nervenheilk, Psychiatrie, u. ger. Psych.*, VIII, 193–196, May.

THOMSON, J. ARTHUR, 1908. Heredity. London, J. Murray.

TIFFANY, NELSON O., 1903. The Tiffanys of America: History and Genealogy. Buffalo, N. Y., 254 pp.

TOMES, JOHN, 1906. A System of Dental Surgery Revised and Enlarged by Chas. S. Tomes and Walter S. Nowell. London, J. & A. Churchill, 1906, 770 pp.

TREASURY OF HUMAN INHERITANCE, Pt. 1–2—Diabetes insipidus, Split-foot, Polydactylism, Brachydactylism, Tuberculosis, Deaf-mutism, Legal Ability. Pt. 3—Angioneurotic Oedema, Hermaphroditism, Deaf-mutism, Insanity, Commercial Ability. Pt. 4—Cleft palate, Hare-lip, Deaf-mutism, Congenital Cataract.

TREDGOLD, A. F., 1908. Mental deficiency (Amentia). London, Bailliere, Tindall & Cox.

TURNER, J. G., 1906. Hereditary Hypoplasia of Enamel. *Trans. Odont. Soc.*, XXXIX, new series, March, 137–151.

TWINING, T. J., 1905. The Twining Family, Descendants of Wm. Twining. Fort Wayne, Ind., 251 pp.

UNNA, P. G., 1883. Ueber des Keratoma Palmare et Plantare Hereditarium. *Viertelj. f. Derm. u. Syph.*, Wien., X, 231–270.

VICE COMMISSION OF CHICAGO, 1911. The Social Evil in Chicago. Chicago, 399 pp.

VIERORDT, KARL HERMANN, 1901. Die angeborenen Herzkrankheiten. In Nothnagel. *Specielle Pathologie and Therapie*, XV, Wien, 1901, Th. 1, Abth. 2.

VIGNES, 1889. Epicanthus hereditaire. *Rec. d'Ophth.*, Paris, pp. 422–425, 3d series, XI.

VIRCHOW, R., 1881. Ueber die Weddas von Ceylon. *Abh. d. königl. Akad. der Wiss. zu Berlin*, 1881.

VOISIN, A., 1865. Contribution à l'histoire des Mariages entre consanguins. *Mém. de la Soc. d'Anthropol. de Paris* (3), II, pp. 433–459, Paris.

WAKLEY, THOMAS 1895. The Influence of Inheritance on the Tendency to have Twins. *Lancet*, London, Nov., 1895, II, pp. 1289–1290.

WALLACE, JOHN H., 1901. Genealogy of the Riggs family, with a number of Cognate Branches descended from the Original Edward through Female Lines and many Biographical Outlines. New York, Vols. I and II.

WEBER, F. PARKES, 1907. Multiple Hereditary Developmental Angiomata (Telangiectases) of the Skin and Mucous Membranes Associated with Recurring Hemorrhages. *Lancet*, 1907, II, 160–162.

WEIL, A. 1884. Ueber die hereditäre Form des Diabetes insipidus. Virchow's *Arch. f. Path. Anat*., etc., Berl., XCV, 70–95.

WHEELER, B. F., 1906. The Varick Family. Mobile, Alabama, 58 pp.

WHITE, CHARLES J., 1896. Dystrophia Unguium et Pilorum Hereditaria. *Jour. Cut. and Gen. Urin. Dis.* N. Y., XIV, 1896, pp. 220–227.

WILSON, E. B., 1900. The Cell in Development and Inheritance. N. Y., xxi + 483 pp.

—, 1902. Mendel's Principles of Heredity and the Maturation of the Germ-cells. *Science*, Dec., p. 991.

—, 1911. Studies on Chromosomes, VII. A Review of the Chromosomes of Nezara; with some more General Considerations. *Journal of Morphology*, XXII, 71–110, Mar., 1911.

—, ERASMUS, 1869. Lecture on Ekzema. *Jour. Cutan. Med.*, London, B. III, 1869, pp. 106–117.

—, WOODROW, 1902. History of the American People. Vols. I–V, New York.

WITHINGTON, C. F., 1885. Consanguineous Marriages; their effect upon offspring. *Mass. Med. Soc.*, Boston, XIII, 453–484.

WOOD, T. B., 1906. Note on the Inheritance of Horns and Face Color in Sheep. *Journ. Agri. Sci.*, v. I, p. 364.

WOODS, FREDERICK ADAMS, 1903. The Correlation between Mental and Moral Qualities. *Pop. Sci. Mo.*, Oct.

—, 1903. Mental and Moral Heredity in Royalty. *Pop. Sci. Mo.*, Aug., 1902, Apr.

—, 1906. Mental and Moral Heredity in Royalty. N. Y., Holt & Co., 312 pp.

—, 1906. Non-inheritance of Sex in Man. *Biometrika*, V, p. 73.

—, 1908. Recent Studies in Human Heredity. *Amer. Nat.*, V. 42, p. 685.

WORTH, CLAUD, 1905. Hereditary Influence in Myopia. *Trans. Ophth. Soc.*, London, XXVI, 141–143.

YULE, G. UDNY, 1902. Mendel's Laws and their Probable Relations to Inter-racial Heredity. *New Phytologist*, I, Nos. 9, 10.

ZAHNISER, KATE M., and CHAS. R., 1906. The Zahnisers: A History of the Family in America. Mercer, Pa., pp. 1–218.

APPENDIX

LIST OF PLACES REFERRED TO, GEOGRAPHICALLY ARRANGED

United States, 30, 31
    Maine: Washington Co., Jonesport, 190
        Hancock Co., Mt. Desert Is., 190
            Swans Is., 190
            Deer Is., 190
            Long Is., 190
    New Hampshire: Hillsboro Co., Milford, 51
    Vermont: Caledonia Co., St. Johnsbury, 57
    Massachusetts, 208
        Berkshire Co., 197
        Bristol Co., New Bedford, 219
        Barnstable Co., Cape Cod, 195, 219
            Falmouth, 43
        Dukes Co., Martha's Vineyard, 182, 188, 190, 192
    Rhode Island, 218
        Newport Co., Block Is., 188, 192
        Washington Co., Point Judith, 195
    Connecticut: Hartford Co., Windsor, 55
        New Haven Co., New Haven, 102, 182
            Wallingford, 57
        Fairfield Co., 182
    New York, 208
        Catskill Mountains, 197
        Ramapo Mountains, 197
        Albany Co., Albany, 237
        Delaware Co., 182
        Westchester Co., Tarrytown, 237
        Kings Co., 83
        Suffolk Co., East Hampton, 182
    New Jersey, Atlantic Co., Hammonton, 217
        Cumberland Co., Vineland, 217
    Pennsylvania, 208, 209, 251
        Allegheny Co., Pittsburgh, 56
        Dauphin Co., Harrisburg, 80
        Sullivan Co., 155, 182
    Maryland, 235

INDEX

NAMES PRINTED IN SMALL CAPS ARE CONSIDERED AS SUBJECTS, THOSE IN ITALICS AS AUTHORS. THE NUMBERS REFER TO PAGES